PROGRESS IN COLLOID & POLYMER SCIENCE

Editors: H.-G. Kilian (Ulm) and G. Lagaly (Kiel)

Volume 77 (1988)

Dispersed Systems

Guest Editors: K. Hummel and J. Schurz (Graz)

Springer-Verlag
Berlin Heidelberg GmbH

ISBN 978-3-662-15132-7   ISBN 978-3-7985-1692-2 (eBook)
DOI 10.1007/978-3-7985-1692-2
ISSN 0340-255-X

© 1988 by Springer-Verlag Berlin Heidelberg
Originally published by Dr. Dietrich Steinkopff Verlag GmbH & Co. KG, Darmstadt in 1988
Softcover reprint of the hardcover 1st edition 1988
Chemistry editor: Heidrun Sauer; Copy edition: James Willis; Production: Holger Frey.

Type-Setting: K+V Fotosatz GmbH, 6124 Beerfelden

Preface

The 33rd meeting of the "Kolloid-Gesellschaft" was held September 1987, in Graz, Austria. Graz holds a long tradition of work in colloid science. R. Zsigmondy, later Nobel prize winner for his work in this field, was a lecturer at Graz University. G.F. Hüttig's investigations into disperse systems and interfaces are well remembered. O. Kratky developed the application of x-ray small angle scattering to colloidal systems and we were honored that Prof. Kratky delivered the opening lecture of the 1987 Conference.

More than 160 active participants attended and the scientific program included 55 lectures and 24 posters. The Conference covered colloids, latices, interface, and surface films, surfactants, membranes, liquid crystalline systems, emulsions, microemulsions, micelles, lipids, polymeric structures, properties of polymers, optical investigation methods, pharmaceutical systems, adsorption processes, clay minerals and related topics. The program was well-balanced and included theoretical and practical aspects of today's colloid science. Scientists from eastern European countries, especially Hungary, Yugoslavia, and Bulgaria, participated along with those from other European and foreign countries, making the meeting a truly international event.

This Progress Volume contains a selection of the papers presented at the 1987 Conference. We hope it will demonstrate the manifold nature and the diversity of modern colloid science as an interdisciplinary field.

Klaus Hummel
Josef Schurz

Contents

Polymer Solutions and Polymers

General

The importance of x-ray small-angle scattering in colloid research

O. Kratky

Key words: Mass determination, shape determination, radial density distribution, persistence length, Porod's relations, scattering power

Introduction

When we ask about possibilities to study the geometric structure of colloidal and macromolecular systems we can mention a number of methods which provide one or two parameters; in addition, three methods exist which lead to much more substantial information:

1) x-ray crystal structure analysis
2) electron microscopy
3) small-angle scattering of x-rays and neutrons

Even though crystal structure analysis exceeds all other methods in terms of resolution, there exists a limitation in its applicability through the requirement for crystals which must be not only of macroscopic size but also largely free of defects: these requirements are currently met only for part of all macromolecular substances.

Electron microscopy has the undoubted advantage of providing immediately comprehensible pictures; however, the method is far from being non-invasive. Moreover, because of the necessity of separating the particles from a solution, it is impossible, for instance, to study biological molecules in their natural surroundings. Considering the limitations of the other two methods, x-ray small angle scattering has certain distinct advantages:

a) x-ray small-angle scattering offers the possibility of studying macromolecules in their natural surroundings, i.e., in aqueous solution.
b) it is practically a non-destructive method.
c) it does not require crystals.
d) it provides information on intermediate states in solution which are likely to be lost during crystallization or in the preparation steps necessary for electron microscopic investigation.

Therefore, x-ray small-angle scattering is an ideal supplement to the other two techniques.

Of all the systems that can be approached by small angle scattering we first describe the analysis of monodisperse solutions of biological macromolecules, because they demonstrate the possibilities of the method best.

Why small-angle scattering? According to the law of reciprocity in optics, diffraction occurs at smaller angles the larger the structural dimensions are. Now, since colloidal particles are huge as compared with the wave length of x-rays (approximately 1 Å), scattering occurs at correspondingly small angles. In a monodisperse solution which is sufficiently dilute so that the average distances between the particles are large as compared with the particle size, the scattering intensities from the individual particles simply add, and the effective scattering curve corresponds to that of one particle averaged over all possible spatial orientations: in this case we speak of *particle scattering*.

A) Parameters

From certain characteristic features of particle scattering a number of parameters can be obtained unambiguously. We shall discuss some of them in the following.

1) Radius of gyration

Just 50 years ago, Guinier [30] showed that the innermost part of any particle scattering curve of monodisperse particles is a Gaussian curve. Allow me then to show a photograph of this pioneer of small-angle scattering: Guinier and the author in discussion at the 2nd International Conference on X-ray Small-Angle Scattering in Graz, 1970 (Fig. 1).

The following relation (1a) describes the Gaussian course of a small-angle scattering curve in its innermost part.

Fig. 1. André Guinier and the author in discussion at the 2nd International Conference on X-Ray Small-Angle Scattering, Graz, 1970

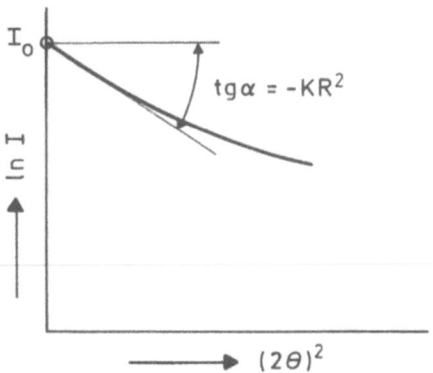

Fig. 2. Schematic course of a particle scattering curve at very small angles

$$I = I_0 \cdot e^{-KR^2(2\theta)^2} \quad K = \frac{1}{3}\left(\frac{2\pi}{\lambda}\right)^2 \tag{1a}$$

$$\ln I = \ln I_0 - KR^2(2\theta)^2 \tag{1b}$$

It contains the scattering angle 2θ, the scattering intensity I and its value at zero angle I_0, and, most importantly, the parameter R, the radius of gyration. It represents the root mean square distance of all electrons within one particle from the common electronic center of gravity, and is, therefore, an illustrative measure for the spatial dimension of particles. Plotting the logarithm of the intensity, $\ln I$ versus $(2\theta)^2$ (Guinier plot) (1b), a curve is obtained (Fig. 2) in which the tangent of the linear course in its innermost part yields the radius of gyration directly. R is a most important parameter, not only as a measure for the size of a particle but also as a very useful indicator for structural changes in a substance. An important field of shape information is opened if we consider the behavior of very anisotropic structures.

2) Rod-like particles [1]

Let us first consider a solution of infinitely thin and infinitely long particles. This system of needles has a scattering function represented by $1/2\theta$. In reality, however, particles always have a finite thickness. This leads to the function I_c, the cross section factor, which depends on the size and shape of the cross section and by which the needle-scattering curve must be multiplied:

$$I = I_c \frac{1}{2\theta} \; ; \quad I_c = I \times (2\theta) \; . \tag{2}$$

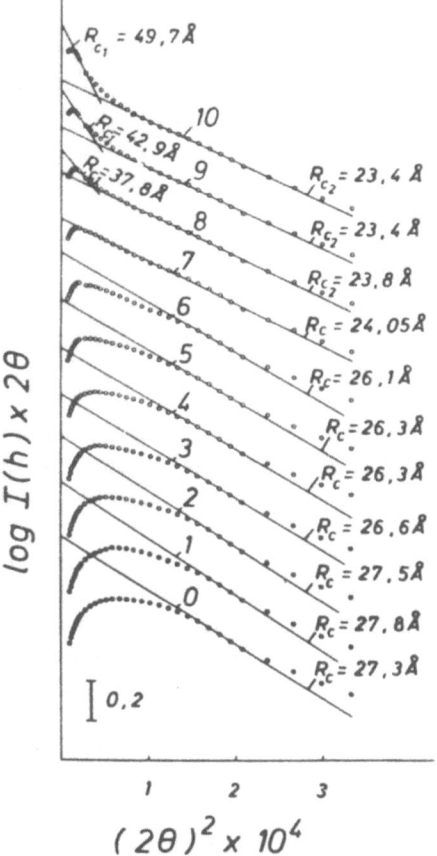

Fig. 3. Changes in cross section factor of malatsynthase with increasing x-ray irradiation time in the Guinier plot [2]

Analogous with the determination of the radius of gyration R from the intensity in the case of corpuscular particles, the radius of gyration of the cross section R_c in the case of elongated particles can be evaluated from the cross section factor I_c by a Guinier plot.

In reality the particles also have a finite length, i.e., the maximum dimensions are lacking in comparison to infinitely long particles. According to the law of reciprocity these distances would scatter to the smallest angles; if they are lacking, a deficit in intensity at the smallest angle will occur, and, consequently, the cross section factor shows a falling tendency in its innermost part.

A very instructive example for these relationships is given by the enzyme malatsynthase, which has been investigated by Zipper and Durchschlag [2]. Figure 3 shows the cross section factor for a series of solutions which have been exposed to increasing x-ray doses, from bottom to top; this treatment leads to an association of the particles. If one neglects the innermost part, the same cross section factor is observed in all cases; hence aggregation takes place in the longitudinal direction.

The original particles are initially short, and therefore, one observes at the beginning a strong decrease of the cross section factor towards the angle zero. With increasing duration of irradiation this effect decreases and the shape approaches that of a long rod. For the longest irradiation period finally, one observes in the innermost parts of the curves an increased cross section factor, which can be easily interpreted in terms of lateral aggregation of two rods.

3) Lamellar particles [1]

An analogous treatment applies to two-dimensional, lamellar particles, only that in this case the square of the scattering angle $(2\theta)^2$ takes the place of 2θ in all the formulas. Since we shall be concerned in the following mainly with the cross section factor, we have limited ourselves to its discussion.

4) Particle volume

A further important parameter which can be obtained from particle scattering is the volume. According to Porod [3] the following equation holds

$$V = \frac{\lambda^3 \cdot I_0}{4\pi \int\limits_0^\infty I \times (2\theta)^2 d(2\theta)} \quad (3)$$

The integral in the denominator is frequently assigned the symbol Q and is called the invariant because it is independent of the degree of dispersion. This quantity will be used repeatedly in the following discussion.

5) Mass determination

The determination of sizes and shapes of dissolved macromolecules is not the only possibility offered by small-angle scattering. In addition if offers the possibility of weighing particles [4]. According to Thomson the ratio of scattering of the single electron to the primary energy is known. Hence one can calculate the ratio of scattering to primary energy for all electrons within one particle of the mass M, and determine the molecular weight from the experimentally observed ratio of I_0 to P_0. The following equation describes this relationship

$$M = \frac{21.0 \cdot I_0 \cdot a^2}{P_0 [z_2 - \bar{v}_2 \varrho_1]^2 d \cdot c} \quad 21.0 = \frac{1}{N_L T_e} \quad (4)$$

The primary energy P_0 is so exceedingly large that it cannot be measured with the same detector as the scattered energy I_0. However, this experimental problem is solved and we shall not discuss this here in any detail. The symbols in Eq. (4) have the following meanings: a — sample-detector distance; d [cm] — sample thickness; c [g/ml] — concentration; z_2 — number of mole electrons per gram of dissolved substance; ϱ_1 — electron density of the solvent; $\bar{v}_2$ — partial specific volume of the dissolved substance; T_e — 7.9×10^{26} (Thomson's constant); N_A — Avogadro's number.

Hundreds of molecular weights of macromolecules have been measured according to Eq. (4). Of course, there are other techniques for the same purpose, especially ultracentrifugation. A unique possibility of the small-angle method, however, is given by the fact that it can determine the mass per unit length for elongated particles, it can, hence, quasi-weigh the slice of a sausage in molecular dimensions (5a); likewise, in the case of lamellar particles the mass per unit area can be determined (5b) [5]. The corresponding relations

$$M_c = \frac{6.68 [I(h) \times h]_0 a^2}{\lambda P_0 [z_2 - \bar{v}_2 \varrho_1]^2 d \cdot c} \quad (5a)$$

$$M_t = \frac{3.34 [I(h) \times h^2]_0 a^2}{\lambda^2 P_0 [z_2 - \bar{v}_2 \varrho_1]^2 d \cdot c} \quad (5b)$$

contain the factors $[I(h) \times h]_0$ and $[I(h) \times h^2]_0$, respectively, wherein

$$h = 4\pi \sin \theta / \lambda \quad . \quad (6)$$

These expressions hence are proportional to the cross section factor $[I_0(2\theta)]_0$ and the thickness factor $[I(2\theta)^2]_0$ at the angle zero.

B) General scattering theory [47]

1) Homogeneous particles

We have derived certain parameters from the scattering curve. In the following we shall attempt to describe the entire shape of the particle as precisely as possible. This requires a quantitative relation such as the following (7a):

$$I(h) = 4\pi \int_0^\infty p(r) \left(\frac{\sin hr}{hr}\right) dr; \quad h = 4\pi \sin\theta/\lambda$$

$$(7a)$$

and by Fourier transformation

$$p(r) = \frac{r^2}{2\pi^2} \int_0^\infty I(h) \times h^2 \frac{\sin hr}{hr} \cdot dh \qquad (7b)$$

The term $\frac{\sin hr}{hr}$ is the scattering of a dumbbell according to Debye, hence the scattering of a body consisting of two point masses separated by the distance r, averaged over all orientations in space. If a structure has the distance distribution function $p(r)$ (which is the number distribution of distances r between any two volume elements within one particle), we have only to perform the above integration in order to obtain the entire scattering curve $I(h)$.

If we know the shape of a body, we can calculate $p(r)$ (this is a merely geometrical task, relatively easy to solve for all tri-axial symmetrical bodies as for instance tri-axial ellipsoids, but it is more difficult for complex particles which are composed of several symmetric parts). We have only to insert the result into Eq. (7a) to obtain the corresponding scattering curve. *However, we intend to go the opposite way: we know the scattering curve and want to evaluate the shape of the body.* The only possibility that remains is the trial and error approach, i.e., to try assumed structures until one finds one that closely approximates the observed distance distribution function $p(r)$ obtained by Fourier transformation (7b) of the measured scattering curve.

Frequently this task can be facilitated by auxilliary information from other sources. An example is given by Immunoglobulin [6]. From the scattering curve a cross section factor is obtained, shown in its Guinier representation in Fig. 4. This plot shows three characteristic properties:

1) the substance is composed from rod-like particles, otherwise no straight section in the Guinier plot of the cross section could be observed (the measurements are shown by the points at the bottom curve).
2) the decay of the cross section factor in the Guinier plot in the innermost part shows that the rods are short.
3) the curve has two branches, which means that there are two cross sections prevailing. In good agreement with this, the biochemical investigations by Edelman [7] have provided evidence for the existence of three subunits. Of the possible configurations following from these facts the one with fully stretched arms (assigned no. 4) leads to the best agreement between experiment (points in curve 5) and theory (curve 4).

Frequently, the power of the trial and error method is amazing. As an example, the comparison of the experimental distance distribution function of the lac repressor to the theoretical curves, which correspond to the models, is shown in Fig. 5 (in collaboration with O. Jardetzky [8]). This model has been formed by 600 small spheres which do not represent subunits but serve for the simulation of the shape only. In the same way a major number of trial and error developments has been carried out by Pilz, a former collaborator of mine. This kind of research requires three distinct factors: 1) diligence and patience, 2) access to a powerful computer, 3) a feeling for the manifold variations, one of which finally leads to success.

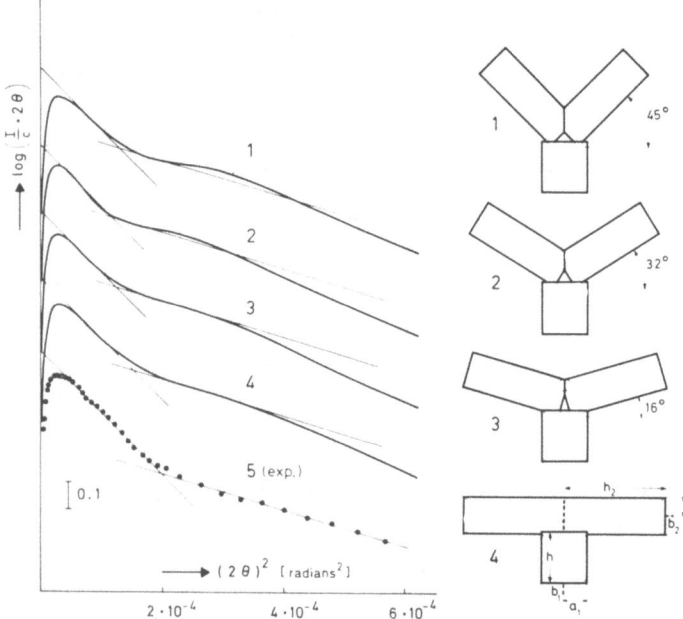

Fig. 4. Guinier plot of the theoretical cross section scattering curves (1−4) of the Y- and X-shaped models shown in the right part of the figure (1−4) of IgG in solution. The experimental curve [5] shows the best fit to model 4, [6]

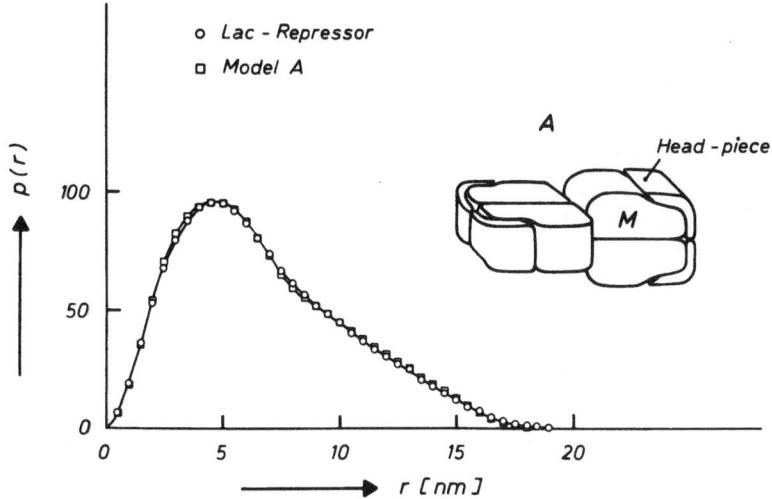

Fig. 5. Comparison between the distance distribution functions obtained from the scattering curve (○) and calculated for Model A (■) [8]

2) Symmetrical particles with strong internal electron density variation

Up to this point the assumption was sufficient that the particles and the subunits, respectively, are internally homogeneous, i.e., they do not contain domains with strongly different electron densities. With internal fluctuations, however, it becomes desirable to determine this electron density distribution. The most favorable cases are given by particles with spherical symmetry. In this case we obtain the radial electron density distribution $\varrho(r)$ from the distance distribution function $p(r)$ by applying a method developed by Glatter [50].

An important application is to plasma lipoproteins. The scattering curve of Low Density Lipoprotein LpB is shown in Fig. 6. This shows the typical appearance of a quasi-spherical particle by the well developed minima and maxima. From the calculated radial electron density distributions at 4° and 37 °C, respectively, it becomes immediately clear that the core of particles undergoes a transition from a highly ordered state at low temperature to a disordered state at high temperature. From the knowledge of the molecular composition an idealized model of the radial cross section can be deduced (Fig. 7). This field of research has been treated in great detail, especially by the groups of Laggner and Luzzati.

For lamellar particles with an internal mirror plane, the electron density distribution vertical to the lamellar plane can be obtained in a similar way. This technique has been applied successfully by Laggner

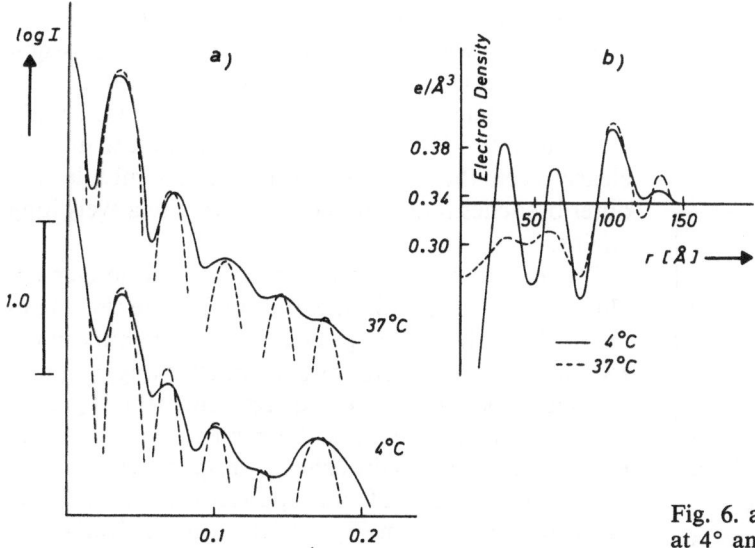

Fig. 6. a) Scattering curves of low-density lipoprotein LpB at 4° and 37 °C. b) Radial electron density distributions at the two temperatures [9]

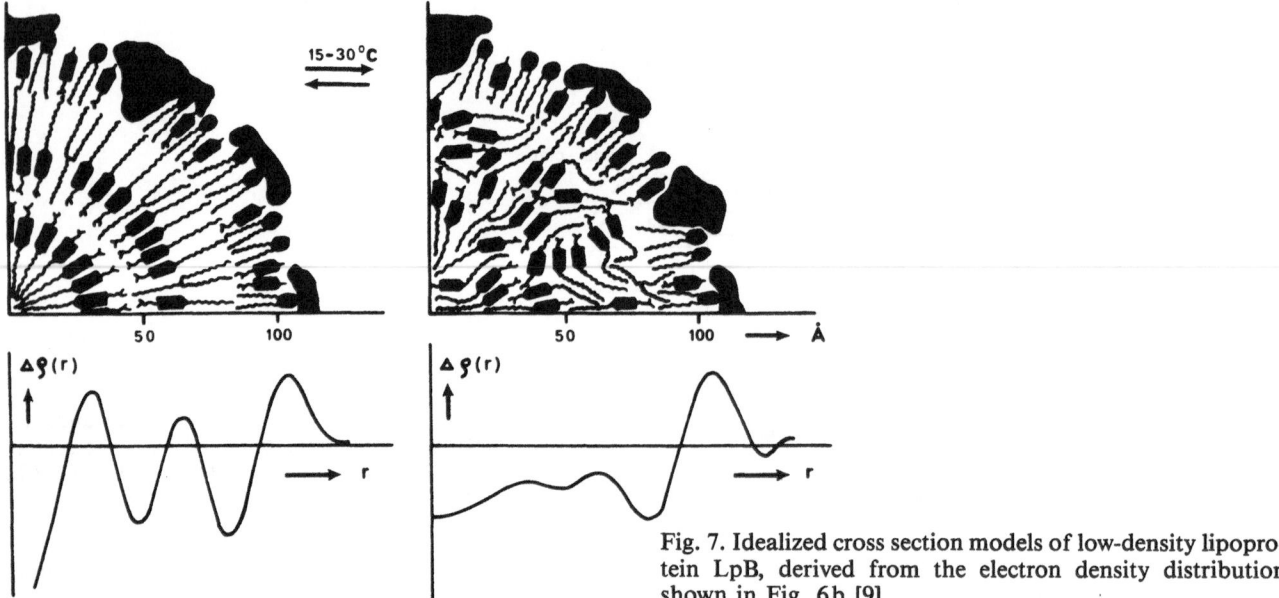

Fig. 7. Idealized cross section models of low-density lipoprotein LpB, derived from the electron density distribution shown in Fig. 6b [9]

and K. Müller [43, 48] in the area of biological membranes.

C) Chain molecules

1) Dilute solutions

In general the previously discussed principles of particle scattering apply largely for chain molecules in solution. The fact that these molecules can constitute coiled and mobile chains necessitates additional consideration; for such objects the term "worm-like chain" has been generally adopted. The schematic scattering curve for this type of chain is shown in Fig. 8a [1, 10]. The innermost part relates to the entire molecule. Since the coil as a whole is isotropic, this innermost part shows an approximately Gaussian shape. The middle part of the scattering functions reflects

larger sections of the chain molecules; these represent a random walk distribution of masses, for which, according to Debye, a $1/(2\theta)^2$ course must be expected [11]. The third, outermost part relates, in keeping with the optical law of reciprocity, to very small sections of the chains which are almost straight. The rod-like section should show a scattering function following a $1/2\theta$ course. Multiplying the entire curve by $(2\theta)^2$ leads to Fig. 8b. The Gaussian branch shows a course decaying towards zero, the $1/(2\theta)^2$ branch gives a constant value and the $1/2\theta$ branch tends towards 2θ and hence is ascending.

According to more detailed calculations, the better the scattering curves approximate the above described ideal behavior the longer the chain becomes. Fig. 9 shows such a series of curves [12]. The number assigned to the individual curves are the ratios between the total chain length and the so-called persistence length, which is half of Kuhn's statistical chain element. It is clearly seen that the kink between the middle and outer branches is only well developed in relatively long chains.

The more stretched a chain is, i.e., the longer the almost rod-like sections are, the smaller are the angles to which the $1/(2\theta)$ range extends, i.e., the kink in the $I \times (2\theta)^2$ plot will move towards smaller angles. From its angular position one can directly infer the degree of coiling or, more precisely, the persistence length and Kuhn's statistical chain element, respectively.

As an example of an application we mention heparin [13] shown in Fig. 10. The kink is clearly visible. The following control can be performed: from the

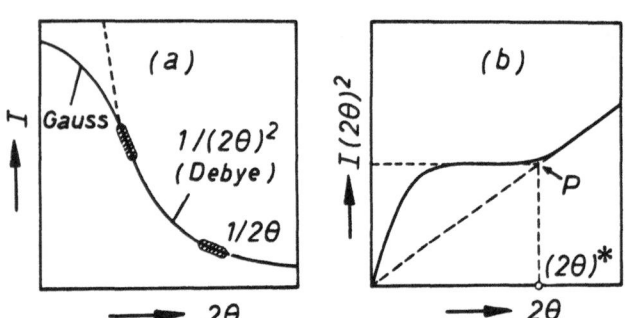

Fig. 8. Schematic representation of a scattering curve for statistically coiled chain molecules

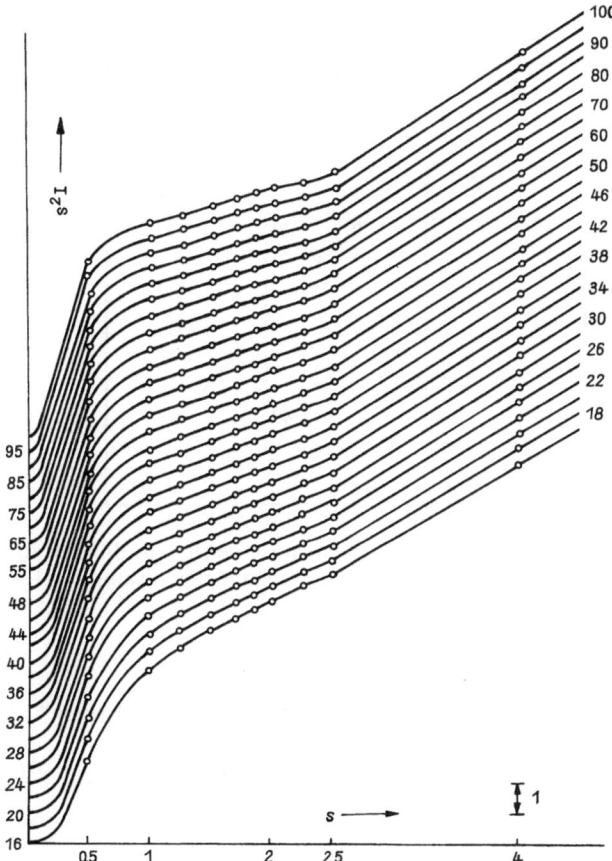

Fig. 9. Scattering curve of chains of $18-100$ persistence lengths in the plot of Is^2 vs s $\left(s = \dfrac{4\pi \sin \theta}{\lambda} \right)$ [12]

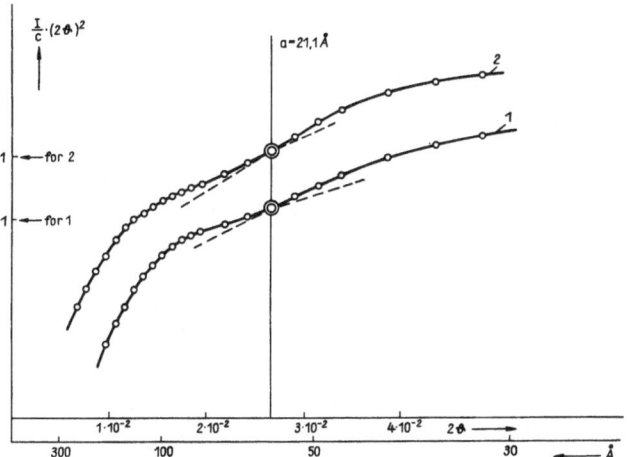

Fig. 10. Plot of $\dfrac{I}{c} \times (2\theta)^2$ vs 2θ for a solution of heparin [13]

molecular weight, which can be determined as described before, and from the known chemical structure, one knows the length of the chain in the fully stretched state. If on the other hand one knows the degree of coiling from the position of the kink, one also knows how big the coil has to be and, consequently, what the radius of gyration should be; this prediction can be checked by a direct analysis of the innermost branch. The agreement in this particular case has been excellent.

The described possibility to determine the degree of coiling in a chain molecule from a single characteristic experimental parameter is unique for small-angle scattering. Quite clearly, one can determine the radius of gyration also by scattering of visible light, provided the particle is of a sufficiently large size, and from this and the molecular weight, the persistence length can be also calculated. However, one has to imply a Gaussian coil, an assumption which is not necessary in case

of the small-angle method, and which, indeed, does not apply in all cases. Another important advantage of the small-angle method in the field of chain molecules: the application of the above described relation (5a) for the mass per unit length leads to information on the configuration of the molecule, i.e., whether it is present as a stretched chain or as a helical structure, or whether lateral associations prevail.

2) Chain molecules in the amorphous condensed state

We now approach the question of what conformation chain molecules obtain in the amorphous condensed state (melt or glass). Obviously, it is impossible to draw immediate conclusions about the shape of individual chains from such a system because the interparticular interferences are dominant. One can find a way out of this dilemma by labeling only a small part of the chains with heavy atoms at both ends through admixture in the melt; in this way a dilute system of marked chains can be obtained. Subtracting the scattering of a sample without labels leaves the scattering of the markers, i.e., of the pairs of atoms which do not interfere in scattering with other pairs because they are sufficiently far form each other. One pair of identical atoms with a mutual distance r gives the scattering function

$$I \sim 1 + \frac{\sin hr}{hr} \tag{8}$$

That is, a damped oscillation, the wave length of which leads to the value of r. In general one will thus obtain an average value of all distances r and can thus draw conclusions regarding the chain coiling.

One can also label each monomer by a marker, as has been performed for instance with polystyrene using iodine labels. In this way one obtains the radius of gyration for the entire marked chain. In many experiments of this kind the Flory hypothesis [14] of mutually independent statistical chains has proven valid also for the condensed state. In essence, this means that even in the bulk polymer the chain dimensions would return to the ideal condition found in solution at the θ point. In particular it has been found that the radius of gyration varies linearly with the square root of the molecular weight as should be expected for a Gaussian chain, i.e., in θ solvent.

3) Neutron scattering experiments

Most convincing are the neutron scattering experiments, wherein the labeling is achieved simply by introduction of D instead of H so that the labeling itself does not imply a source of errors through alteration of the structure. This possibility is based on the fact, that hydrogen and deuterium differ strongly in their scattering of neutrons [41, 46].

D) Semi-crystalline and air-swollen high polymers

1) General considerations

A much more complex situation is encountered in the case of semi-crystalline samples in a concentrated state. At this point we shall recall that in 1938, two separate paths developed in small-angle scattering research. Parallel to the studies of dilute systems, concentrated and initially completely compact solid natural fibers were considered [15]. In this latter case the condition no longer holds that the distances between the particles are large compared with the particle size, and consequently interparticular interference effects cannot be neglected.

Hosemann [16] has noted that interference effects are less important for increasingly polydisperse systems. This does not invalidate the task of collecting as much unambiguous information on each individual case as possible.

In this area much is owed to the fundamental insight by Porod.

2) Porod's relations [3]

Dividing the invariant Q by the primary energy, Eq. (9a)

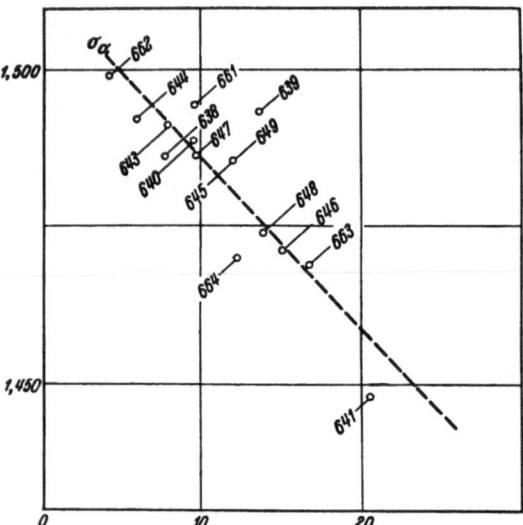

Fig. 11. Connection between mass density (ordinate) and scattering power (abscissa) for various rayon samples [18]

$$Q/P_0 = \frac{Kd}{a^2}\,\overline{(\Delta\varrho)^2} \qquad K = \frac{i_0}{4\pi}\lambda^3 N_L^2 = 22.8\times10^{20}\lambda^3$$

$$(9\,\mathrm{a})$$

$$\overline{(\Delta\varrho)^2} = (\varrho_2 - \varrho_1)^2\, w_2 w_1\ , \tag{9b}$$

$$Q/P_0 = \frac{Kd}{a^2}\, w_2(1-w_2)(\varrho_2-\varrho_1)^2 \tag{9c}$$

leads to a parameter, which is fundamental for the material structure of the system, i.e., the average square of the electron density fluctuations $(\Delta\varrho)^2$, frequently termed the "scattering power" of an object. For a two-phase system with volume fractions w_1 and w_2 the scattering power is given by Eq. (9b) which can be inserted into the first formula. Knowing two of the three unknowns on the righthand side of the Eq. (9c) leads to the evaluation of the third unknown. If for instance the electron densities ϱ_1 and ϱ_2 are known from the material composition, the volume fractions w_1 and $w_2 = 1 - w_1$ can be calculated. In the case of very condensed systems, this method allows one to obtain the volume fractions of the voids, i.e., the holes. Figure 11 shows that, according to the work of Hermans on different samples of solid cellulose with decreasing macroscopic density (i.e., increased volume fractions of voids), plotted at the ordinate, the scattering power (abscissa) increases.

A further important finding was obtained by Porod [3], which states that the outermost part of any scattering curve from systems containing sharp phase

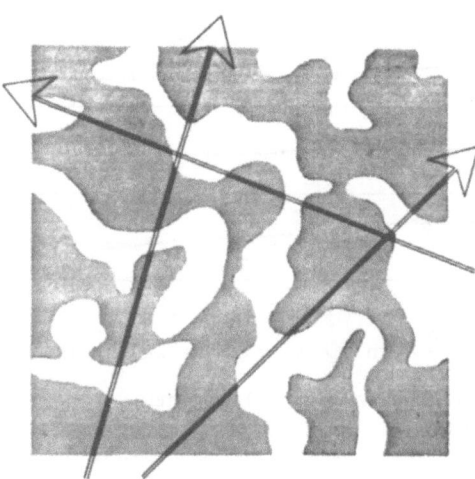

Fig. 12. Sponge-like colloid with schematic representation of the "intersection length"

boundaries should decrease according to Eq. (10a) with the inverse fourth power of the scattering angle

$$\lim I \sim \frac{1}{(2\theta)^4} \quad \lim I \times (2\theta)^4 = K \tag{10a}$$

$$O_s = \frac{2\pi^2}{\lambda} w_1 w_2 \frac{K}{Q} \tag{10b}$$

$$\overline{l_1} = \frac{4 w_1}{O_s} ; \quad \overline{l_2} = \frac{4 w_2}{O_s} \tag{10c}$$

This general relation is called Porod's law. Finally, Porod showed that the constant term K in the case of a two-phase system relates to the interfacial area of the two phases, and thus leads to a specific inner surface of a sample O_s (Eq. 10b).

This parameter can be related to the so-called intersection length, which characterizes the degree of dispersion in a sponge-like colloid (Eq. 10c), where no particle size can be defined (Fig. 12). If, in a model as depicted in Fig. 12, arrows are shot through the system in all possible directions, and an average is taken over all length $\overline{l_1}$ and $\overline{l_2}$ passing through a single phase, one indeed obtains a characteristic parameter for the dispersity of the two phases. In this case, the simple relationship (Eq. 10c) holds for the average values of $\overline{l_1}$ and $\overline{l_2}$.

These parameters are particularly relevant to solid colloidal systems.

We now return to the discussion of special condensed systems.

3) Air-swollen cellulose

The link between dilute and condensed systems, as far as the behavior in small-angle scattering is concerned, is given by objects consisting of highly anisotropic particles, for instance, rods. The same holds also for lamellar elements. If these are present in disordered arrangement it can be easily shown that interparticle interferences are practically negligible [17]. An illustrative example is regenerated cellulose. Upon regeneration this appears in a strongly swollen situation and consists of extended supramolecular associates frequently termed micelles. According to a treatment proposed first by Hermans [18] such samples can be transformed into an air-swollen state, through exchange of water with an organic solvent and subsequent evaporation; by this treatment the rod-shaped micelles largely lose their mutual connection and reach a disordered state.

Figure 13 shows the Guinier plots of the cross section factors for different samples as a function of the degree q of air swelling.

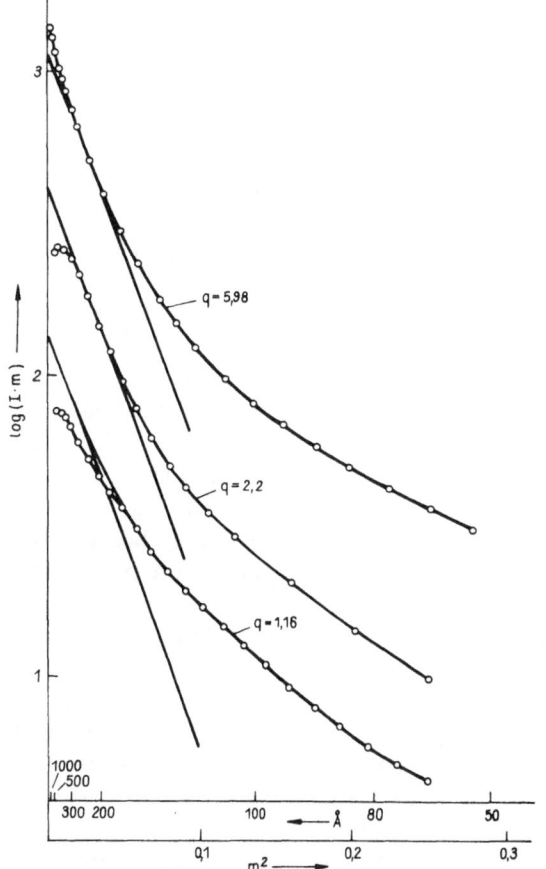

Fig. 13. Cross section factor of three air-swollen samples of regenerated cellulose in the Guinier plot [19]

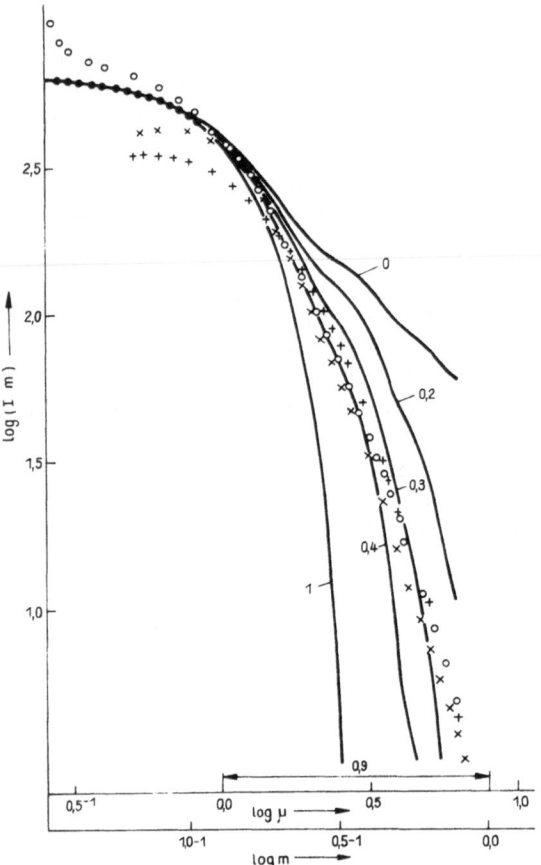

Fig. 14. Comparison between theoretical cross section curves for different axial ratios and experimental cross section curves of air-swollen regenerated cellulose with the following degrees of swelling: 5.98 (top curve); 2.2 (medium curve); 1.16 (bottom curve) [19]

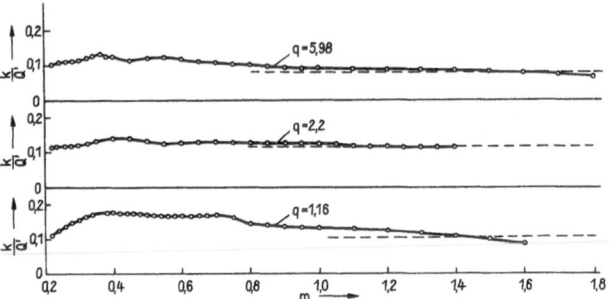

Fig. 15. Plot of $\dfrac{K}{Q'} \equiv Im^4/Q'$ vs $m(m/a \approx 2\theta)$ for the tail end of the scattering curve: ($\circ\,\circ\,\circ$) measurements, ($-\,-\,-$) calculated values [19]

Within the range of degrees of swelling corresponding to volume occupancies of 16 and 30 volume percent, one observes constant slopes from which the radius of gyration of the cross section can be calculated. With very small degrees of swelling, however, the shapes of the curves change, simply because the dense packing forces the units into a more parallel organization.

From the higher degrees of swelling a radius of gyration $R_c = 47$ Å has been obtained. Furthermore one can obtain from the shape of the cross section factor in the log-log plot the ratio of the secondary axes of about 0.4 (Fig. 14). A very simple calculation then leads to the absolute values for the secondary axes of 60 and 150 Å, respectively.

The outer part of the scattering curve corresponds well to Porod's law (Fig. 15) and one can calculate from this an inner surface, which agrees well with the

expected values from the cross section of the rods, so that this model can be considered as very well founded; the same follows from considerations of absolute intensity which shall not be discussed in any detail here.

In essence, these are quite condensed systems, which, can be treated in the data evaluation similar to dilute systems inspite of the fact that they do not necessarily imply strong polydispersity. Indeed, the reason for the reduced interference effect is given by the elongated structure of the unit in combination with the orientational disorder.

It seems surprising that this technique of air swelling has not found more wide spread application. Considering the elaborate procedures for electron microscopic investigations one could also apply these simple preparative steps for the samples in x-ray small-angle scattering experiments.

The approach has its limitations as we have seen in the example of the curves for the cross section factors, since too small a degree of swelling forces the micellar domains into parallel organization. Through this the system finally approaches the state of a stack of parallel-oriented lamellae which has also been treated quantitatively; this approach yields an average value for the thickness of the single lamella of 60 Å. Thus a good agreement between the results obtained from the treatment as a dilute system and that obtained from the dense system is achieved.

The basic models in these considerations of cellulose are finally always micellar networks of crystalline domains which are linked by fringes as is schematically shown in Fig. 16.

4) Neutron small-angle scattering

It shall also be mentioned that neutron small-angle scattering offers an elegant possibility of describing

Fig. 16. Scheme of a micellar network [20]

the semi-crystalline nature of cellulose. There one starts from partially deuterated samples which are obtained by treatment of the fibers with D_2O vapor. In this case the degree of hydrogen-deuterium exchange depends on the accessibility of the various domains. This makes periodicities of micellar and amorphous regions more visible than in the case of x-ray scattering, for which the contrast is too small. While native cellulose fibers (Ramie) do not show long periods in the direction of the fibers even with prolonged exposure to D_2O vapor, one observes strong meridional reflections at 165 and 193 Å with regenerated

celluloses (Fortisan and Rayon). These reflections can be recognized in Fig. 17b as intensity contours halfway between the center and the upper and lower edges of the picture (Fischer et al. [21]). In these studies one has also analyzed the intensity distribution of the layer lines in the horizontal direction with the aim of defining the radius of gyration of the lateral dimension of the micellar particles. The obtained value of 17 Å agrees with other ultrastructural studies but it is not clear yet how this result fits with the values which have been obtained from air-swollen fibers. In any case these studies must be continued.

5) Chain folding

For synthetic high polymers, a model such as has proven successful in the analysis of cellulose usually does not lead to success because here the chains are generally folded in the condensed state, as has been found by the important studies carried out independently by Fischer [22], Keller [23], and Till [24]. These folded chains also interconnect adjacent crystalline domains as shown by the schematic model in Fig. 18 which originates from a work of Zachmann [25]. This scheme makes it clear that in this case the quantitative treatment is based on lamellar systems, for the characterization of which suitable parameters must be determined. This is a rather difficult field with many facets which cannot be treated here in any detail.

For wholly condensed, solid celluloses and for synthetic high polymers it is possible through Porod's

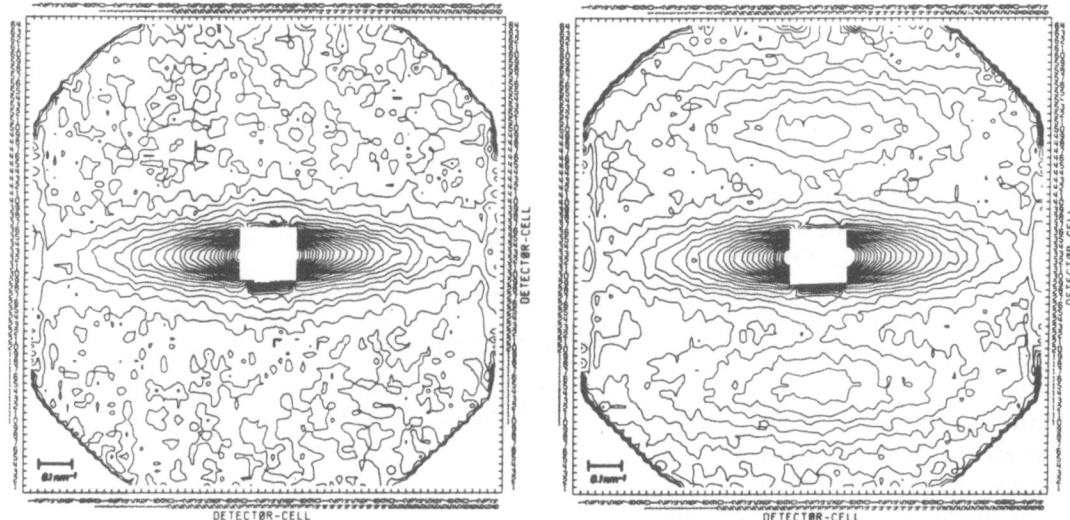

Fig. 17. Two-dimensional intensity distribution of the neutron scattering pattern of regenerated cellulose (Fortisan; logarithmic scale of equiintensity curves): (*a*) untreated; (*b*) deuterated for 5 hours. Fiber direction vertical [21]

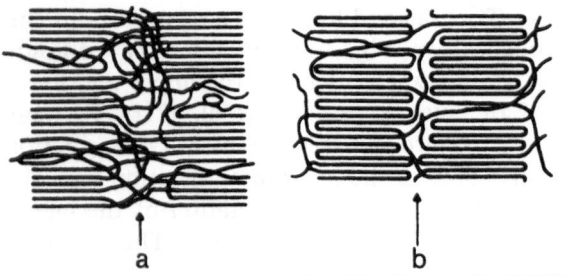

Fig. 18. Models of supramolecular structures in the case of folded chains [25]

treatment to obtain important technological information. In recent times this has been carried out especially by Schurz et al. [26]. The relations discussed above for two-phase systems have been extended to three phases: crystalline substances, amorphous substances and voids [27].

E) Notes on small-angle scattering of inorganic substances

This field encompasses natural and synthetic materials, among others metals, alloys, semi-conductors, glasses, and catalysts. The small-angle method always determines the structural inhomogeneities. These can be of great interest, particularly for mechanical properties, catalytic activity, or electric conductivity.

In contrast to the organic substances, which have been discussed so far, the interesting inorganic substances mostly consist of relatively heavy atoms. Even if one uses the lower wavelength of molybdenum radiation the transparent thicknesses are so small that their very preparation can already change the properties of the material as compared with the bulk form. Here, no doubt two approaches are promising: one is neutron scattering which is not the central theme of this lecture; the other is the use of even more energetic radiation such as white tungsten radiation.

Another complication in polycrystalline materials is double Bragg reflection, which cannot be distinguished from the scattering effect proper. One possible way to circumvent this problem is again to use neutron scattering with a suitably large wave length. If the wavelength is more than twice as large as the lattice period no Bragg reflection can be observed.

Due to the large variety of different objects, a simple unifying treatment according to a certain scheme is impossible. In particular the idea of particle scattering is applicable only in the minority of cases and can be used to describe the size and shape of the inhomogeneities. In the large majority of cases nonuni-

form sizes and large concentrations prevail. Very often, however, the main task of such investigations is not the quantitative description of the structural details but the studies of relative changes following different compositions and different pretreatments. A detailed interpretation is very difficult and frequently impossible.

In the framework of this presentation a further discussion of this subject is dispensible because for inorganic substances the field of small-angle scattering tends to move strongly in the direction of neutron scattering, and as far as metals are concerned, this research is carried out by metallurgists who consider this research as part of their discipline. Interested readers are referred to an instructive paper by Kostorz ([50] pp 467).

F) Notes on the small-angle camera

The slit cameras used until the mid-1950s were unsatisfactory because of their parasitic scattering. The suggestion made at that time for a collimation system is shown in Fig. 19. It consists of a U-shaped body B_1 on which a flat bridge B_2 is mounted. The essential feature is the precise coincidence of the upper plane B_1 of the U-shaped body and the lower plane B_2 of the bridge. If the x-ray beam enters the system through slit E, propagates along the surface B_1, and proceeds below B_2, no parasitic scattering can arise in the plane of registration (PR) above the upper edge of the primary beam stop (PS). This can be understood by the sections shown in Fig. 20 [28], vertical to the long dimension of the focus, and parallel with the length of the collimation system. A picture of the camera mounted in front of the x-ray tube is shown in Fig. 21 [29]. (The camera is produced in Graz by Messrs. A. Paar KG. About 400 laboratories worldwide use it.)

The most important goal at present for the small-angle method is the development of methods to follow

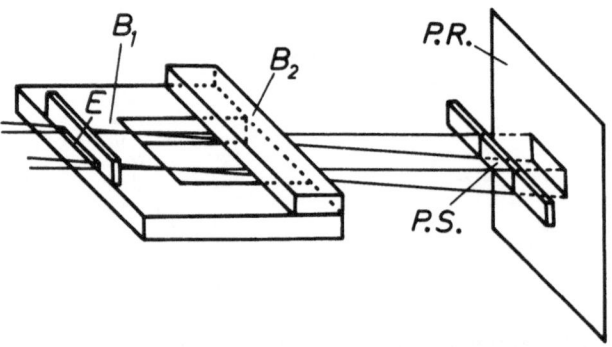

Fig. 19. Scheme of the U-shaped collimation system [28]

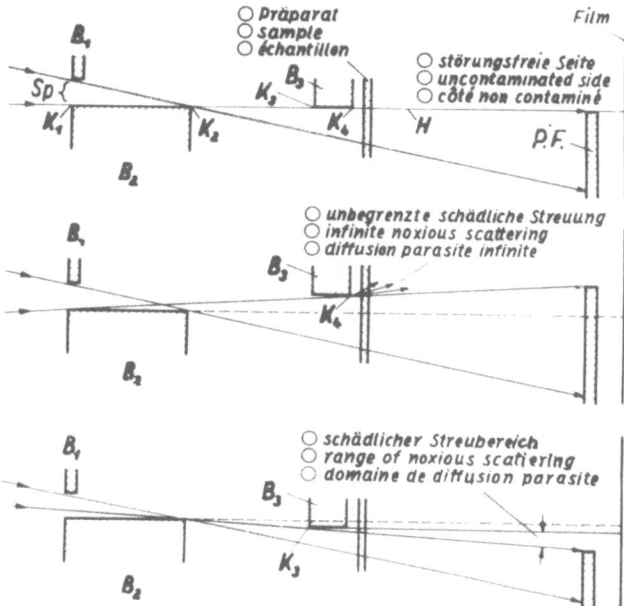

Fig. 20. Sections through U-shaped collimation system, vertical to the long dimension of the focus and parallel with the length of the camera. The dimensions in the vertical direction are greatly enlarged. The bridge B_3 is correct in position in the top arrangement, it is too high in the middle and too low in the bottom arrangement

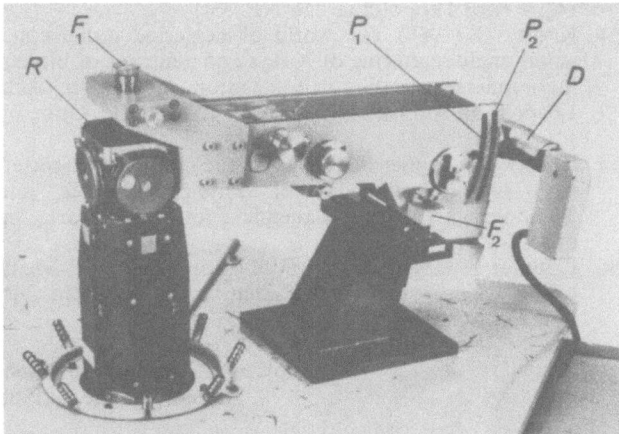

Fig. 21. Camera mounted in front of the x-ray tube

the progress of macromolecular reactions. This has become feasible with position-sensitive detectors and high-intensity x-ray generators (the price of a 12 kW rotating anode is approximately the same as that of an x-ray four-cycle goniometer), which have reduced measuring times to minutes. An even further acceleration to results in microseconds has come about with the use of synchrotron radiation [56]; this, however, is presently available only at a few places in the world.

We see that the method is on its way to cinematography of reactions with medium velocity.

Closely related to x-ray small-angle method is small-angle neutron scattering, which has inherited much, namely the scattering theory, including the method of measuring the distance of labels. Unfortunately, the few available neutron sources with sufficient high flux restrict the extension of the method.

References

I. Original papers

This is not a complete bibliography. It contains the selected examples and special theoretical results. Information on the general theory is included in "Monographs and Review Articles".

1. Kratky O, Porod G (1948) Acta Physica Austriaca 2:133
   Porod G (1948) Acta Physica Austriaca 2:255
2. Zipper P, Durchschlag H (1980) Radiat Environ Biophys 18:99
   (1980) Z Naturforsch 35 c:890
3. Porod G (1951) Kolloid Z 124:83; 125:51
4. Kratky O, Porod G, Kahovec L (1951) Z Elektrochem 55:53
   Kratky O (1964) Z analyt Chem 201:161
5. Kratky O, Porod G (1953) In: Die Physik der Hochpolymeren (ed by HA Stuart), II. Springer, Berlin
6. Pilz I, Puchwein G, Kratky O, Herbst M, Haager O, Gall WE, Edelmann GM (1970) Biochemistry (NY) 9:211
7. Edelman GM, Gall WE (1969) Ann Rev Biochem 38:415
8. Pilz I, Goral K, Kratky O, Bray RP, Wade-Jardetzky NG, Jardetzky O (1980) Biochemistry 19:4087
9. Laggner P, Degovics G, Müller KW, Glatter O, Kratky O, Kostner G, Holasek A (1977) Hoppe-Seyler's Z Physiol Chem 358:771
10. Kratky O, Porod G (1949) Rec Trav Chim Pays-Bas 68:1106
    Kratky O (1960) Makromol Chem 35 a:12
11. Debye P (1944) J Appl Phys 15:338; (1946) ibid 17:392
12. Heine S, Kratky O, Roppert J (1962) Makromol Chem 56:150
13. Stivala SS, Herbst M, Kratky O, Pilz I (1968) Arch Biochem Biophys 127:795
14. Flory PJ (1953) Principles of Polymer Chemistry, Cornell University Press, Ithaca and London, ps 399
    Tonelli AE (1970) J Chem Phys 53:4334
15. Kratky O (1938) Naturwiss 26:94; (1942) 30:542
    Kratky O, Sekora A (1943) Naturwiss 31:46
16. Hosemann R (1949) Z Physik 127:16; (1950) ibid 128:1, 128:465
17. Kratky O, Porod G (1956) Z physik Chem 7:236
18. Hermans PH, Heikens D, Weidinger A (1959) J Polymer Sci 35:145
19. Kratky O, Miholic G (1963) Mh Chem 94:151; J Polymer Sci C2:449
20. Baule B, Kratky O, Treer R (1941) Z physik Chem B 50:255
21. Fischer EW, Herchenröder P, Manley RSJ, Stamm M (1978) Macromolecules 11:213
22. Fischer EW (1957) Z Naturforsch 12a:753
23. Keller A (1957) Phil Mag 2:1171

24. Till PH (1957) J Polymer Sci 24:301
25. Zachmann HG (1974) Angew Chem 86:283
26. Schurz J, Jánosi A, Wrentschur E, Krässig H, Schmidt H (1982) Celloid Polym Sci 260:205
    Schurz J, Jánosi A (1982) Holzforschung 36:307
    Jánosi A, Schurz J, Matin N (1985) Cellulose Chem Technol 19:487
    Lenz J, Schurz J, Wrentschur E, Geymayer W (1986) Angew Makromol Chemie 138:1
27. Jánosi A (1986) Z phys Chem – Condensed Matter 63:375, 383
28. Kratky O (1954) Z Elektrochem 58:49; (1958) 62:66; (1960) Makromol Chem 35a:12
29. Kratky O, Stabinger H (1984) Colloid & Polymer Sci 262:345

II. Monographs and Review Articles (chronological order)

30. Guinier A, Fournet G (1955) Small-Angle Scattering of X-Rays. New York: Wiley Sons and London: Chapman & Hall
31. Beeman WW, Kaesberg P, Anderegg JW, Webb MB (1957) Size of Particles and Lattice Defects. In: Handbuch der Physik. Springer, Berlin, 32:322–389
32. Kratky O (1963) X-Ray Small Angle Scattering with Substances of Biological Interest in Diluted Solution. In: Progress in Biophysics, Vol. 13, Pergamon Press, Oxford, London, New York, Paris, pp 105–173
33. Luzzati V (1963) Small Angle X-Ray Scattering on an absolute Scale. In: X-Ray Optics and X-Ray Microanalysis. Academic Press, New York, p 123
34. Kratky O (1966) Possibilities of X-Ray Small Angle Analysis in the Investigation of Dissolved and Solid High Polymer Substances. Pure and Appl Chem 12:483–523
35. Brumberger H (ed) (1967) Proceedings of the Conference held at Syracuse University, June 1965, on "Small-Angle X-Ray Scattering". Gordon & Breach, New York, London, Paris
36. Alexander LE (1969) X-Ray Diffraction Methods in Polymer Science, Wiley-Interscience, pp 280–356. Wiley & Sons, New York, London, Sydney, Toronto
37. Kratky O, Pilz I (1972) Recent Advances and Applications of Diffuse X-Ray Small-Angle Scattering on Biopolymers in Dilute Solutions. Quart Rev Biophysics 5:481–537
38. Pilz I (1973) Small-Angle X-Ray Scattering, In: Physical Principles and Techniques of Protein Chemistry, Part C, pp 141–243. Academic Press, New York, London
39. Pessen H, Kumosinsky TF, Timasheff SN (1973) Small-Angle X-Ray Scattering. Methods Enzymol 27:151–209
40. Engelman DM, Moore PB (1975) Determination of Quarternary Structure by Small Angle Neutron Scattering. Ann Rev Biophysics Bioengin 4:219–241
41. Jacrot B (1976) The study of biological structures by neutron scattering from solution. Rep Progr Phys 39:911–953
42. Kratky O, Pilz I (1978) A Comparison of X-Ray Small-Angle Scattering Results to Crystal Structure Analysis and other Physical Techniques in the Field of Biological Macromolecules. Quart Rev Biophysics 11:39–70
43. Laggner P, Müller KW (1978) The structure of Serum Lipoproteins as Analysed by X-Ray Small-Angle Scattering. Quart Rev Biophysics 11:371–425
44. Stuhrmann HB, Miller A (1978) Small-Angle Scattering of Biological Structures. J Appl Cryst 11:325–345
45. Pilz I, Glatter O, Kratky O (1979) Small-Angle X-Ray Scattering. Methods Enzymol 61:148–249
46. Stuhrmann B (1979) Neutronenstreuung an Biopolymeren. Chemie unserer Zeit 13:11–22
47. Damaschun G, Müller JJ, Bielka H (1979) Scattering Studies of Ribosomes and Ribosomal Components. Methods Enzymol 59:706–750
48. Laggner P (1982) Lipoproteins and Membranes, see [47]
49. Kratky O (1982) Diffuse Kleinwinkelstreuung von makromolekularen Lösungen. In: Biophysik, 2 ed, pp 69–78. Springer, Berlin, Heidelberg, New York
50. Glatter O, Kratky O (eds) (1982) Small-Angle X-Ray Scattering. Academic Press, London. (Overview on the entire field of X-Ray Small Angle Scattering presented in 15 chapters, on 515 pages, by 14 authors)
51. Hendrix J (1982) Position Sensitive X-Ray Detectors. In: Synchrotron Radiation in Biology (Stuhrmann HB (ed)), pp 285–319. Academic Press, New York
52. Kratky O (1983) Diffuse Small-Angle Scattering of Macromolecules in Solution. In: Biophysics, pp 65–74. Springer, Berlin, Heidelberg, New York, Tokyo
53. Kratky O (1983) Die Welt der vernachlässigten Dimensionen und die Kleinwinkelstreuung der Röntgenstrahlen und Neutronen an biologischen Makromolekülen. Nova Acta Leopoldina 256 NF, 55:1–72
54. Kratky O (1984) The world of neglected dimensions, small-angle scattering of X-rays and neutrons of biological macromolecules. (Translation of Nova Acta Leopoldina 256 NF, 1–72, 1983). A Paar KG, Graz, pp 1–103
55. Kratky O, Laggner P (1987) X-Ray Small-Angle Scattering. In: Encyclopedia of Physical Science and Technology, Vol 14. Academic Press, New York, pp 693–742
56. Laggner P (1987) X-Ray Studies on Biological Membranes Using Synchrotron Radiation. In: Topics in Current Chemistry 145:173

Received February 26, 1988;
accepted June 1, 1988

Author's address:

Dipl.-Ing. Dr. Dr. h.c. mult. O. Kratky
Drosselweg 15
A-8010 Graz, Österreich

Progress in Colloid & Polymer Science   Progr Colloid & Polymer Sci 77:15–25 (1988)

The α*-law in colloid and polymer science

R. Hosemann

Gruppe Parakristallforschung, Bundesanstalt für Materialforschung und -prüfung, Berlin, F.R.G.

Key words: α*-law, bearing netplanes, catalysts, equilibrium state, free enthalpy, microparacrystals, radial potential, tangential potential

W. Graham stated in 1861 that all colloids are not crystalline. On the one side, colloid and polymer science includes stereochemistry but presently gives no direct insight into the mutual arrangements of the atoms or molecules in "solids" or "liquids" at the present time. Only expressions such as "nodules", "globules", "entanglements", "micelles", "meander cubes", "liquid crystals", "amorphous phases", etc. are used. The paracrystal theory on the other hand leads to a direct explanation of x-ray and neutron diffraction patterns of colloids. This theory is a fusion of the ideas of M. v. Laue, P. P. Ewald, P. Debye, and J. D. Landau and it enhances the concept of solid and liquid state physics. Its application to catalysts, biopolymers, fibers, synthetic polymers, glasses, and melts, led to an empirical relation of a new kind of equilibrium state of colloids: $\sqrt{\bar{N}}g = a^*$; $a^* = 0.15 \pm 0.05$. $\bar{N}$ is the number of net planes and g the relative distance fluctuation of lattice bricks in the adjacent net planes. This equilibrium state can be explained, if one adds to the free enthalpy a new summand ΔG_p, the paracrystalline enthalpy term. It is given for cubic microparacrystals with N^3-lattice cells by $\Delta G_p = \frac{3}{2}N^4 A_0 g^2$, where A_0 is the coefficient of the atomic tangential potential. The generalized enthalpy has for $g \neq 0$ a minimum at $N = (a^*/g)^2$ with

$$a^{*2} = \beta(1 + \sqrt{1 - 4g^2/\beta}); \quad \beta = \Delta G_v/4A_0 \text{ and } \frac{\sigma}{d\Delta G_v}.$$

This a-relation is based on thermodynamics. It depends somewhat on g, but mainly on the quotient of the volume enthalpy ΔG_v and A_0, hence, on the radial atomic potential and tangential potential.

I. The first example of a microparacrystal

Since the discovery of x-ray diffraction in 1912 [1] it is known that most minerals, salts, and metals, are built up of crystals. During the 20th century the atomistic structure of matter has opened up, and now we know that in the crystalline state, atoms are arranged in straight lines, rows, and columns and thereby create "three-dimensional lattices". Following the terminology of W. H. and W. L. Bragg [2], the straight rows of columns are called "netplanes", and their mutual distances d define the angle 2ϑ by which the x-rays (or neutrons) with wave length λ are "reflected" and related to

$$2d\sin\vartheta = h\lambda \quad \text{or} \quad b = \frac{2\sin\vartheta}{\lambda} = h/d \ . \tag{1}$$

These are the well known Bragg-Laue-Equations whereby h is an integer. Since 1912, physicists, chemists, mineralogists, metallurgists, and crystallographers have been applying and developing x-ray diffraction methods worldwide. An up-to-date laboratory without x-ray equipment and adequate analysis is almost inconceivable today.

Since 1930, the importance of diffraction analysis of noncrystalline materials (plastics, elastics, biopolymers, pyrolytics, liquids, and catalysts) has been recognized in many laboratories because x-ray scattering provides new and interesting information about their non-crystalline structures and properties. However, the lack of an adequate diffraction theory has hindered further development.

The theory of paracrystals [3], published in 1950, can be found in many textbooks but it has rarely been applied in its full three-dimensional form. (The term, "paracrystal", was introduced for biopolymers and liquid crystals by Rinne [15].) This probably results from the fact that three-dimensional convolution- and Fourier-integrals in their mathematical language are

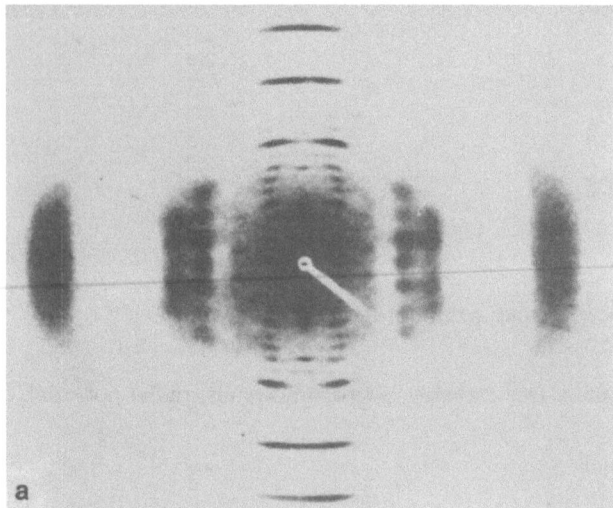

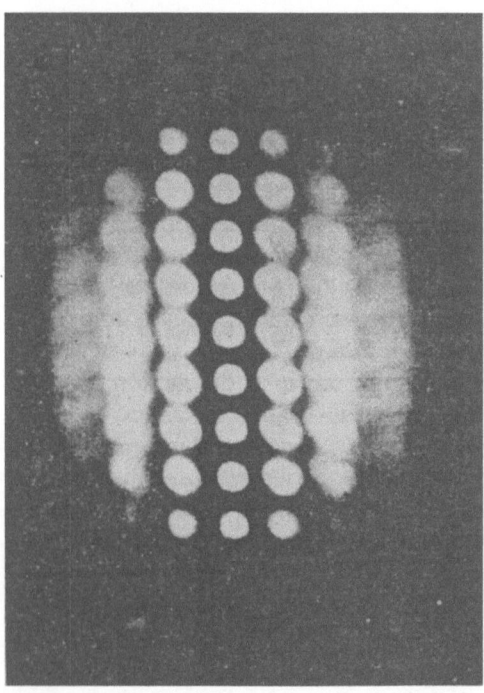

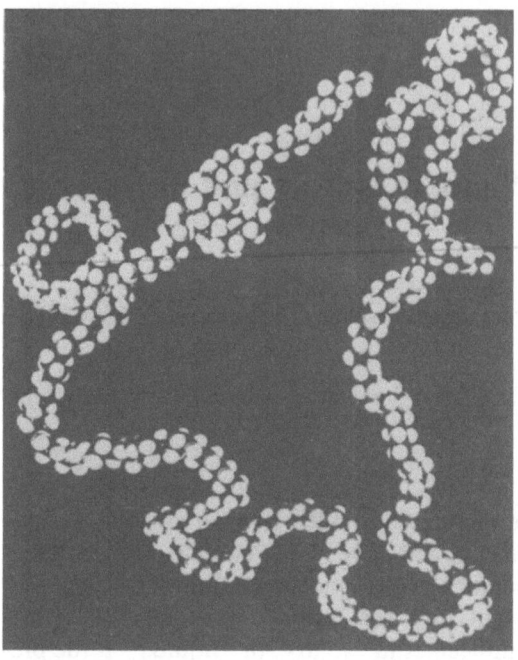

Fig. 2. Model of a long-chained PE molecul [5] where the crystalline conformation is destroyed

Fig. 1. Small-angle diffraction of β-feather kerating [4] of the seagull quill (a); two-dimensional paracrystalline point model (b) and its Fraunhofer pattern (c). The fibrous character of the vertical rows determines the macroscopic properties of the fiber

difficult to understand *). Nine statistical fluctuations Δa_{ik} of the lattice cell vectors $\underline{a}_1$, $\underline{a}_2$, $\underline{a}_3$, which become zero in the case of ideal crystals are of fundamental importance. Δa_{12} for instance is the statistical variance of $\underline{a}_1$ in the direction of $\underline{a}_2$. One of the first paracrystals was detected in the β-feather keratin of the seagull quill whose small angle diffraction pattern is shown in Fig. 1a, as published in 1951 by R. S. Bear and H. J. Rugo [4]. The quill axis is the vertical direction $\underline{a}_2$. All reflections from this ordinate have the same width and that become more and more diffuse at increasing distances from the ordinate. Figure 1b shows a two-dimensional point model and Fig. 1c the Fraunhofer-pattern which is similar to Fig. 1a with regard to the type of paracrystalline distortion. The points represent lattice cells positioned on equidistant, horizontal straight lines. The horizontal lattice vectors $\underline{a}_1$ vary statistically by Δa_{11} but $\Delta a_{12} = 0$. The vertical rows of neighboring points are, on the other hand, statistically curved with a certain Δa_{21} value while $\Delta a_{22} = 0$. The fibrous character of the β-keratin is now manifested on the atomic scale by the unique concept of paracrystallinity.

The most relevant parameters of the paracrystal theory are the nine relative distance fluctuations

―――――――

*) This explains the abundance of misinterpretations in literature discussed further on.

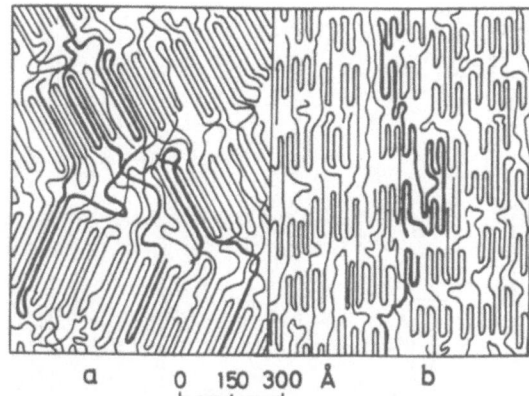

Fig. 3. Rubber-like network of chains in polymers (schematic). (a) The "knots" are microparacrystals in which the chains are parallel and densely packed in large domains of width ~400 Å. (b) After stretching, microparacrystals of reduced sizes of ~100 Å have been formed. The volume occupied by the microparacrystals is conventionally called crystalline phase. The regions between the microparacrystals in both a & b are called amorphous phases

$$g_{ik} = \Delta a_{ik}/a_k \qquad (i, k = 1, 2, 3) , \qquad (2)$$

of the lattice vector a_i in the direction of a_k divided by $\underline{a}_k$.

It is noteworthy that the so called "fiber-texture", depicted in Fig. 1, is typical of most biopolymers and also synthetic polymers consisting of long molecules. Figure 2 illustrates the model by H. A. Stuart of polyethylene having $-(CH_2-CH_2 \ldots CH_2)$-long chains [5]. In a condensed state such chains are packed within domains exhibiting parallel-positioned chain segments (Fig. 3). These aggregations do not develop large crystallites, as in the case of small molecules, but are statistically distorted along their chain axes (see Fig. 1 b). Lateral direction x-ray diffraction analysis yields g_{ik}-values of a few percent and domain sizes of 5–100 nm, which we identify as "microparacrystals".

II. The liquid-like character of synthetic polymers

A further source of paracrystalline distortions is the kinks (Fig. 4a) [6] or caterpillars (Fig. 4b) [7], embedded within the microparacrystals. They are induced by 120°-twists of the CH_2-groups and cause slips of chain segments along the fiber direction. During the stretching procedure of the fibers, the microstructure changes drastically: Adjacent chains glide along each other. The larger domains (Fig. 3a) are more and more destroyed and reduce their mean size (Fig. 3b). Their chain segments join other single ones and are aggregated onto other domains to

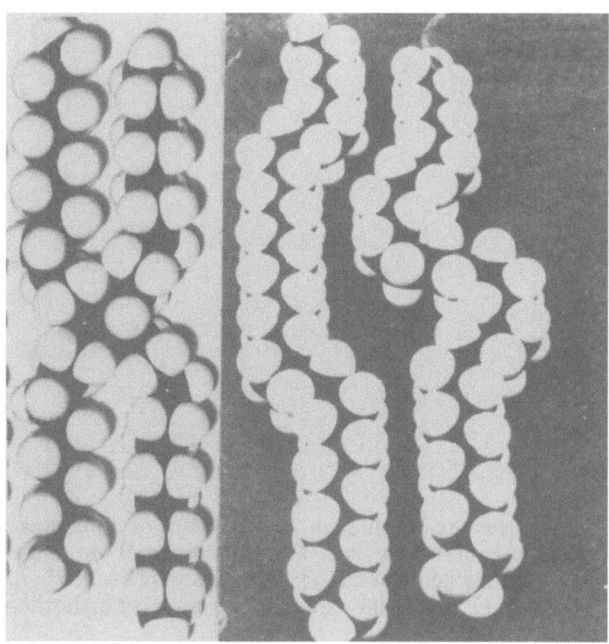

Fig. 4. Three consecutive kinks [6] (a) and two separated kinks (b) form a Reneker caterpillar [7] which is able to creep along the chain. If γ-radiated, two kinks that are neighbors of adjacent molecules join the cross link without remarkable changes in their conformations (c)

become larger units again. The "solid" polymer hence has a liquid-like mobility. The long chain molecules traverse higher ordered domains (see Fig. 3) and build up a rubber-like network. In natural rubber the knots between the chain molecules consist of one homopolar bonded monomer whereas polymers have large knots composed of 50–1000 parallel-aligned chain segments. During stretching the sample suffers plastic deformation. The microparacrystals build up the "crystalline phase" and the network in between them is called the "amorphous phase". The next section explains some special macroscopic properties of "plastics", "elastics", "crystalline", "amorphous", or "fibrous" polymers. Even microparacrystals in higher-ordered domains of polyethylene are full of distortions which can be quantitatively defined by the theory of paracrystals.

Figure 5 shows a sketch of the cross section of such a polyethylene microparacrystal perpendicular to the chain. Single C_2H_4-groups of chains and about six rather large gaps (10% of the chain cross section) are distinguishable. These gaps, resulting from tilts, kinks, and caterpillars, are only one of the sources of paracrystalline distortion. The microparacrystals have grain boundaries like the crystals in polycrystalline material. As an example, in the so called single-

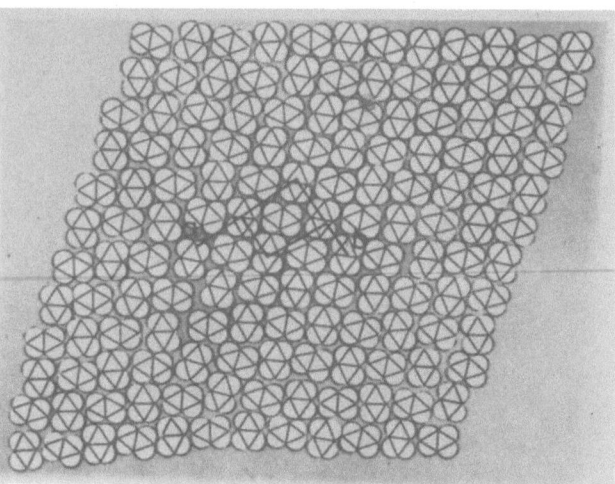

Fig. 5. Cross section through the PE chains of a microparacrystal in equilibrium state. The 110 netplanes are the surfaces (bearing netplanes, *BN*). The gaps were evoked by the kinks. (a) and (b) are the axes of a face-centered orthorhombic lattice cell

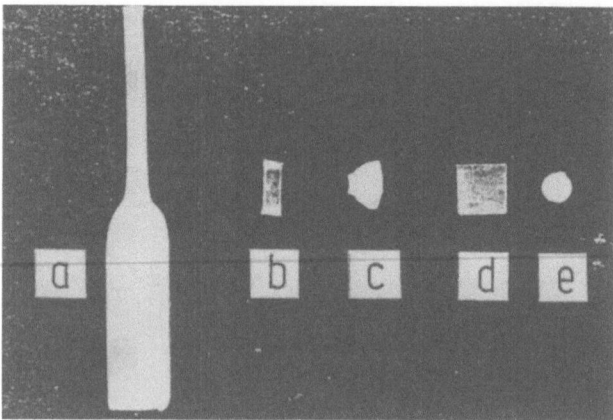

Fig. 6. A partly stretched PE rod (a) and a piece of the stretched part (b). After melting it tries to regain its former shape (c). While swimming in a glycerine bath the melt regains the cross section it had before stretching and retains it in the molten state for a certain time. This can be appreciably extended if γ-irradiated after stretching (d). After some time it finally reaches the liquid state and becomes a droplet (e)

crystals of polyethylene, twist boundaries combine the 400 Å large microparacrystals in a well defined way. They do not lose their mutual orientation over hundreds of entities [8]. In melt-crystallized and drawn material the grain boundaries, along the fiber direction, consist more or less of distorted chains with the mean microparacrystal length. The topology of microparacrystals within this "amorphous phase" can be directly obtained from small angle scattering (SMAS). The "polycrystalline" material then satisfies a "two-phase-model".

A final example may illustrate the role of microparacrystals in synthetic fibers which are arranged as knots of a three-dimensional rubber-like network: a linear polyethylene rod of $4 \times 4 \text{ mm}^2$ cross section is carefully stretched 10 times (Fig. 6), after which a piece of the stretched region (Fig. 6a) is cut off (Fig. 6b). After floating in glycerine at some degrees above the melting point the sample shrinks (Fig. 6c) and finally becomes a platelet with the same cross section ($4 \times 4 \text{ mm}^2$) as the original rod (Fig. 6d). In Fig. 6e the sample finally becomes a droplet [9]. The sample's memory of the unstretched state (Fig. 6d) is remarkably strengthened if one irradiates the *stretched* sample with γ-rays. This seems to contradict the conventional point of view which postulates that cross links are built into the amorphous phase relating to its higher mobility and which would fix the network in the streched state. Actually, the *knots* are strengthened by cross links because microparacrystals, in contrast to crystals, have a large inner mobility on account

of the Reneker-caterpillars (Fig. 4b) and there is the advantage of chains touching each other. Adjacent kinks need change their conformations only slightly to attain the structure of a cross link (Fig. 4c). The memory of the melt originates in the paracrystalline knots.

These experiments clearly show that molten material can have a well defined shape. Here the word "amorphous" ($\mathring{\alpha}\mu o\varrho\varphi o\varsigma$ which means "shapeless" in Greek) is not appropriate because the final shape of the melt is determined more or less by stable knots of microparacrystals. It may be mentioned that molten metals also consist of ~10 Å large microparacrystals with lifetimes of only 10^{-11} s [10].

It is a paradox that this mathematically-based language is not used to a greater extent in the vast field of "non-crystalline" matter. Instead of paracrystals it is currently more conventional to speak of nodules, globules, entanglements, twist-dispiration loops, micelles, quasi-crystallites, meander-cubes, short and middle-range order, mesophases, disclinations, etc. The flow and deformation behavior of polymers is explained by rheological phenomena, and their structure by an amorphous and crystalline phase. The main reason for the late use of the paracrystal theory is apparently the unfamiliar mathematics [11]. One exception, however, is the field of promoted catalysts wherein the existence of microparacrystals has been fully accepted since 1981. Details are discussed later in this article.

Considerations on microparacrystals in synthetic polymers and biopolymers offer an adequate and fundamental way to understand their macroscopic properties quantitatively. We try to explain briefly the background of the mathematics applied here, i.e., the combination of the Laue conditions (Eq. 1) with P. P. Ewald's concept [12] of reciprocal b-space that leads directly to the microparacrystal equations (see Table).

III. The $a*$ relation

The table contains the essential equations describing microparacrystals. The unit vectors of the direction of the scattered and primary x-ray-beam are $\underline{s}$ and $\underline{s_0}$. By a fixed term $\underline{s_0}$ the endpoint of $\underline{s}$ describes the "Ewald-sphere" in a three-dimensional reciprocal space that is expanded by the reciprocal vector $\underline{b} = (\underline{s} - \underline{s_0})/\lambda$ (see Eq. 1). Ewald's intensity function $I(\underline{h})$ [12] contains the volume v of a lattice cell of the crystal, the structure factor $f^2(\underline{h})$ and the convolution product of the lattice peak factor $Z(\underline{h})$ with the shape factor $S^2(\underline{h})$:

$$I(b) = \frac{1}{v} f^2 Z S^2 \; ; \quad Z(\underline{b}) S^2(\underline{b}) = \int Z(\underline{c}) S^2(\underline{b} - \underline{c}) dc^3 \tag{3}$$

This formula also applies to microparacrystals, if the lattice peak function of Ewald is replaced by a function which, according to the Table, depends on the Fourier transform $F_k(\underline{b})$ of the three coordination statistics $H_k(\underline{x})$. These coordination statistics create a three-dimensional convolution polynom relating to the results of Landau [14]. It combines Laue's idea [1] of a three-dimensional lattice with P. Debye's postulate wherein all bricks of a liquid should have a priori, the same distance statistics to that of their neighbors [13]. If, $H_k(\underline{z})$ is the probability of finding a vector $\underline{z}$ between an atom center and its nearest neighbor in the k direction, then the probability for a vector y between the next atoms in the k direction is again given by the same coordination statistics $H_k(\underline{y})$. If, according to Debye, no statistical correlation exists between the two distances $\underline{z}$ and $\underline{y}$, then the probability of finding a combination $\underline{x} = \underline{z} + \underline{y}$ is given by $H_k(\underline{y}) H_k(\underline{z})$. In order to estimate the statistics $H_k(\underline{x})$ from the first atom to the next but one, we have to integrate over all positions of the intermediate atom and using $z = x - y$ derive:

$$H_k = \overset{1}{H_k H_k} = \int H_k(\underline{y}) H_k(\underline{x} - \underline{y}) dy^3 \; . \tag{4}$$

Table 1. Fundamentals of the generalized crystallography

M. von Laue (1912):
$$\underline{a_k} \cdot \underline{b} = h_k \; ; \quad b = (\underline{s} - \underline{s_0})/\lambda$$

P. P. Ewald (1940):
$$I(\underline{h}) = \frac{1}{v} f^2 Z(\underline{h}) S^2(\underline{h}) \; ; \quad Z(\underline{h}) = \sum_{hkl} \frac{1}{v} P(\underline{h} - \underline{h}_{hkl})$$

P. Debye (1930):
The "a priory probability function"

J. D. Landau (1937):
$$H(y) H(z) \xrightarrow{y + z = x} \overset{1}{H(x)} = \int H(y) H(\underline{x} - \underline{y}) dy^3$$

$$z(\underline{x}) = \sum_{pqr} P(\underline{x} - \underline{x}_{pqr}) \rightarrow \sum_{pqr} \overset{p}{H_1} \overset{q}{H_2} \overset{r}{H_3}$$

F. Rinne (1933):
"Paracrystals and Bioparacrystals"

R. Hosemann (1950):
$$F_k(\underline{b}) = \mathscr{F} H(\underline{x}) \; ; \quad \mathscr{F} = \int e^{-2\pi i(bx)} d\infty_x \; .$$

$$Z(\underline{b}) = \mathscr{F} z(\underline{x}) = \prod_k \mathrm{Re} \frac{1 + F_k(\underline{b})}{1 - F_k(\underline{b})} \; ; \quad F_k = |F_k| e^{-2\pi i \underline{b} \cdot \underline{a_k}}$$

$$Z(\underline{b}) = \prod_k^3 \frac{1 - |F_k|^2}{(1 - |F_k|)^2 + 4|F_k| \sin^2 \pi \underline{a_k} \cdot \underline{b}} \; ;$$

$$|F_k| = e^{-2\pi^2 g_{ki}^2 h_i^2}$$

$$g_{ki}^2 = 1/a_i^4 \int H_k(\underline{x} - \underline{a_k}, \underline{a_i})^2 dx^3 \; ;$$

$$\delta b_k \cong 1/N_k + \pi^2 g_{ki}^2 h_i^2$$

The symbol $\overset{\frown}{p}$ stands for the three-dimensional convolution and $\overset{p}{H}$ for a p-fold convolution product of H.

The function $z(\underline{x})$ in the Table is the p-sum of all p-fold convolution integrals of H_1 folded q times with H_2, and r times with H_3. Its Fourier transform easily can be calculated because the Fourier transform of a convolution product is the product of the two Fourier transforms (see Table). Folding this lattice factor $Z(\underline{b})$ with the shape factor S^2 produces paracrystalline reflections whose integral widths δb_k increase quadratically with the order of reflection h (see the formula at the bottom of the Table).

In Fig. 7 the measured integral widths of the reflections (002), (004) and (006) of the DuPont fiber PRD 49 are plotted against the squared value of the order h. According to the equation a straight line can be drawn through the three measured values. From its slope one obtains the relative distance fluctuation $g = 0.02$ of atoms between their neighboring netplanes

(002), and from the intercept one finds the mean number of netplanes $\bar{N} = 59$.

Figure 8 shows a plot of the measured $\bar{N}$ and g values for different classes of substances [17] demonstrating that all of them have finite sizes that increase with decreasing distortions. The plot of $\sqrt{N}$ against $1/g$ leads empirically to the relation

$$\sqrt{N} \cdot g = a^* \; ; \quad a^* = 0.15 \pm 0.05 \tag{5}$$

which is $0.02 \cdot \sqrt{59} = 0.153$ in the case of Fig. 7. There is no doubt that this a^* relation has a definite physical meaning. The coin model of Fig. 9a illustrates this concept in more detail.

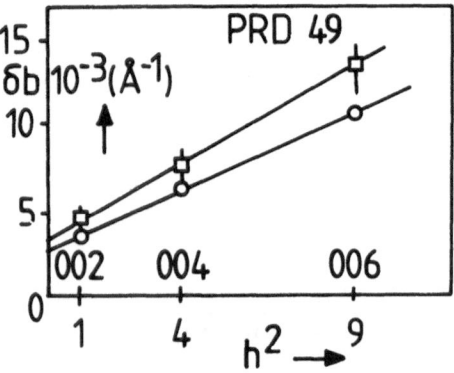

Fig. 7. $\delta b - h^2$ plot of the DuPont fiber PRD 49

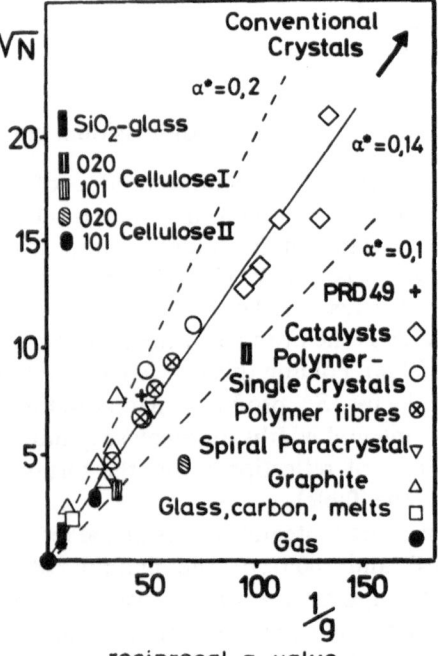

Fig. 8. Presentation of the a^* law $\sqrt{N} = a^* \cdot 1/g$ for a large variety of compounds

IV. The thermodynamic equilibrium state

As mentioned above, ideas of Laue [1] and Debye [13] are combined in a more general theory in order to understand the state of condensed matter better and quantitatively. The two-dimensional coin model of Fig. 9 shows all essential details of such a state, depicting 160 smaller coins with diameters of $D = 16$ mm mixed statistically with nine coins of 19 mm diameter situated in cubic cells of 13×13 atoms. All coins inside the lattice touch each other. Their distances from center to center range from 16 to 19 mm. According to Debye *no statistical correlations* exist between these *distances*. The g value of this paracrystalline lattice is given by

$$g = \sqrt{\gamma(1-\gamma)} \, (D_2 - D_1)/\bar{D} = 4.5\% \; , \tag{6}$$

since $\gamma = 0.06$ and $\bar{D} = 16.4$ mm. $N = 13$, hence $a^* = 0.155$ in accordance with Eq. (5). The crystalline-like lines and rows are statistically curved because of their tight packing. The *disturbed directions* between adjacent coins *produce a correction term* of the convolution polynom $z(\underline{x})$: The distances between neighboring atoms fluctuate, according to Eq. (2) by $\Delta a = ag$, and to Eq. (4) for n^{th} neighbors in a row or line by $\sqrt{n} \, ag$. The netplanes are statistically roughened [18, 19]. The binding angles between the atoms are changed. This gives rise to a new kind of lattice energy because the tangential potentials $A r_i^2$ of the lattice bricks are affected, where r_i is the tangential deviation of the adjacent atom in the roughened netplane [20]. In the nth netplane the deviation angle has the value $\sqrt{n} \, g$. On this account the tangential potential of a cubic paracrystal with N^3 lattice cells is given by

$$\Delta G_p = \sum_{n=m}^{m+N} |n| A_0 g^2 N^2 \; ; \quad |n| A_0 g^2 = A r_i^2 \tag{7}$$

This sum has a minimum for $m = -N/2$, the lattice cells in the center of a microparacrystal, hence, have a minimum of distortion. Brämer [21] was the first to mention that paracrystalline lattices are not homogeneous.

Moreover, we should take into account that the neighbors within each netplane fluctuate relatively with $2ng^2d^2$. Thus, for one netplane with N^2-lattice cells we obtain an energy $N^2 A_0 2ng^2$ where $A_0 = A d^2 \cdot n$ has to be integrated from $-N/2$ to $+N/2$. In this way we directly attain a new term ΔG_p for the free enthalpy ΔG [22]:

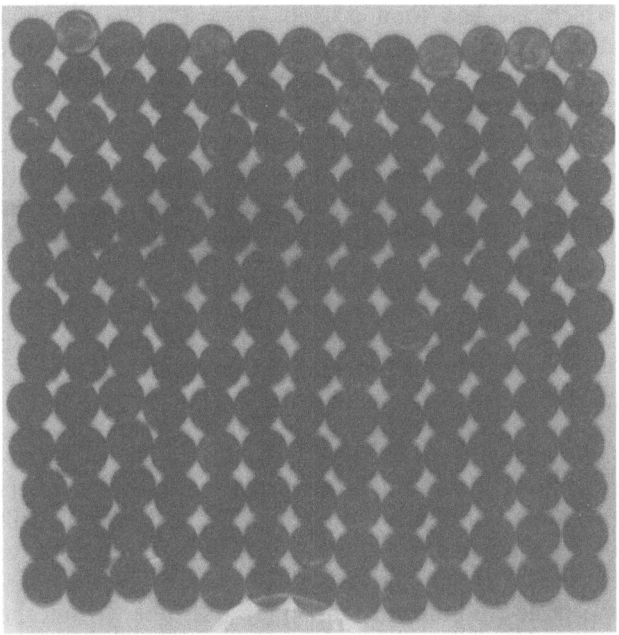

a

N-3 N-2 N-1   N   N+1 N+2 N+3

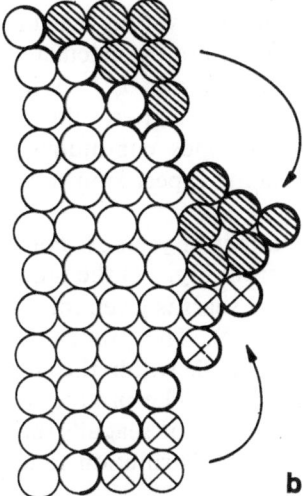

b

Fig. 9. (a) Coin model of a paracrystal in equilibrium state. In it nine larger and 160 smaller coins are mixed statistically and touch each other along the bearing netplanes 10 and 01 and build up surfaces with the same roughness. (b) At the right corner of the 10 surfaces six and three coins, respectively, are shifted to build up 11 and $1\bar{1}$ surfaces with larger tangential energies that destroy the equilibrium

$$\Delta G_p = 3/2 N^4 A_0 g^2 \; ;$$

$$\Delta G = 6 N^2 \sigma/d - N^3 \Delta G_v + \Delta G_p \qquad (8)$$

The factor 3 takes into account the existence of three families of netplanes with the largest packing density

in a primitive cubic lattice. Netplanes 10 and 01 of Fig. 9 are called bearing netplanes because their bricks are directly in touch with the bricks of the neighboring bearing netplanes. These netplanes are also the surfaces of all microparacrystals in a thermodynamic equilibrium state. If we consider creating (11) and $(1\bar{1})$ surfaces at the right side of Fig. 9a, we have to move six coins from the upper right and three coins from the lower right side to the center (see Fig. 9b). All nine ng^2 values increase and the free enthalpy loses its minimum value; Wulff's rule loses its significance.

In crystal nuclei that have reached a certain critical size, N_1 begins to grow automatically due to the slower increase in surface free energy $\Delta G_e = 6 N^2 \sigma/d$, in competition with the negative volume enthalpy $-N^3 \Delta G_v$. The value of ΔG finally decreases (Eq. 10) until the new positive term ΔG_p stops decreasing due to its N^4 dependency. The quantity ΔG has now reached its minimum. Only one maximum $N_1 = 4\sigma/d\Delta G$ exists for crystals ($g = 0$). In Fig. 10 it is assumed that $\sigma/d\Delta G = \frac{1}{2}$. Then $N_1 = 2$ is the critical germ size at which the crystal automatically begins to grow due to the decrease of ΔG. For $g \neq 0$ two extremes appear in Eq. (8)

$$N_{1,2} = \beta/g^2 (1 \pm \sqrt{1 - 4g^2/\beta} \; ; \quad \beta = \Delta G_v/4 A_0 \; . \quad (9)$$

One automatically obtains the value of N_2 and the a^* relation for the positive sign

$$a^{*2} = N_2 g^2 = \beta (1 + \sqrt{1 - 4g^2/\beta}) \; . \quad (10)$$

At $g = 0.075 N_1 = N_2$ and both extremes join at a turning point $N = N_1 = N_2$. From Eq. (9) it follows that

$$4g^2/\beta = 1 \; ; \qquad \beta = \Delta G_v/4 A_0 = 0.022 = a^{*2} \; ,$$
$$a^* = 0.15 \; , \qquad N_1 = N_2 = (a^*/g)^2 = 4 \; ; \quad (11)$$
$$\frac{A_0}{\Delta G_v} = 11.4 \; .$$

The a^* relation is no longer an empirically determined relation but is fixed in the equations of a generalized thermodynamics of colloid science. One obtains from Eq. (9) and Fig. 10 for $g = 0.042$:

$$N_2 - N_1 = 2(\beta/g^2) \sqrt{1 - 4g^2/\beta} = 12 - 2 = 10 \; . \quad (14)$$

This leads to the unique solution $\beta/g^2 = 7.5$, hence $\beta = 0.013$, $a^{*2} = 0.022$ and

$$a^* = \sqrt{N_2} g = 0.15 \; ; \quad \frac{A_0}{\Delta G_v} = 19 \; . \quad (15)$$

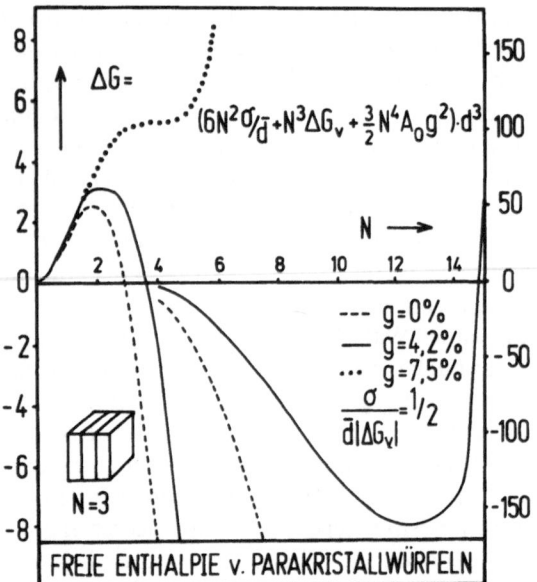

Fig. 10. Free enthalpy of microparacrystals with different g values as a function of the number N of lattice planes [22]. $g = 0$ for crystals; $g = 0.042$ for paracrystals; $g = 0.075$ for melts

Again we obtain the a^* relation which is a fundamental law. The change in $A_0/\Delta G_v$ with g is connected with the fact, that the tangential potential, contrary to the assumption made above, does not have a quadratic course.

The coin model of Fig. 9 represents the platonic idea of this state: According to Debye, all bricks (coins) have the *same interatomic distance* statistics and the coins are in Laue-like lattice positions without any statistical correlations between their single shapes. Correlations between the single rows and lines exist only with respect to the *direction of the distances*. Their directions are therefore distorted by angles $\sqrt{n}\,g$ that increase with the root of the number n of netplanes removed from the center of the microparacrystal. The roughness of the netplanes reaches the highest value on *all* surfaces, that consist of bearing netplanes (see Fig. 9b). The lattice strains being connected with these paracrystalline distortions demonstrate the physical reason for the existence of colloids. Equation (7) explains how the summation of all these correlated displacements $\sqrt{n}\,g$ leads to the endotherm summand ΔG_p of free enthalpy. One has to consider, on the other hand, their influence on the scattered intensity, because it is based on the Fourier transform $Z(\underline{b})$ of the convolution polynom $z(\underline{x})$, which – according to Eq. (4) – excludes all statistical correlation effects. The answer is given in the smallness of a^*. These $\sqrt{n}\,g$ displacements produce shifting angles

which are less than $a^* \, 360°/2\pi = 8°$. The netplane distances of the roughened netplanes are diminished hereby from d to $d \cos a^*$, less than 1%. The diffraction intensity depends on the d distances which lie within experimental error. The theory was called into question many years ago because of the problem with the "ideal paracrystal". The empirically detected a^* relation, developed with the help of this theory, has now solved the problem.

V. Size distribution of microparacrystals in the thermostable-equilibrium state

It is a demonstration of the reality of microparacrystals described in the theory, that this theory predicts fact which until now were unknown. An example will be demonstrated for the size distribution $H(N)$ of an ensemble of microparacrystals (mPCs) in the thermostable-equilibrium state [23]. Assuming that the probability $E(N)$ of further growth is given by the fraction of displacements $\bar{r}_i$ within the surface netplanes which are smaller than a certain critical value β^*, then $E(N)$ decreases monotonically with increasing N because the mean value $\bar{r}_i$ within the surface increases with N. Figure 11 shows the calculated dependence of E on N/M, where $M = (\beta^*/g)^2$. With increasing number N of netplanes and a given value of g the probability of growth becomes continuously smaller as does the value of the life expectation function $E^*(n)$.

$E^*(n)$ decreases with the number of years N. For a new born seagull (well known from Fig. 1a) the probability of reaching the age of five years is given for example by $E^*(1)\,E^*(2)\,E^*(3)\,E^*(4)$. The probability that this bird at the age of four will die within the next year is demonstrated by $(1 - E^*(5))$. The product of both functions defines the percentage of seagulls that reach their fifth birthday. The size distribution function $H(N)$ is given by analogy:

$$H(N) = E(1)\,E(2)\,E(3)\ldots E(N-1)(1-E(N)) \quad (12)$$

The function $E(1)\,E(2)\ldots E(N-1)$ decreases continuously with N whereas $(1-E(N))$ increases monotonically with N. As a matter of fact, the product of both functions leads to a frequency distribution $H(N)$ with a maximum and a mean number $\bar{N}$ at values which depend on g. Figure 12 shows $H(N)$ for $\bar{N} = 11$ and $\bar{N} = 62$.

Subsequent to development of these results, high resolution transmission electron diagrams of the Du-Pont fiber PRD49 [24, 25] were received, in which the

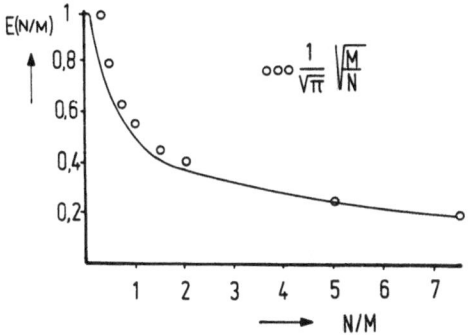

Fig. 11. Growth probability $E(N)$ calculated for different N values [23]

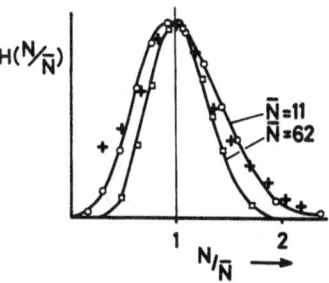

Fig. 12. Equilibrium size distribution for two different g values with $\bar{N} = 62$ and $\bar{N} = 11$ (————) [23]. The crosses indicate the experimentally obtained $H(N/\bar{N})$ values [26] of $\bar{N} = 12$ of Fig. 13

Fig. 13. Transmission electron micrograph of the DuPont-fiber PRD 49 [24, 25]

lattices of the 110- and 002-netplanes were obtained in the shape of an arrray of almost parallel fringes, which were curved in rare cases. The lattice fringes of Fig. 13 yield to a value of $\bar{N} = 12$ and a distribution function $H(N)$ marked by crosses in Fig. 12. Within the statistical error of experiment they agree with those previously calculated from the paracrystal theory [23] for $\bar{N} = 11$. This theory predicts new phenomena unknown in solid state physics.

VI. Microparacrystals in promoted catalysts

In this connection a first acceptance must be mentioned regarding promoted catalysts. Borghard and Boudart of Stanford University [27] and Puxley et al. of British Gas Corporation, London, and Harwell [28], fully confirmed our results on promoted ammonia catalysts [29, 30], published in 1966. Both departments started systematically to improve the technological efficiency of the ammonia, nickel-alumina and methanol catalysts on the basis of the paracrystal theory. The development of promoted catalysts is fascinating and demonstrates the importance of adequate diffraction analysis. Haber and Mittasch found at the beginning of the 20th century that reactions on suitable solid faces can produce NH_3 from N_2 and H_2 gas at $450\,°C$ and $220\,atm$. Bosch and coworkers [31] found that pure Fe molten with Al_2O_3 and small amounts of potash and lime yields mixed spinels $Fe_{3-x}Al_xO_4$ which, after reduction at $400\,°C$, produce the industrially important promoted ammonia catalyst. The role of the Al_2O_3 and K_2O promoters remained unknown. Brill [32] showed that x-ray reflections from a promoted ammonia catalysts have the same spatial positions as in body-centered aFe. Ignoring the line broadening caused by distortions, he concluded that the catalysts consisted of pure metallic aFe crystals with the promoters distributed outside the crystals. This opinion is still accepted. Ertl stated that the "promoter Al_2O_3 builds up a rigid frame work" during reduction which prevents the sintering of aFe. The Al_2O_3 builds up large crystals to a certain amount whereas the rest forms a thin layer at one part of the aFe crystal [33]. With respect to paracrystallinity he argues that "this possibility cannot clearly be ruled out" and that "it is difficult to reconcile it with the fact that an appreciable amount of alumina is certainly on the surface" [34]. On the other hand a paracrystal analysis made in 1966 found that [29] $Fe_{3-x}Al_xO_4$ consists of a face-centered-cubic lattice of (large) O^{2-}-anions. The Fe^{3+} and *statistically* distributed Al^{3+} ions occupied the octahedral sites, whereas the Fe^{2+} ions occupied the tetrahedral sites [35]. During reduction the O^{2-} anions bind Al^{3+} and Fe^{2+} cations to form $FeAl_2O_4$ groups which are familiar as the structure of hercynite ($FeAl_2O_4$). These groups replace seven Fe atoms at the positions 000 for Fe^{2+}, 100 and $\bar{1}00$ for Al^{3+} and at $\frac{111}{222}$, $\frac{\bar{1}\bar{1}1}{222}$, $\frac{\bar{1}1\bar{1}}{222}$, $\frac{1\bar{1}\bar{1}}{222}$ for O^{2-} (see Fig. 14). These endotactically built-in groups cannot change their statistical distributions within the aFe lattice because of their sizes. The large O^{2-} ions – similar to the larger coins shown in Fig. 9 – produce paracrystalline distortions in the 110-bearing netplanes BN of the body-centered aFe lattice, but they do not affect its lattice constant because (due to the small three cations) they have the same volume as the seven Fe-atoms replaced. Because of their roughness the paracrystalline surfaces have special adsorption potentials which are unknown in crystals. In the industrially important low-pressure methanol catalyst the Cu microparacrystals play the predominant role for CO adsorption. This catalyst converts $CO+2H_2O$ into CH_3OH+O_2 molecules. A calculation by means of a generalized Hückel theory [36] leads to the adsorption potentials drawn in Fig. 15. If the CO compound approaches the Cu surface at position a, then crystals lead to the potential $(-\cdot-\cdot-)$ and paracrystals to $(---)$. The minimum of the potential is shifted from 2.1 Å to 2.6 Å, changing its value from -25 to -50 kJ/mol. The number of binding electrons between C and O diminishes from 1.4 to 1.1. These changes produce the specific chemical surfac activity of promoted catalysts [36]. Another characteristic peculiarity of the paracrystalline catalyst is the thermostability of its large, uncovered inner surfaces.

The key for studying non-crystalline structures is an adequate application of the diffraction phenomena. Alexander in 1969 stated [37] that in the field of lattice distortions the literature is full of uncertainties and controversies. We too cite an abundance of dubious formalisms of the last decade [38]. A more detailed paper demonstrates the separation of paracrystalline distortions from others [39].

The elucidation of the paracrystal theory still meets various difficulties. Many scientists who are generally concerned with specialized problems often reject generalized theories which do not have a direct relation to their own practical application. For example, scientists engaged with liquids or glasses will rarely be stimulated by a theory which involves lattice cells. In a technological world we should not ignore the admonition of Aristoteles "that detailed studies should

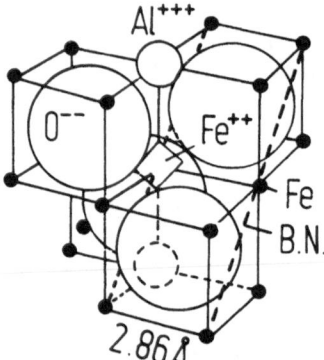

Fig. 14. A $FeAl_2O_4$ molecule endotactically included in the aFe lattice. The bearing netplane BN $(---)$ is disturbed by the larger O^{2-} anions

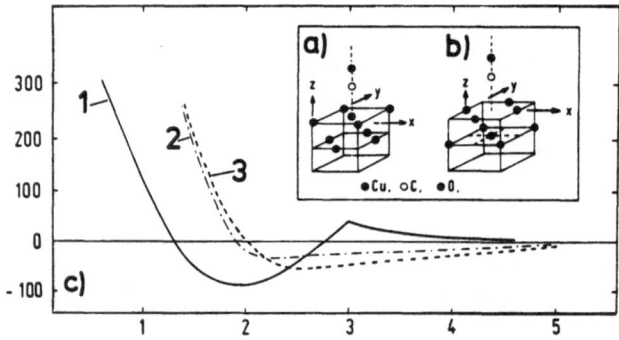

Fig. 15. Adsorption potential of microparacrystals of the methanol catalyst [35] assuming that CO units approach the Cu surface along (a) curve 2 and 3, or along (b) curve 1. Curve 3 demonstrates the advantageous properties of microparacrystals in comparison with crystals, curve 2

be supplemented by the conviction of the importance of an embracing science". Finally it should be mentioned that 40 years after his discovery of x-ray diffraction, Laue was still active enough in 1952 to establish a laboratory in Berlin where interferences within microparacrystals could be investigated. He recognized "a new science" dealing with "bodies intermediate" between the liquid and crystalline state [40].

Acknowledgement

The author thanks Dr. J. Stephenson for his excellent assistance in preparing the manuscript.

References

1. Laue M, Friedrich F, Knipping P (1912) Sitzungsber Bayer Akad Wiss Math Phys Kl 303:363
2. Bragg WH, Bragg WL (1913) Proc Roy Soc A88:428
3. Hosemann R (1950) Z Phys 1 and 465
4. Bear RS, Rugo HJ (1951) Ann NY Acad Sci 53:627

5. Stuart HA (1952) Die Struktur des freien Moleküls. Springer-Verlag, Heidelberg, p 230
6. Pechhold WR, Blasenbrey W (1967) Kolloid-Z u Z Polymere 216:237
7. Renecker DH (1962) J Polymer Sci 59:39
8. Hosemann R, Wilke W, Baltá-Calleja FJ (1966) Acta Cryst 21:118
9. Hosemann R, Loboda-Čačković J, Čačković H (1972) Z Naturforsch 27 a:478
10. Steffen B (1978) Progr Colloid & Polymer Sci 65:133
11. Hosemann R, Bagchi SN (1962) Direct Analysis of Diffraction by Matter. North-Holland Publ Comp, Amsterdam
12. Ewald PP (1940) Proc Phys Soc (London) 52:167
13. Debye P (1927) Phys Z 28:135
14. Landau LD, Lifshitz W (1938) Statistical Physics. Oxford
15. Rinne F (1933) Trans Act Faraday Soc 292:1016
16. Hindeleh AM, Hosemann R (1982) Polymer Commun 23:1101
17. Hosemann R (1982) Physica Scripta P 1:142
18. Hosemann R, Hentschel MP, Baltá-Calleja FJ, Lopez Cabarcos E, Hindeleh AM (1985) Exp Technik der Physik 33:135
19. Hosemann R, Hentschel MP, Baltá-Calleja FJ, Lopez Cabarcos E, Hindeleh AM (1985) J Phys C: Solid State Phys 18:961
20. Hosemann R (1985) J de Physique 46:C8−379
21. Brämer R (1950) J Appl Phys 21:595
22. Hosemann R (1982) Colloid & Polymer Sci 260:864
23. Hosemann R, Schmidt W, Lange A, Hentschel M (1981) Colloid & Polymer Sci 259:1161
24. Dobb MG, Hindeleh AM, Johnson DJ, Saville BP (1975) Nature 253:189
25. Dobb MG, Johnson DJ, Saville BP (1977) J Polymer Sci, Polymer Symp Ed 58:237
26. Hindeleh AM, Hosemann R (1982) Polymer Commun 23:1101
27. Borghard WS, Boudart M (1983) J Catal 80:194
28. Puxley DC, Wright CJ, Windsor CG (1982) J Catal 78:257
29. Hosemann R, Preisinger A, Vogel W (1966) Ber Bunsenges Phys Chem 70:796
30. Ludwiczek H, Preisinger A, Fischer A, Hosemann R, Schönfeld A, Vogel W (1978) J Catal 51:326
31. Bosch C, Mittasch A, Stern G, Wolf H (1910) DRP 249447, DRP 258146
32. Brill R, Richter EL, Ruch R (1967) Angew Chem Int Ed Engl 6:882
33. Ertl G (1983) Nachr Chem Techn Lab 31 (3):178
34. Ertl G, Prigge D, Weiss M (1983) J Catal 79:357
35. Fischer A, Hosemann R, Vogel W, Koutecky J, Pohl J, Ralek M (1981) Proc 7. International Congress on Catalysis, Tokyo 1980. Elsevier Sci Publ Comp, Amsterdam
36. Hosemann R, Fischer A, Ralek M (1980) Phys Bl 36:334
37. Alexander LE (1969) X-ray diffraction methods in polymer science, Chapt. VII: Lattice distortions and crystallite size. Wiley-Interscience, New York
38. Hosemann R, Lange A, Hentschel MP (1985) Acta Cryst A41:434
39. Hosemann R, Vogel W, Weick D (1981) Acta Cryst A37:85
40. Laue M v (1960) Röntgenstrahl-Interferenzen, 3. Aufl. Akad Verl Ges, Frankfurt/Main

Received February 2, 1988;
accepted June 9, 1988

Author's address:

R. Hosemann
Gruppe Parakristallforschung
Bundesanstalt für Materialforschung und -prüfung
Unter den Eichen 44−46
D-1000 Berlin 45

Progress in Colloid & Polymer Science   Progr Colloid & Polymer Sci 77:26–39 (1988)

The theory of surface tension components and the equation of state approach

J. K. Spelt and A. W. Neumann[1]

Alcan International Limited Kingston Research and Development Centre, Kingston, Ontario, Canada, [1]Department of Mechanical Engineering, University of Toronto, Ontario, Canada

Abstract: There are at present two main approaches to the calculation of solid surface tensions from contact angles: the theory of surface tension components and the equation of state approach. These are compared on the basis of their abilities to predict both the outcome of a specially designed contact-angle experiment and the engulfing behavior of microscopic particles at advancing solidification fronts. Each of these experiments provides an independent test of the validity of the two approaches. The equation of state passes both tests. The theory of surface tension components is shown to be inadequate. A theoretical assessment of the Fowkes equation for surface tension components reveals that the "dispersion components" of solid or liquid surface tension are, in fact, meaningless functions of the corresponding total solid and liquid surface tensions. More generally, it is found that contact angles can never be used to measure surface tension components. This is a consequence of Young's equation and the existence of an equation of state relating the solid-liquid surface tension to the total surface tension of the liquid and solid. Equation of state results are compared with direct measurements of solid surface tension made with an Israelachvili machine and recently reported in the literature.

Key words: Solid surface tension, contact angles, surface tension components, equation of state

Introduction

The determination of solid and solid-liquid surface tensions is of importance in a wide range of problems in pure and applied science. Because it is not possible to measure directly surface tensions involving a solid phase, there exist, at present, many indirect approaches for obtaining these values. These various methods are often in considerable disagreement, both quantitatively and theoretically. The problem persists because most of these approaches have not been tested objectively through the prediction of physical phenomena which could be independently observed and thus used to validate the various theories.

The present work addresses this issue by making reference to recently completed experimental and theoretical work pertaining to the following two theories:
1. The Fowkes theory of surface tension components [1].
2. The equation of state approach to interfacial tensions [2, 3].

These two approaches for evaluating solid surface tension are briefly outlined below.

1. The theory of surface tension components was pioneered by Fowkes [1] who proposed that surface tension should be considered in terms of components, each due to a particular kind of intermolecular force. Thus, a given organic liquid may have discrete surface tension components attributable in London dispersion forces, dipole-dipole (Keesom) forces, induction (Debye) forces, and hydrogen-bonding forces. Such surface tension components, although not thermodynamically defined, are, nevertheless, regarded by Fowkes as unique physical properties of the material. Liquid-solid interactions are considered only between those surface tension components which arise from the same types of forces. Therefore, at a water-Teflon interface, for example, because only dispersion forces are present in the Teflon, the large polar and hydrogen-bonding forces in the water will not act across the interface to affect the interfacial tension directly. In this approach the solid-liquid interfacial tension is a function of the types and relative magnitudes of the intermolecular forces in the solid and the liquid.

The Fowkes equation and methodology is used to measure only the dispersion component of the surface tension of a solid (or a liquid). There are no widely accepted methods for determining nondispersion components of surface tension and, therefore, little consensus exists regarding the magnitudes of total solid surface tensions.

2. The equation of state approach is based on thermodynamic arguments [3] which lead to the conclusion that the solid-liquid interfacial tension is only a function of the total solid and liquid surface tension. Unlike the Fowkes approach, the types and relative magnitudes of the intermolecular forces in either phase are not considered to be directly relevant. The interfacial tension is believed to be completely defined by the total surface tensions of the separate phases.

In both of these theories, solid surface tensions are evaluated using contact angle measurements. From Young's equation

$$\gamma_{SL} = \gamma_{SV} - \gamma_{LV} \cos \theta \qquad (1)$$

it is seen that if γ_{LV} and γ_{SV}, the liquid-vapor and solid-vapor surface tensions, respectively, are fixed for a series of solids and liquids, then the contact angle is directly related to γ_{SL}, the solid-liquid interfacial tension. This consept forms the basis for a direct experimental comparison of the predictions of the two theories.

Consider first two different pure liquids that are chosen to have equal overall surface tensions, as measured by, for example, the Wilhelmy plate technique. These same liquids are, however, also selected to have widely disparate compositions of intermolecular forces. In other words, one liquid may be an alkane (a liquid which has only dispersion forces) whereas the other may be characterized by a large dipole moment and perhaps significant hydrogen bonding. According to the theory of surface tension components, the contact angles of these two liquids on a single solid surface should differ in proportion to the differences in the makeup of the intermolecular forces. In contrast, the equation of state approach predicts that the contact angles will be equal since the total liquid and solid surface tensions are constant. This simple experiment provides a direct test of the basic premise of each of the two theories, and, moreover, it is independent of the specific form of any Fowkes-type equation or of any particular equation of state. In the following section, contact angle measurements with pairs of liquids of the type described above are reported and discussed from the points of view of the theory of surface tension components and the equation of state approach [4].

A second and completely independent test of the two theories is described in the third part of the paper. Small particles, initially submerged in the liquid phase of a solidifying matrix material, are observed as they encounter the advancing solid-liquid interface. In conjunction with a simple model, either theory is capable of predicting whether a particle should be engulfed or rejected, i.e., pushed by the advancing solidification front.

In the context of these experimental results, the final section considers a number of theoretical aspects of the equation of state approach and the theory of surface tension components.

Contact angle experiment

Liquids

In order to interpret the significance of the final contact-angle results correctly, it is desirable to characterize each liquid in terms of the relative magnitudes of dispersion and nondispersion forces. In keeping with the goals of the experiment, this should be done independently of the methodology of Fowkes. A semiquantitative, relative assessment of the magnitudes of non-dispersion forces may be achieved by comparing either the dipole moments or the empirical solubility parameters of the various liquids.

The solubility parameter concept has been used successfully as a practical tool in a wide variety of areas [5]. Solubility between two liquids is predicted on the basis of the degree of matching encountered among three components of the solubility parameter; viz., the dispersion (δ_d), polar (δ_p), and hydrogen-bonding (δ_h) components. This is analogous to the familiar adage that "like dissolves like", with the three empirically determined components defining the molecular character of a liquid. Although such solubility parameters lack a rigorous theoretical basis, they continue to be widely used in practice. Therefore, it is possible, with some confidence, to assume that at least a semiquantitative measure of the relative importance of dispersion forces can be obtained through the consideration of the relative magnitudes of published solubility parameters.

The first three liquids in Table 1 were selected to be significantly more "dispersive" than the remaining four liquids, which are characterized by much larger dipole moments and by relatively smaller dispersion components of the solubility parameters. In addition, the prediction of the Burrell "hydrogen-bonding"

Table 1. Liquid properties

Liquid	Dipole moment[a] Debye	Surface tension[b] mJ/m^2	$\dfrac{\delta_d}{\delta_T} \times 100$[c]
Pentadecane	0.0	$28.93 - (0.08531)\,T$	100
Dibenzylamine	0.97	$42.14 - (0.1054)\,T$	71
1-Methyl-naphthalene	0.23	$41.82 - (0.1188)\,T$	79
Benzaldehyde	2.77	$43.24 - (0.1195)\,T$	60
Ethyl caprylate	1.68	$29.12 - (0.1018)\,T$	50
Heptaldehyde	2.58	$28.50 - (0.0766)\,T$	47
Methyl salicylate	2.23	$41.84 - (0.1201)\,T$	44

a: McClellan [7]
b: Measured by the Wilhelmy plate technique [8] with an uncertainty of ± 0.15 mJ/m^2
c: Fraction of total solubility parameter (δ_T) attributed to dispersion forces at 25 °C. Calculated using the solubility parameters in [6] and the correlations suggested in [5] (for details see [8])

Table 2. Predictions of the Beerbower correlation of liquid surface tension, γ_{LV}, with solubility parameter components (δ_d, δ_p, and δ_h). Comparison of predicted and measured surface tensions (25 °C) and the percentage of the total predicted surface tension due to the dispersion solubility parameter component (δ_d)

Liquid	Beerbower γ_{LV} mJ/m^2	Measured γ_{LV} mJ/m^2	Percentage of Beerbower γ_{LV} due to δ_d
Pentadecane	29.6	26.8	100
Dibenzylamine	41.0	39.5	94
1-Methylnaphthalene	38.8	38.8	97
Benzaldehyde	39.7	40.2	76
Ethyl caprylate	32.5	26.6	58
Heptaldehyde	31.4	26.6	56
Methyl salicylate	39.2	38.8	56

classification [6], which is another empirical aid for the prediction of solubility, is "moderate" for the last four liquids and "poor" for the rest. This is not meant to imply the actual existence of hydrogen-bonding in our systems, but serves in the present context to indicate that independent experimental observation has established significant differences in the character of the intermolecular forces.

Taken together, the information in Table 1 indicates that dispersion forces are responsible for significantly larger fractions of the total intermolecular binding energy in pentadecane, dibenzylamine, and 1-methylnaphthalene than in the last four liquids. For the purpose of the present investigation, the exact magnitudes of such differences are unimportant. It is only necessary to establish that there are appreciable dif-

ferences in the relative degrees of "dispersion" in the first three liquids as compared with the last four liquids.

Beerbower [9] has developed a correlation between liquid surface tension and the dispersion, polar, and hydrogen-bonding components of the solubility parameter. Table 2 compares the predictions of the Beerbower correlation with the measured surface tensions for seven liquids. The third column of this table lists the percentage of the total predicted surface tension which is due to the dispersion component of the solubility parameter. Beerbower [9] noted, that this dispersive fraction bears no relation to the "dispersion component of surface tension" as presented in the Fowkes theory. Table 2 provides another indication that pentadecane, dibenzylamine, and 1-methylnaphthalene are characterized by substantially higher fractions of dispersion forces than are the other four liquids. In order to minimize the potential for vapor adsorption, the seven liquids were chosen to have relatively high boiling points, the lowest being that of heptaldehyde at 153 °C.

Solid surfaces

The acquisition of thermodynamically significant contact-angle data is largely dependent on the quality of the substrate surface. The effects of roughness and heterogeneity can easily overshadow the influence of interfacial energetics [10]. It is, therefore, important in a study of this type to produce solid surfaces of sufficient quality to ensure that the observed contact angles accurately reflect the interaction between the solid and liquid surface tensions as given by Young's equation, Eq. (1).

Contact angle measurements were performed on two surfaces. The first was heat-pressed Teflon FEP (Dupont), a surface which is exceptionally smooth and homogeneous. At 24 °C the surface gave rise to advancing and receding contact angles of 52° and 49°, respectively, for hexadecane, and 46° and 43°, respectively, for undecane. The method of preparation of this surface is summarized in ref. 11, with greater detail supplied in ref. 8.

In the present work, two FEP samples (designated A and B) were employed, each having been prepared in a different way and having a unique thermal history. The surface tensions of the samples were slightly different.

A second solid surface, which was used for only one pair of liquids, was siliconized glass (details of the dimethyldichlorosilane treatment are given in ref. 10). The advancing contact angle with water was 105° while the receding angle was between 95° and 100°.

Apparatus and procedure

The present sessile-drop contact angles were measured using Axisymmetric Drop Shape Analysis (ADSA) [12]. The approach is unique because it does not depend on the location of specific points or features of the drop shape, and it is generally applicable to all axisymmetric liquid-fluid interfaces; i.e., sessile or pendant drops and contact angles greater or less than 90°. The technique has an uncertainty of less than ±0.4° (standard error of the mean for three samples) [9, 10]. Above all it is objective, being independent of the skill and experience of the operator, and is relatively straightforward in terms of the apparatus and the details of its use.

Drops are photographed in the horizontal plane and approximately 40 coordinate points are selected arbitrarily from the profile utilizing either a manual digitizing tablet or a digital-image analyzer. This array of points is then fitted, by the method of least squares, to the Laplace equation which yields both the contact angle and the liquid surface tension (for details see [8, 11]).

Table 1 lists the measured surface tension-temperature relations for each of the seven liquids. By controlling the temperature of the contact angle experiment, it was possible to match the total surface tensions of the various liquids more exactly. All of the experiments were, therefore, performed in a temperature-

controlled chamber. Advancing contact angles were produced by growing sessile drops through a small hole in the center of the solid substrate with a motorized syringe drive [11]. A single experiment consisted of photographing a series of successively larger drops (usually three) on a given substrate. Individual drops were photographed three times, refocusing the microscope each time, and the contact angles obtained from these photographs were averaged to provide a mean contact angle for the drop. The mean contact angle for a single experiment, usually consisting of three different sized drops, was simply the average of the individual mean contact angles for each drop. In other words, the contact angle reported for a single experiment, of one liquid on a given substrate, was the average of (usually) nine ADSA results, comprising three replications for three distinct drops. In some cases, only two different drops were photographed in a given experiment.

Results and discussion

The results of the contact angle experiments for the solid substrates Teflon FEP sample A, FEP sample B, and siliconized glass are reported in Tables 3, 4, and 5. The data are grouped in pairs according to the matched surface tensions of the liquids used. The first liquid in each pair is the one which is completely or

Table 3. Grand average advancing contact angles of all available experiments on substrate FEP A

Liquids[a]	Grand[b] average contact angle	Error[c] limits (±)	Contact[d] angle difference	Expt. temp. °C	Liquid surface tension mJ/m^2	No. of expts./drops
M	72.6	1.4		24	39.0	2/6
MS	72.8	0.7	−0.2	24	39.0	2/6
D	75.4	0.6		3	41.8	1/3
B	73.4	0.4	+2.0	3	42.9	2/6
D	75.4	0.6		3	41.8	1/3
MS	72.9	0.6	+2.5	3	41.5	2/5
P	52.4	0.7		39	25.6	2/5
H	53.2	0.3	−0.8	39	25.5	2/5
P	53.6	0.3		14	27.7	1/5
EC	53.0	0.4	+0.6	14	27.7	1/5

a: B − Benzaldehyde, D − Dibenzylamine (predominantly dispersive), EC − Ethyl caprylate, H − Heptaldehyde, M − 1-methylnaphthalene (predominantly dispersive), MS − Methyl salicylate, P − Pentadecane (totally dispersive)

b: Grand average contact angle − The average of the mean contact angles (the average contact angle for a single experiment) for the number of experiments indicated

c: Error limits − The "worst possible case" values required to encompass all mean contact angles from the grand average contact angle. In cases where only one experiment was performed, the 95% Student-*t* confidence limits on the mean contact angle are quoted

d: Contact angle difference − The contact angle of the predominantly dispersive liquid (the first one listed in each pair) minus that of the significantly nondispersive liquid

Table 4. Grand average advancing contact angles of all available experiments on substrate FEB B

liquids[a]	Grand[b] average contact angle	Error[c] limits (±)	Contact[d] angle difference	Expt. temp. °C	Liquid surface tension mJ/m^2	No. of expts./drops
D	72.4	3.2		3	41.8	4/12
MS	69.3	0.6	+3.1	3	41.5	4/12
P	49.1	0.5		39	25.6	1/3
H	50.3	0.4	−1.2	39	25.5	1/3

a−d: See footnotes of Table 3

Table 5. Grand average advancing contact angles of all available experiments on the siliconized glass substrate

Liquids[a]	Grand[b] average contact angle	Error[c] limits (±)	Contact[d] angle difference	Expt. temp. °C	Liquid surface tension mJ/m^2	No. of expts./drops
M	58.3	1.1		24	39.0	2/5
MS	61.0	0.3	−2.7	24	39.0	2/5

a−d: See footnotes of Table 3

overwhelmingly composed of London dispersion forces, and the second liquid has relatively large non-dispersion forces. The fourth column in these tables, the contact angle difference, is defined as the contact angle of the first liquid minus that of the second liquid. In four of the eight cases, this difference exceeds the range of the combined error limits, indicating the possible influence of some systematic factor, distinct from the random error which contributes to the error limits. The average of the eight contact angle differences is +0.4°..The following is a brief discussion of the possible explanations for these results, in terms of both the theory of surface tension components and the equation of state.

The Fowkes equation for solid-liquid interfacial tensions,

$$\gamma_{SL} = \gamma_S + \gamma_L - 2\sqrt{\gamma_S^d \gamma_L^d} \tag{2}$$

is strictly applicable only to situations in which at least one phase is a saturated hydrocarbon (n-alkane, paraffin wax, etc.) because this ensures that only dispersion forces are operative within that phase. Here, γ_S and γ_L are, respectively, the solid and liquid surface tensions (neglecting, as is customary, the equilibrium spreading pressures); γ_{SL} is the solid-liquid interfacial tension, and γ_S^d and γ_L^d are the dispersion components of the solid and liquid surface tension, respectively. The presence of the two different dispersion components under the square root sign in Eq. (2) should not be construed as implying the ability to treat the interaction of nondispersion forces in the two phases in a similar fashion.

Equation (2) predicts that if γ_S and γ_L are fixed, then γ_{SL} will vary inversely with γ_L^d. With respect to the contact angle experiments, the "dispersive" liquid in each pair should, therefore, have the smaller contact angle on a surface which is interacting only through dispersion forces. As was mentioned above, however, the average contact angle difference for the eight liquid pairs was +0.4°, indicating that the opposite trend was prevalent. On average, the "dispersive" liquid had a slightly larger contact angle than the more "nondispersive" liquid, directly contradicting the predictions of Eq. (2). In two of the four cases where the contact angle difference exceeds the error limits (dibenzylamine with benzaldehyde and with methyl salicylate on FEP A) the difference is positive (contrary to Eq. (2)), whereas in the other two cases it is indeed negative (pentadecane and heptaldehyde on FEP B, and 1-methylnaphthalene and methyl salicylate on siliconized glass).

Equation (2) may be combined with the Young equation, Eq. (1) to yield,

$$2\sqrt{\gamma_S^d} = \sqrt{\frac{\gamma_L}{\gamma_L^d}} \sqrt{\gamma_L} (1 + \cos\theta) . \tag{3}$$

For a given pair of liquids (denoted "1" and "2") on a single substrate, the lefthand side of Eq. (3) is constant so that

$$\frac{\gamma_{L_2}^d}{\gamma_{L_2}} = \frac{\gamma_{L_1}^d}{\gamma_{L_1}} \frac{\gamma_{L_2}}{\gamma_{L_1}} \left(\frac{1+\cos\theta_2}{1+\cos\theta_1}\right)^2 . \qquad (4)$$

Considering liquid "1" to be the dispersive liquid (the first liquid listed in each pair in Tables 3, 4, and 5) and assuming for these liquids that the dispersion fraction is that listed in the last column of Table 2, Eq. (4) may be used to calculate the implied dispersion fraction of liquid "2" (the lefthand side of Eq. (4)). Note that Eq. (4) can also be used to give the ratio of the dispersive fractions of liquids "1" and "2", without regard to the Beerbower correlation.

Equation (4) is applicable to situations where the solid substrate comprises only London dispersion forces. This condition is satisfied by the use of both Teflon FEP and siliconized glass that has a surface consisting of methyl groups and thus behaves as a saturated hydrocarbon.

Table 6 presents the results of these calculations for the nondispersive liquids on the three substrates. The last column in this table is the Fowkes-theory prediction of the dispersive fraction of the "2" liquid (the "nondispersive" one) within each liquid pair. In all cases, this dispersive fraction is very close to 1.00, indicating that Eq. (2) predicts that the "2" liquids are just as dispersive as are the "1" liquids. This is clearly

Table 6. Fowkes equation predictions of the fraction of the total surface tension due to dispersion forces for the non-dispersive ("2") liquids in each pair

Liquids[a]	Expt. temp. °C	γ_L^d/γ_L (Beerbower) (Table 2)	$\gamma_{L_2}^d/\gamma_{L_2}$ (Eq. 4)
FEP A			
M (1)	24	0.97	
MS (2)	24	0.56	0.97
D (1)	3	0.94	
B (2)	3	0.76	1.02
D (1)	3	0.94	
MS (2)	3	0.56	1.00
P (1)	39	1.00	
H (2)	39	0.56	0.98
P (1)	14	1.00	
EC (2)	14	0.58	1.01
FEP B			
D (1)	3	0.94	
MS (2)	3	0.56	1.01
P (1)	39	1.00	
H (2)	39	0.56	0.98
Siliconized glass			
M (1)	24	0.97	
MS (2)	24	0.56	0.92

a: See footnote of Table 3

contrary to the predictions of the Beerbower correlation and, in general, to the expectations based on solubility parameters, molecular structure, and molecular properties. Within the context of the Fowkes approach, it therefore needs to be asked why the apparently significant differences in the relative magnitudes of dispersion forces have not manifested themselves in concomitant differences in the observed contact angles measured for each pair of liquids.

One possible answer is that the "2" liquids are, in effect, almost completely dispersive in behavior with the nondispersive contributions being too small to affect the contact angle noticeably. As was mentioned above, however, this is in considerable disagreement with the independent measures of the relative importance of nondispersion forces; viz., solubility parameters, the Beerbower correlation, and dipole moments. Nevertheless, if the "2" liquids are indeed accepted as being almost completely dispersive, then the conclusions remain significant within the context of the Fowkes approach. If liquids as apparently nondispersive as the "2" liquids of Table 6 are found to behave as if they were overwhelmingly dispersive (as do the "1" liquids), it in effect means that in many practical situations it is unnecessary to evaluate the non-dispersion components of surface tension. The wetting behavior is then well modeled on the basis of a single component of surface tension (dispersion) that is equal to the total surface tension. In other words, the equation of state approach is applicable.

Alternatively, if it is agreed that there is a significant difference between the liquids with regard to the types and relative magnitudes of the intermolecular forces, then it must be concluded that Eq. (2) has failed to detect these differences in terms of surface tension components.

Therefore, within the context of the theory of surface tension components there are primarily two possible responses to the observed contact angles. The first may be to repudiate the independent measures of a liquid's "dispersive" character such as solubility parameters, dipole moments, and molecular structure. The argument would then be that the liquids in each polar-nonpolar pair are equally dispersive and hence the contact angles should indeed be the same. The second possible response would acknowledge the differences in the liquids within each pair and would conclude that the Fowkes equation does not correctly identify the anticipated non-dispersion components of surface tension.

In contrast to the above dichotomy, the present contact-angle results are fully consistent with the equation of state approach to interfacial tensions. Recall that, since the total liquid surface tension is constant within a given pair of liquids, the contact angles are predicted (by an equation of state) to be equal on a single solid substrate. This does appear to be largely the case, although explanations must be found for the small contact angle differences which persist. There are several possibilities for these discrepancies:

1. Vapor adsorption leading to small equilibrium spreading pressures.
2. Liquid contamination.
3. Nonmaximal advancing contact angles.

With respect to the last of these points, dynamic contact angle experiments have demonstrated that a static "advancing" contact angle may be as much as 3° less than the true Young contact angle encountered at very low three-phase line velocities [13].

Regarding the first possible explanation, although it is customary to neglect vapor adsorption (and, hence, equilibrium spreading pressures) in all practical contact-angle measurements, it is important to appreciate that this is an approximation, albeit a good one, for most low-energy solids [14]. It is widely accepted that surfaces such as Teflon and polyethylene contain a small fraction of hydrophilic sites which may lead to spreading pressures of over $3 \, \text{mJ/m}^2$ with octane and approximately $2 \, \text{mJ/m}^2$ for water on powdered Teflon [18]. The manner in which these powder measurements relate to solid surface of Teflon is uncertain, but it seems plausible to expect, under certain circumstances, equilibrium spreading pressures of the order of $1 \, \text{mJ/m}^2$.

Table 7 provides an estimate of the equilibrium spreading pressure required to make the contact angles equal in each liquid pair. The observed contact angle difference, $\Delta\theta$, between two liquids in a pair may be considered in two distinct parts, one due to the small differences in the liquid surface tension ($\Delta\theta_\gamma$), and the other to vapor adsorption ($\Delta\theta_\pi$). We define $\Delta\theta$ as $\theta_1 - \theta_2$, and "1" and "2" denote, respectively, the dispersive and nondispersive liquids in a given pair. The equation of state of Neumann et al. [2, 10, 15–18] was used to calculate these contact angle differences as described in [4].

Of the eight hypothetical equilibrium spreading pressures listed in Table 7, six are less than or equal to $1 \, \text{mJ/m}^2$. As demonstrated by Good [14], it is not unreasonable to assume that such spreading pressure can occur on surfaces of Teflon FEP and siliconized glass. It is, therefore, concluded that the observed differences in the contact angles within a given liquid pair may reasonably be attributed to experimental error and specifically to vapor adsorption. The present contact-angle data are consistent with the predictions of

Table 7. Equilibrium spreading pressures required to make the contact angles equal in each liquid pair. Calculations based on the Neumann equation of state

Liquid[a] pair	Expt. temp. °C	$\Delta\theta$[b] (deg)	$\Delta\theta_\gamma$[c] (deg)	$\Delta\theta_\pi$[d] (deg)	π_e[e] mJ/m^2
FEP A					
M + MS	24	−0.2	0.0	−0.2	0.1
D + B	3	+2.0	−1.4	+3.4	1.4
D + MS	3	+2.5	+0.4	+2.1	0.9
P + H	39	−0.8	+0.2	−1.0	0.3
P + EC	14	+0.6	0.0	+0.6	0.1
FEP B					
D + MS	3	+3.1	+0.4	+2.7	1.2
P + H	39	−1.2	+0.2	−1.4	0.4
Siliconized Glass					
M + MS	39	−2.7	0.0	−2.7	1.0

a: See footnote in Table 3. The first liquid in each pair is the "1" (dispersive) liquid and the second is the "2" liquid
b: $\Delta\theta = \theta_1 - \theta_2$
c: $\Delta\theta_\gamma$ = Contact angle difference due to the difference between γ_{L_1} and γ_{L_2}
d: $\Delta\theta_\pi$ = Contact angle difference due to π_e
e: π_e = Equilibrium spreading pressure responsible for $\Delta\theta_\pi$

the equation of state approach and provide an independent experimental verification of this theory.

Attraction and repulsion of solid particles by solidification fronts

In the previous section the equation of state theory was seen to be in accord with the measured contact angles whereas the Fowkes theory of surface tension components led to a number of discrepancies. A second experiment which provides an objective assessment of these two theories is the observation of the pushing or engulfment of microscopic particles at solidification fronts.

When a microscopic particle (diameter $10–200 \, \mu$), embedded in the liquid phase of a solidifying matrix material encounters the advancing solidification front, the particle may either be immediately engulfed by the solid phase or it may be pushed through the melt by the advancing front. The behavior of such a particle may be predicted by the size of the net change in the free energy of adhesion, which is the first step in the process of engulfment.

The free energy of adhesion is given by

$$\Delta F^{adh} = \gamma_{PS} - \gamma_{PL} - \gamma_{SL} \tag{5}$$

where γ_{PS}, γ_{PL}, and γ_{SL} are, respectively, the particle-solid, particle-liquid and solid-liquid interfacial tensions at the melt temperature. If the free energy change is positive, the engulfment process is thermodynamically unfavorable and particle rejection will occur at the solidification front. A negative change in the free energy would lead to particle engulfment, since the overall system energy is thereby decreased. Equation (5), together with experimental observations of particle behavior in solidifying melts, provides an excellent, independent test of the validity of equations for interfacial tensions. Omenyi et al. recognized this and used contact angle measurements and the equation of state of Neumann et al. [2] to calculate the interfacial tensions in Eq. (5) and to predict engulfment or rejection for a range of particle materials in different organic melts such as naphthalene and biphenyl [19, 20]. The agreement between direct experimental observations and the predictions of the equation of state was excellent. Note that in order to calculate the particle-solid interfacial tension in Eq. (5) it was necessary to use the equation of state of Neumann et al. [2] in a "generic" sense. In other words, the equation of state, "f", was used to evaluate

$$\gamma_{PS} = f(\gamma_{PV}, \gamma_{SV}) \qquad (6)$$

where γ_{PV} and γ_{SV} are the particle and solid surface tensions obtained from the equation of state and contact angle measurements. In spite of the fact that this equation of state was empirically derived from solid-liquid contact-angle data, the success of the freezing front predictions [19−22] lends confidence to this usage. It is interesting to note that the freezing front experiment has since been developed into a technique for measuring the surface tension of solid particles [23, 24] that has been applied to a wide variety of situations [25−27].

In the present work, it is of interest to apply the Fowkes theory of surface tension components to the evaluation of Eq. (5) and to compare the results with the independent experimental observations of particle rejection or engulfment [20].

The Fowkes equation may be combined with Young's equation, Eq. (1), to yield

$$\cos \theta = -1 + 2\sqrt{\gamma_S^d \gamma_L^d}/\gamma_L . \qquad (7)$$

Note that Eq. (7) provides only the means to evaluate the dispersion component of the solid or liquid surface tension.

In order to obtain the surface tensions of the particle materials as well as those of the solid phases of the various matrix materials, Omenyi performed temperature dependent contact-angle measurements with

glycerol [19, 20]. The restrictions inherent in the use of the Fowkes equation, Eq. (7), however, permit the use of glycerol (a liquid with significant polar and hydrogen-bonding elements) only in the measurement of contact angles on totally dispersive solids. Of the nine matrix materials for which data exist: naphthalene, biphenyl, o-terphenyl, thymol, salol, 2-phenylphenol, pinacol, benzophenone, and bibenzyl [20], the first three satisfy this requirement best. The validity of this assertion within the framework of the Fowkes approach is probably best established by comparing the structure and solubility parameters [6] of these materials with those of some other large organic molecules which are acknowledged to be overwhelmingly dispersive [1, p. 48]: tricresyl phosphate, 1-bromonaphthalene, and trichlorobiphenyl. Such comparisons are detailed in [28] and demonstrate that the dispersive approximation is at least as good for most of the matrix materials as that made by Fowkes [1].

Attempts were also made to evaluate the dispersive surface tension component of solid o-terphenyl by following the Fowkes methodology and measuring the contact-angles of totally dispersive liquids. The problem, however, is that o-terphenyl has a relatively large surface tension which makes it impossible to use a saturated hydrocarbon as the test liquid. In such cases, the common solution among those employing the Fowkes approach is to use one of the higher surface tension liquids: 1-methylnaphthalene. diiodomethane, or 1-bromonaphthalene, and assume that the nondispersive surface tension components are negligible. Unfortunately, o-terphenyl is soluble in all of these liquids, making contact angle measurements impossible.

Not all of the particle materials employed by Omenyi et al. were completely "dispersive" [20]. Teflon, siliconized glass, and polystyrene certainly fall into this category, but other materials have, to some extent, some nondispersion contributions.

In terms of the Fowkes equation, Eq. (2), the free energy of adhesion is expressed as

$$\Delta F^{adh} = 2 [\sqrt{\gamma_P \gamma_L} + \sqrt{\gamma_S \gamma_L} - \sqrt{\gamma_P \gamma_S} - \gamma_L] \qquad (8)$$

where the assumption has been made that both the matrix and the particle materials are completely "dispersive". Here, γ_S and γ_P are the surface tensions of the matrix solid phase and the particle, respectively, at the melting point of the matrix with liquid surface tension γ_L. The solid surface tensions, γ_S, are determined by using the available temperature-dependent glycerol contact-angle data together with Eq. (7). For glycerol at 20 °C, $\gamma_L^d/\gamma_L = 0.58^{(1)}$. This ratio was

Table 8. Theoretical predictions and microscopic observations of particle behavior at solidification fronts; ΔF_I^{adh} = free energy of adhesion in mJ/m² calculated from equation of state; ΔF_{II}^{adh} = free energy of adhesion in mJ/m² calculated from Fowkes' theory; R = rejection; E = engulfment

		Naphthalene	Biphenyl	Thymol	o-Terphenyl	Salol	2-Phenylphenol	Pinacol
Acetal	ΔF_I^{adh}	2.58	1.99	0.198	1.12	0.204	0.013	0.047
	ΔF_{II}^{adh}	1.56	− 0.181	− 5.66	− 1.90	− 5.41	− 5.47	− 5.33
	Observation	R	R	R	R	R	R	R
Nylon-6	ΔF_I^{adh}	1.98	1.76	0.186	1.00	0.189	0.011	0.045
	ΔF_{II}^{adh}	1.42	− 0.168	− 5.39	− 1.77	− 5.14	− 5.06	− 5.12
	Observation	R	R	R	R	R	−	E
Nylon-6,6	ΔF_I^{adh}	2.12	1.67	0.177	0.922	0.174	0.009	0.043
	ΔF_{II}^{adh}	1.29	− 0.154	− 5.02	− 1.62	− 4.74	− 4.57	− 4.81
	Observation	R	R	R	R	E	−	E
Nylon-12	ΔF_I^{adh}	1.57	1.28	0.147	0.631	0.125	0.003	0.037
	ΔF_{II}^{adh}	1.12	− 0.135	− 4.50	− 1.41	− 4.19	− 3.86	− 4.36
	Observation	R	R	R	R	R	E	R
Nylon-6,10	ΔF_I^{adh}	0.720	0.735	0.109	0.263	0.070	− 0.003	0.029
	ΔF_{II}^{adh}	0.967	− 0.118	− 4.05	− 1.22	− 3.69	− 3.27	− 3.97
	Observation	R	R	R	R	R	E	R
Nylon-6,12	ΔF_I^{adh}	− 0.279	0.038	0.056	− 0.236	− 0.013	− 0.014	0.018
	ΔF_{II}^{adh}	0.667	− 0.084	− 3.06	− 0.822	− 2.56	− 1.97	− 3.06
	Observation	R	R	R	R	E	E	E
Polystyrene	ΔF_I^{adh}	− 1.42	− 0.757	− 0.002	− 0.797	− 0.105	− 0.025	0.006
	ΔF_{II}^{adh}	0.187	− 0.031	− 1.67	− 0.250	− 1.02	− 0.082	− 1.88
	Observation	E	E	R	E	E	E	R
Teflon	ΔF_I^{adh}	− 5.10	− 3.21	− 0.186	− 2.51	− 0.380	− 0.059	− 0.030
	ΔF_{II}^{adh}	− 0.962	0.098	1.91	1.20	2.99	4.66	1.28
	Observation	E	E	E	E	E	E	R
Siliconized glass	ΔF_I^{adh}	− 6.58	− 4.05	− 0.231	− 3.00	− 0.437	− 0.068	− 0.038
	ΔF_{II}^{adh}	− 1.47	0.145	2.86	1.64	3.85	6.06	1.98
	Observation	E	E	E	E	E	E	E

Additional data:
Teflon in benzophenone: ΔF_I^{adh} = − 3.2 mJ/m²; ΔF_{II}^{adh} = 2.6 mJ/m²; observation: E
Teflon in bibenzyl: ΔF_I^{adh} − 0.96 mJ/m²; ΔF_{II}^{adh} = 0.03 mJ/m²; observation: E

assumed to be independent of temperature up to the melting point of naphthalene (80 °C).

Table 8 contains the estimates of the free energy of adhesion as calculated with Eq. (8), together with the experimental observation of particle engulfment (*E*) or rejection (*R*) [28]. Also included in Table 8 are the ΔF^{adh} predictions as calculated using the equation of state of Neumann et al. [2]. Recall that a negative value of the free energy corresponds to a prediction of particle engulfment while a positive value is a prediction of rejection.

The results listed in Table 8 indicate that in the vast majority of cases the Fowkes approach does not correctly predict particle behaviour at solidification fronts. This is generally true, even for those matrix and particle materials which are completely or overwhelmingly dispersive; viz. polystyrene, Teflon, siliconized glass in naphthalene, biphenyl, and o-terphenyl.

In contrast, in almost every case the equation of state predicts the observed behavior correctly. In only two instances, both involving Nylon-6,12, are there discrepancies which cannot be attributed to errors associated with the extremely small values of ΔF^{adh}. The origin of these two anomalies is unknown. The present results reinforce doubts about the basic soundness of the Fowkes approach which have arisen in connection with the contact-angle measurements described previously.

Theoretical assessment

The two experiments described above cast into doubt both the legitimacy and the necessity ·of dividing surface tensions into components in order to predict solid and solid-liquid interfacial tensions [1].

On the other hand, these same contact-angle and freezing-front data clearly support the equation of state approach [2], whereby the solid-liquid interfacial tension is thought to be a function only of the total solid and liquid surface tensions, irrespective of the types and relative magnitudes of the intermolecular forces present within each phase.

As far as the determination of solid surface tensions from contact angles is concerned, one might well discard the approach of surface tension components altogether in favor of an equation of state approach. Nevertheless, there are three points which merit further consideration. First, the contact-angle experiment does not involve hydrogen bonding between the solids and liquids, and proponents of the Fowkes approach might argue that the contact angle of hydrogen-bonding systems may not follow the same pattern. While it is possible to perform such measurements, the present purpose is to consider the theoretical aspects of the problem relevant to this question. Second, there is the question of whether the apparent breakdown of the Fowkes approach could have been predicted and, if so, where the fallacies are. Finally, what is the correct interpretation of the quantities known as "dispersion components of the interfacial tension"?

These questions are best approached from an historical perspective.

Historical perspective

The calculation of solid surface tension γ_{SV}, from the contact angle θ, of a liquid of surface tension γ_{LV}, starts with Young's equation, Eq. (1). Of the four quantities in Young's equation, γ_{LV} and θ are readily measurable. In order to determine γ_{SV}, further information is necessary. Conceptually, one obvious approach is to seek one more relation between the parameters in Eq. (1), such as an equation of state, possibly of the form

$$\gamma_{SL} = f(\gamma_{LV}, \gamma_{SV}) . \tag{9}$$

The simultaneous solution of Eqs. (1) and (9) would solve the problem. Note that if the commonly used assumption of negligible liquid vapor adsorption is applied, then Eq. (1) and Eq. (9) may be written in terms of γ_L and γ_S, rather than of γ_{LV} and γ_{SV}.

An old equation of state for solid-liquid interfacial tensions was contributed by Rayleigh, and later Good et al. (see review [8])

$$\gamma_{SL} = \gamma_S + \gamma_L - 2\sqrt{\gamma_s \gamma_L} \tag{10}$$

Table 9. Solid surface tension (mJ/m²) of n-hexatriacontane at 20°C. γ_S calculated using Eq. (11) and γ_S^{ES} calculated with the equation of state of Neumann et al. [2]. Contact angle data (degrees) from [10]

Liquid	γ_{LV}	θ	γ_S	γ_S^{ES}
Water	72.8	104.6	10.2	19.8
Glycerol	63.4	95.4	13.0	20.0
Thiodiglycol	54.0	86.3	15.3	19.8
Ethylene glycol	47.7	79.2	16.8	19.8
Hexadecane	27.6	46	19.8	20.1
Tetradecane	26.7	41	20.6	20.7
Dodecane	25.4	38	20.3	20.4
Decane	23.9	28	21.2	21.2
Nonane	22.9	25	20.8	20.8

where γ_S is the solid surface tension (equal to γ_{SV} if adsorption is neglected), and γ_L (or equivalently γ_{LV}) is the surface tension of the liquid. Using these equivalents and combining Eq. (10) with Eq. (1) gives

$$\gamma_S = \tfrac{1}{4}\gamma_L(1+\cos\theta)^2 . \tag{11}$$

Early investigations by Good et al. [29] showed that Eq. (11) yields consistent value of γ_S when γ_S and γ_L are both relatively small (e.g., liquid alkanes on Teflon or paraffin wax). Contact-angle data for liquids of higher surface tension, however, lead to values of solid surface tension which become progressively smaller as the liquid surface tension increases. The fourth column of Table 9 illustrates this for the contact angles of a wide range of liquids on solid hexatriacontane. On the assumption that the surface tension of the hexatriacontane should be approximately constant for all of these liquids, it is evident that Eq. (10) is inadequate as an equation of state.

Good et al. proposed that Eq. (10) could be modified so that it yields constant values of γ_S for all γ_L by incorporating an adjustable parameter, Φ, the Good interaction parameter [29]. Eq. (10) is then written as,

$$\gamma_{SL} = \gamma_S + \gamma_L - 2\Phi\sqrt{\gamma_S \gamma_L} \tag{12}$$

and Eq. (11) becomes

$$\gamma_S = \frac{1}{4}\frac{\gamma_L^2}{\Phi^2 \gamma_L}(1+\cos\theta)^2 \tag{13}$$

with the understanding (from experimental observation) that Φ approaches unity whenever γ_S and γ_L are both relatively small. It should be noted, however, that Good et al. believed that it is not the magnitudes of the solid and liquid surface tensions which govern the condition $\Phi = 1$, but rather the similarity in the types of intermolecular forces in the solid and liquid [29]. These two points of view are often coincidental, since as surface tension decreases below, say, $30 \, \text{mJ/m}^2$, it is generally found that London dispersion forces are predominant in both solids and liquids.

At this point three options present themselves for the further development of Eqs. (10) or (12):

1. Attempts can be made to calculate Φ in terms of statistical mechanics, which is the very complex path followed by Good et al. [29].
2. As described previously, Fowkes' arguments start with the conviction that the total surface tension can be decomposed into surface tension components

$$\gamma = \gamma^d + \gamma^h + \ldots \qquad (14)$$

where γ^d and γ^h are considered as unique physical properties called the dispersion and hydrogen-bonding components of surface tension.

In the context of Eq. (10) this latter approach interprets the decrease in γ_S with increasing γ_L (cf. Table 9) as a reflection of the decrease in the relative importance of dispersion forces in the higher surface tension liquids found in Table 9. Equation (10) is thus thought to be deficient because it does not take into account the nature of the intermolecular forces present in these various liquids. This reasoning led to the Fowkes equation

$$\gamma_{SL} = \gamma_S + \gamma_L - 2\sqrt{\gamma_S^d \gamma_L^d} \qquad (15)$$

where γ_S^d and γ_L^d are the dispersion components of the solid and liquid phases. Equation (15) is supposed to be valid only for cases where at least one phase is completely "dispersive". Considering Table 9, such would be the case with water, since $\gamma_L^d < \gamma_L$ that γ_{SL} would be larger than the value obtained from Eq. (10); hence, in view of Young's equation, γ_{SV} would also be larger, potentially equal to the values obtained with low surface tension liquids, which coincidentally, are also the liquids having dispersion forces only. Combining Eq. (15) with Young's equation, Eq. (1), yields

$$\gamma_S^d = \frac{1}{4} \frac{\gamma_L^2}{\gamma_L^d} (1 + \cos \theta)^2 \ . \qquad (16)$$

Within the context of the surface tension components approach, γ_S^d can be determined from the liquid surface tension γ_L, its dispersive component γ_L^d, and the contact angle θ.

3. The approach of Neumann et al. [2] begins with the question of the existence of an equation of state in the form of Eq. (9). For a three phase, two-component system consisting of a pure liquid and an inert solid which does not dissolve in the liquid and has zero vapor pressure, the three Gibbs-Duhem equations [3] for the solid-vapor, the solid-liquid and the liquid-vapor interfaces can be written as

$$d\gamma_{SV} = -s_{SV} dT + \Gamma_{SV} d\mu_L \qquad (17\,\text{a})$$

$$d\gamma_{SL} = -s_{SL} dT + \Gamma_{SL} d\mu_L \qquad (17\,\text{b})$$

$$d\gamma_{LV} = -s_{LV} dT + \Gamma_{LV} d\mu_L \qquad (17\,\text{c})$$

where s denotes the specific surface entropy, T the temperature, Γ the specific interfacial masses and μ_L the chemical potential of the liquid. Since the three equations (17 a, b, c) contain on their righthand sides, only the two independent properties T and μ_L, it is apparent that the lefthand sides of the equations are not independent, i.e., that a relation of the form of Eq. (9) exists.

An explicit equation of state was formulated [2] empirically from contact-angle data, utilizing the experimental fact that $\Phi = \Phi(\gamma_{SL})$. It should be noted that the final product of this approach is not an analytical equation, but rather a computer program [2, 17] or, alternatively, a set of tables [18] which permit the determination of γ_S and γ_{SL} from contact angles. It may also be noted that the use of Φ in the development of the equation of state was solely a matter of convenience. Since the approach utilizes curve-fitting procedures, the selection of Φ as a correlating variable, while convenient, was essentially arbitrary.

Consequences for the Fowkes approach of the existence of an equation of state

The existence of an equation of state, generally written as Eq. (9), which is firmly based on thermodynamics, has far-reaching consequences.

Most important, an equation such as Eq. (15) would also have to satisfy Eq. (9), but this is apparently impossible, since the dispersion components of the surface tensions would have to be functions of the total surface tensions. This would preclude the possibility of γ^d being a unique function of intermolecular forces, as stipulated by the Fowkes approach. This impossibility carries over into contact angles. Combining Eqs. (1) and (9) yields

$$\cos \theta = \frac{f(\gamma_S, \gamma_L) - \gamma_S}{\gamma_L} \qquad (18)$$

so that

$$\cos \theta = g(\gamma_S, \gamma_L) , \qquad (19)$$

indicating that the contact angle is completely determined by the total surface tensions γ_S and γ_L. This implies that surface tension components cannot be determined from contact angles. Clearly, any arbitrary solid-liquid-vapor system with the same total surface tensions γ_S and γ_L must have the same contact angle, θ. This is true regardless of the relative magnitudes and types of intermolecular forces, including hydrogen bonding. The results in Tables 3–5 constitute experimental confirmation of this thermodynamic result.

Overall, it is apparent that while Eq. (14) might be correct, Eqs. (15) and (16) are in conflict with the existence of an equation of state, i.e., with Eq. (9). Any attempt to develop a theory of surface tension components would have to come to terms with Eq. (9). Particularly, there is no apparent way of using contact angles to measure surface-tension components.

Reinterpretation of the parameter γ^d

While it is apparent from the above that γ^d parameters as obtained from the Fowkes appraoch cannot be dispersion components of surface tension, the question remains as to whether they may not carry some meaning nevertheless. For a liquid which, in Fowkes' sense, is capable only of interaction by means of dispersion forces, Eq. (15) becomes

$$\gamma_{SL} = \gamma_S + \gamma_L - 2\sqrt{\gamma_S^d \gamma_L} . \qquad (15a)$$

Comparing Eq. (15a) with Eq. (12) yields

$$\gamma_S^d = \Phi^2 \gamma_S . \qquad (20)$$

Considering Eq. (12) as the definition of Φ, and using Eq. (9), we see that

$$\gamma_S^d = \zeta(\gamma_S, \gamma_L) \qquad (21)$$

that is, γ_S^d would have to be some function of γ_S and γ_L. Similarly, for a solid which in Fowkes' sense is capable of interaction only by means of dispersion forces, one obtains

$$\gamma_L^d = \Phi^2 \gamma_L \qquad (20a)$$

and

$$\gamma_L^d = \eta(\gamma_L, \gamma_S) . \qquad (21a)$$

Not only is there no room for a direct dependence of γ_S^d on intermolecular forces in Eq. (21); there is also a dependence of γ_S^d on γ_L which destroys Fowkes' stipulation that γ_S^d is a property of the solid phase alone. It becomes apparent from Eqs. (20) and (20a) that γ^d is a correction factor playing virtually the same role as Φ. Because the equation of state approach of Neumann et al. [2] utilizes an empirical relation between Φ and γ_{SL}, it is possible to calculate Φ and hence γ_S^d or γ_L^d values from that approach; this is illustrated in [8, 30]. This demonstrates quite clearly that γ_S^d and γ_L^d are adjustable parameters defined by both γ_L and γ_S.

Generalizations of the Fowkes approach

Given that γ_S^d and γ_L^d are not physical properties, but rather artifacts of Raleigh-Good equation, it is obvious that the considerable number of extensions and generalizations of the Fowkes approach are also deficient. For example, the Owens and Wendt [31] parameters, γ_S^d and γ_S^h (termed the dispersion and polar components of the solid surface tension) can be shown to be essentially meaningless functions of γ_S [30]. Moreover, since γ_{SL} is a known function of the total solid and liquid surface tensions, [2, 3] in many instances there is little incentive to attempt an evaluation and treatment of surface tension components. Total solid surface tensions can be determined from a single contact-angle measurement using the equation of state approach [2].

Equations of state

The experimental and theoretical results presented thus far lead one to conclude that the theory of surface tension components, based on the work of Fowkes, is incorrect. This is not to say that surface tension components do not exist, only that they cannot be determined using present-day methods.

Turning then to equation of state approaches and in particular to that due to Neumann et al. [2], how confident can we be in its accuracy? As was mentioned in the introduction to this chapter, most approaches to the determination of solid and solid-liquid interfacial tensions have not been verified using experimental results which are independent of the equation being tested. In a recent paper [16], five distinct experimental methods for the determination of solid surface tension were considered in light of the equation of state of Neumann et al. [2]. The five methods, most of which utilize small solid particles, are (1) adhesion, (2) phagocytosis, (3) sedimentation volumes, (4) solidification front, and (5) contact angles. It was demonstrated that the first three of these techniques yield consistent values of solid surface tension if the thermodynamic models utilize equations of state f, with symmetry; that is,

$$\gamma_{12} = f(\gamma_1, \gamma_2) = f(\gamma_2, \gamma_1) \tag{29}$$

and have zero minimum interfacial tension, γ_{12}, i.e.,

$$\gamma_{12} = f(\gamma_1, \gamma_2) = 0 \tag{30}$$

when $\gamma_1 = \gamma_2$.

In Eqs. (29) and (30) γ_1 and γ_2 are generalized solid or liquid surface tensions, or both.

Experimental methods (4) and (5), however, impose additional requirements for consistency of results. The equation of state of Neumann et al. [2] has been found to satisfy these conditions as well as those expressed by Eqs. (29) and (30). In other words, using the equation of state of Neumann et al. [2], all five methods yield the same value of solid surface tension for a given material. Of all available tests, this is probably the most comprehensive.

Another completely independent test of the equation of state of Neumann et al. [2] has come to light recently in the work of Claesson et al. [32]. These authors reported direct measurements of the interaction between molecularly smooth mica surfaces covered with monolayers of dimethyldioctadecylammonium (DDOA). The forces of interaction were

recorded with the apparatus of Israelachvili [33] and were used to measure solid surface tension directly. Water contact angles were also obtained, thereby allowing a comparison between the equation of state of Neumann et al. [2] and the direct γ_S measurements. This comparison is given in Table 10.

The excellent agreement between these direct measurements of surface tension and the values predicted by the equation of state of Neumann et al. [2] provides an important new confirmation of the latter approach. It should be emphasized that these results were obtained using only the total surface tension of water, completely neglecting any attempt to account for hydrogen bonding or the polar components of surface tension. This further supports our contention that contact angles are functions only of γ_S and γ_L.

Conclusions

The contact angles of pairs of liquids on a single solid substrate were found to be equal when the surface tensions of the liquids were the same, regardless of the relative magnitudes of the dispersion forces contained in each of the two liquids. This observation is in direct conflict with the theory of surface tension components. It is, however, consistent with the expectations of the equation of state approach, in which the solid-liquid interfacial tension is only a function of the total solid and liquid surface tensions.

The theory of surface tension components cannot be used to predict the outcome of engulfing experiments involving microscopic particles at advancing solidification fronts. The equation of state of Neumann et al. passes this independent test [2].

It has also been shown that the existence of an equation of state, which has been demonstrated thermodynamically, precludes the use of contact angles as a means of determining surface tension components.

Finally, the quantities γ_L^d and γ_S^d derived from the Fowkes equation are not "dispersion components of surface tension", but rather adjustable parameters which are functions of the total surface tensions γ_L and γ_S. The magnitudes and functional characteristics of the variables γ_L^d and γ_S^d are artifacts of the behavior of the simple Rayleigh-Good equation of state which breaks down and requires corrective terms as the total surface tensions of the solid and liquid increase.

Table 10. Direct measurement of γ_{SV} and γ_{SL} using Israelachvili machine [32] in comparison with the surface tension predictions of the equation of state of Neumann et al. [2], based on the contact angles

	Direct force measurement Claesson et al. [32]	Eq. of state Neumann et al. [2]
γ_{SV} (mJ/m^2)	27	26.0[a]
γ_{SL} (mJ/m^2)	32[b]	31.0[a]
	29[c]	
	34[d]	
	Ave = 31.7	

a: Calculated with equation of state [2], $\gamma_{LV} = 72$ mJ/m^2, $\theta = 94°$ (5)
b: Calculated with Young's equation, $\gamma_{SV} = 27$ mJ/m^2 (direct measurement in air), $\theta = 94°$, $\gamma_{LV} = 72$ mJ/m^2
c: Based on extrapolated force law measured as DDOA surfaces are brought together
d: Based on direct force measurement as DDOA surfaces are separated

Acknowledgements

This work was supported by the Natural Science and Engineering Research Council of Canada under grant A 8278.

References

1. Fowkes FM (1964) Ind Eng Chem Dec 40−52
2. Neumann AW, Good RJ, Hope CJ, Sejpal M (1974) J Colloid Interface Sci 49:291−304
3. Ward CA, Neumann AW (1974) J Colloid Interface Sci 49:286−290
4. Spelt JK, Absolom DR, Neumann AW (1986) Langmuir 2:620−625
5. Barton AFM (1975) Chem Rev 75:731−753
6. Barton AFM (1983) CRC Handbook of Solubility Parameters and Other Cohesion Parameters, CRC Press
7. McClellan AL (1963) Tables of Experimental Dipole Moments, WH Freemann and Co.
8. Spelt JK (1985) Solid Surface Tension: The Equation of State Approach and the Theory of Surface Tension Components, Ph. D. Thesis, University of Toronto, Toronto
9. Beerbower A (1971) J Colloid Interface Sci 35:126−132
10. Neumann AW (1974) Adv Colloid Interface Sci 4:105−191
11. Spelt JK, Rotenberg Y, Absolom DR, Neumann AW (1987) Colloids and Surfaces 127−137
12. Rotenberg Y, Boruvka L, Neumann AW (1983) J Colloid Interface Sci 93:169−183
13. Cain JB, Francis DW, Venter RD, Neumann AW (1983) J Colloid Interface Sci 94:123−130
14. Good RJ (1975) In: Adsorption at Interfaces (Mittal KL, ed.), ACS. Symposium Ser No 8, Washington
15. Neumann AW, Spelt JK, Smith RP, Francis DW, Rotenberg Y, Absolom DR (1984) J Colloid Interface Sci 102:278−284
16. Smith RP, Absolom DR, Spelt JK, Neumann AW (1986) J Colloid Interface Sci 110:521−532
17. Taylor CPS (1984) J Colloid Interface Sci 100:589−594
18. Neumann AW, Absolom DR, Francis DW, Van Oss CJ (1980) Sep Purif Methods 9:69−163
19. Omenyi SN, Neumann AW (1976) J Appl Phys 47:3956−3962
20. Omenyi SN, Neumann AW, Van Oss CJ (1981) J Appl Phys 52:789−795
21. Omenyi SN, Neumann AW, Martin WW, Lespinard GM, Smith RP (1981) J Appl Phys 52:796−802
22. Neumann AW, Omenyi SN, Van Oss CJ (1982) J Phys Chem 86:1267−1270
23. Omenyi SN, Smith RP, Neumann AW (1980) J Colloid Interface Sci 75:117−125
24. Smith RP, Omenyi SN, Neumann AW (1983) In: Physicochemical Aspects of Polymer Surfaces (Mittal KL, ed), Plenum, New York. Vol 1, pp 155−171
25. Spelt JK, Absolom DR, Zingg W, Van Oss CJ, Neumann AW (1982) Cell Biophysics 4:117−131
26. Smith RP (1984) Applied Surface Thermodynamics for the Interaction of Small Particles with an Advancing Solidification Front, Ph. D. Thesis, University of Toronto, Toronto
27. Soulard MR, Vargha-Butler EI, Hamza HA, Neumann AW (1983) Chem Eng Comm 21:329−344
28. Spelt JK, Smith RP, Neumann AW (1987) Colloids and Surfaces 28:85−92
29. Good RJ (1977) J Colloid Interface Sci 59:398−419
30. Spelt JK, Neumann AW (1987) Langmuir 3:588−591
31. Owens DK, Wendt RC (1969) J Appl Polymer Sci 13:1741−1747
32. Claesson PM, Blom CE, Herder PC, Ninham BW (1986) J Colloid Interface Sci 114:234−242
33. Israelachvili JN, Adams GE (1978) J Chem Soc Faraday Trans 74:975−1001

Received February 12, 1988;
accepted July 7, 1988

Authors' address:

J. K. Spelt
Alcan International Ltd.
Kingston Research and Development Centre
Kingston, Ontario
Canada

Progress in Colloid & Polymer Science

Progr Colloid & Polymer Sci 77:40–48 (1988)

Characterization of polymer surfaces by means of electrokinetic measurements

H.-J. Jacobasch and J. Schurz[1]

Institut für Technologie der Polymere, Akademie der Wissenschaften der DDR, Dresden, [1]Institut für Physikalische Chemie, Karl-Franzens-Universität Graz

Abstract: First, open questions as well as advantages of zetapotential-measurements are discussed. With the help of experiments it is shown that the methods of streaming potential-streaming current and electro-osmosis measurements yield reproducible results that agree. A new measuring system (EKM) is described. It is shown that with its help and with newly developed measuring cells the surface of polymers can be investigated. Several applications to fibers and flat polymers (foils) are demonstrated, and it is shown how this method can be used to investigate the pK of dissociable groups, the adsorption potential of ions, and preferential adsorption.

Key words: Electrokinetic measurements, polymer surface, adsorption of surfactant anions and cations

1. Introduction

The measurement of the zetapotential is a classic method of colloid chemistry. Although it has been used for many decades, some of its features are still controversial. In Table 1 we have listed a few "pros" and "contras" of this method. The advantages of the method often lead to its uncritical use or to the neglect of open questions. Our groups working in Dresden and in Graz therefore started a systematic investigation of the possibilities of this method. We wanted to develop an unambiguous and reproducible measuring technique for organic polymers, and to arrive at unequivocal correlations between zetapotential and other interfacial properties.

2. Measuring method

We concentrated on measuring techniques in which the solid phase is fixed. That is, we used streaming potential-streaming current and electro-osmosis. In these methods, a flowing liquid gives rise to electrical potential difference or electric current through a single capillary or a bundle of capillaries, or an electrical potential difference produces material flow. According to the deductions of Helmholtz and Smoluchowski, the zetapotential ζ is given

$$\zeta = \frac{I_s}{\Delta p} \cdot \frac{L}{Q} \cdot \frac{\eta}{\varepsilon \cdot \varepsilon_0} \tag{1}$$

$$\zeta = \frac{U_s}{\Delta p} \cdot \frac{L}{Q} \cdot \frac{\eta}{\varepsilon \cdot \varepsilon_0} \cdot \frac{1}{R} \tag{2}$$

$$\zeta = \frac{D}{I} \cdot \frac{L}{Q} \cdot \frac{\eta}{\varepsilon \cdot \varepsilon_0} \cdot \frac{1}{R} \tag{3}$$

Table 1. Open questions and advantages of electrokinetic methods

Open questions	Advantages
Disagreement between results obtained in different laboratories	Easy applicability, relatively small expenditures for apparatus and time
Dependence of the results on the method applied	Applicability for different shapes of solid materials
Strong influence of geometrical factors (shape of particles, pore geometry)	Correlation of zetapotential with numerous technologically and biologically relevant parameters
Dubios theoretical model	Correlation of the zetapotential with numerous material parameters
Position of the shearing plane	
Influence of surface conductivity	
Often complex influences of various material parameters	

$$\zeta = \frac{D}{U} \cdot \frac{L}{Q} \cdot \frac{\eta}{\varepsilon \cdot \varepsilon_0} \tag{4}$$

I_s streaming current
U_s streaming potential
D volume flow
I electrical current
U voltage
L, Q length and cross section of capillary
η, ε viscosity and permittivity (relative dielectric constant)
ε_0 influence constant
R electrical resistance

The open questions for both methods are mainly
– dubious assumptions about the geometry of the capillaries (L and Q)
– the comparability of zetapotentials obtained by either streaming potential-streaming current or by electro-osmosis. The relations (5) and (6) should hold

$$\frac{I_s}{\Delta p} = \frac{D}{U} \tag{5}$$

and

$$\frac{U_s}{\Delta p} = \frac{D}{I} \tag{6}$$

Table 2. Capillary bundle models

Authors	Equation		
Fairbrother and Mastin [11]	$L/Q = \kappa \cdot R$		(7)
	κ	spec. electric conductivity	
	R	electric resistance	
Biefer and Mason [9]	$\dfrac{L}{Q} = \dfrac{L_M}{Q_M}(1 - a \cdot d)^{5/2}$		(8)
	L_M, Q_M	length, cross section of measuring cell	
	a	spec. hydrodynamic volume	
	d	packing density	
Chang and Robertson [10]	$\dfrac{L}{Q} = \dfrac{L_M}{Q_M} \cdot e^{-B \cdot d}$		(9)
	B	empirical constant	
Happel [7], Ciriacks [8]	$\zeta = f$ (streaming parameters)		
Schurz and Erk [1]		empirically corrected Happel-Ciriacks model	

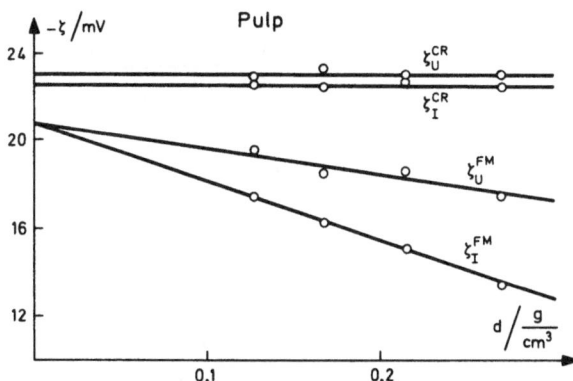

Fig. 1. Dependence of zetapotential in 10^{-4} N KCl solution of cellulose fibers on packing density determined by different methods. ζ_U^{CR}: streaming potential (Eq. 9); ζ_I^{CR}: streaming current (Eq. 9); ζ_U^{FM}: streaming potential (Eq. 7); ζ_I^{FM}: streaming current (Eq. 7)

– independence of the zetapotential of the pressure gradient or the voltage applied (it could be that strong mechanical or electric forces will place the shear plane in the double layer at another site as compared with that caused by weak forces)
– the importance of the influence of surface conductivity on zetapotential calculated according to Eqs. (1)–(4).

Erk, Schausberger and Bauböck [1–5] have studied the applicability of models for the determination of L/Q as described in literature for capillaries consisting of fibrous material. Table 2 gives a listing of the most important capillary-bundle models. As a criterion for the applicability of a pore model, the independence of the calculated zetapotential of the packing density of the diaphragm is taken, since the quotient L/Q will vary with packing density (Figs. 1–3). Figures 1–3

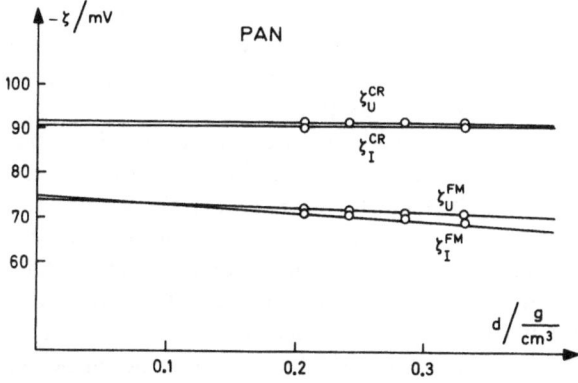

Fig. 2. Dependence of zetapotential in 10^{-4} N KCl solution of PAN fibers on packing density determined by different methods (see Fig. 1)

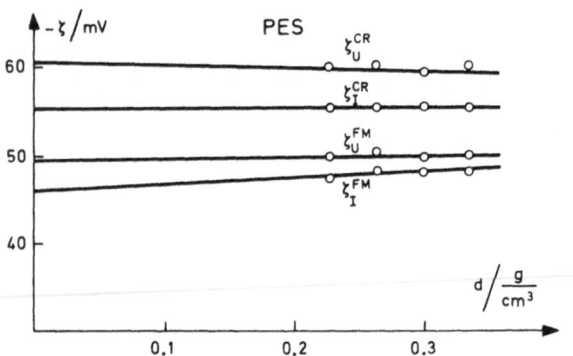

Fig. 3. Dependence of zetapotential in 10^{-4} N KCl solution of polyester fibers on packing density determined by different methods (see Fig. 1)

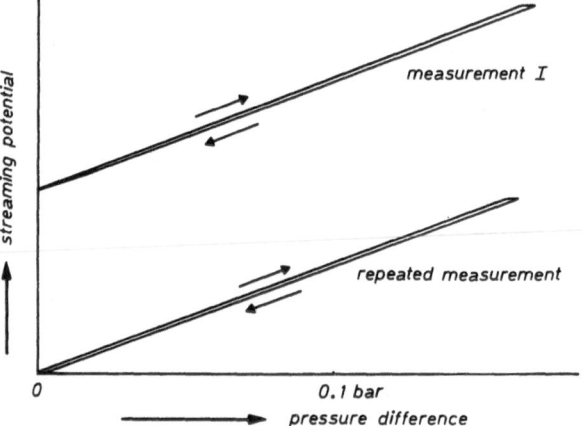

Fig. 4. Dependence of streaming potential on pressure for PAN fibers in 10^{-4} N KCl solution with different flow directions

show the dependence of zetapotential on the packing density for various fibers in 10^{-4} N KCl-solution, calculated according to Chang and Robertson [10] and Fairbrother and Mastin [11] (cf. 3). The evaluation according to Chang and Robertson yields a zetapotential independent of the packing density for all fibers, whereas the evaluation according to Fairbrother and Mastin gives a packing density independent zetapotential only for the hydrophobic fibers. The deviations of the last method are due to the surface conductivity of the cellulose fibers, and to errors in the determination of plug resistance. The differences between the zetapotentials obtained by the two evaluation methods disappear when the measurements are performed at a high electrolyte concentration and the resistance is measured by means of an impedance bridge. The empirically corrected Happel-Ciriacks model, as introduced by Schurz and Erk [1], leads to zetapotentials that are independent of packing density, but requires considerable mathematical effort. The uncorrected Happel-Ciriacks model and the method by Biefer and Mason are not applicable to all fibrous materials [10]. Therefore it is possible to determine the quotient L/Q in Eqs. (1)–(4) in a reproducible way. Analogous to fibers, measurements with plate-shaped polymers yielded the same results [6].

Next we shall consider the determination of the parameters $U_s/\Delta p$, $I_s/\Delta p$, D/U, and D/I. All experimental results published so far [12–14] show, that the value of this quotient does not depend on the pressure difference or the voltage applied or the current. Our own results agree with these findings [4]. As an example for a great number of measurements, Figs. 4 and 5 show the dependence of the streaming potential on pressure, or of volume flow on voltage for polyacrylonitrile fibers in 10^{-4} N KCl. A strictly linear relation is obtained. This means, that independent of the

pressure or the voltage applied, changes in these parameters will always result in equal changes in the streaming potential or the volume flow. We believe this indicates that the position of the shearing plane is well defined and is independent of the applied mechanical or electric force. Therefore, measurements of either streaming potential-streaming current or electro-osmosis must necessarily lead to identical values for the zetapotential. Figure 6 presents some experimental results that show this is actually the case. On the one hand, we conclude that the zetapotential of the investigated polymers is an unambiguous and reproducible parameter. On the other hand these results have led the way to the development of an instrument which allows an unambiguous determination of zetapotential. For the measurement of streaming potential-streaming current it is, according to Eqs. (1) and (2),

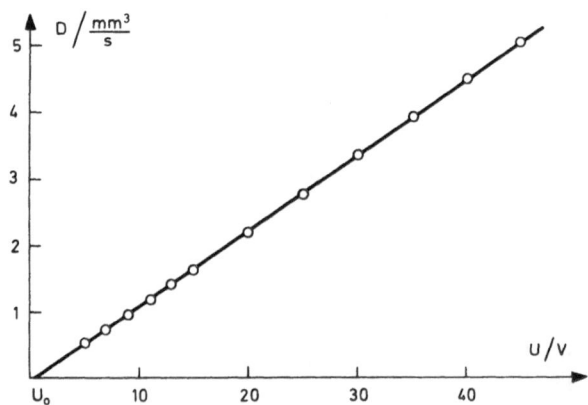

Fig. 5. Dependence of electro-osmotic volume flow D on externally applied voltage U for a PAN fiber plug in 10^{-4} N KCl solution

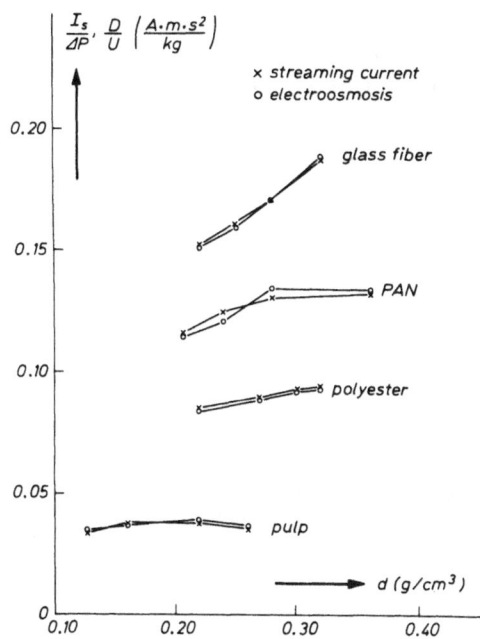

Fig. 6. Comparison of streaming current and electro-osmosis for fiber plugs of different packing density

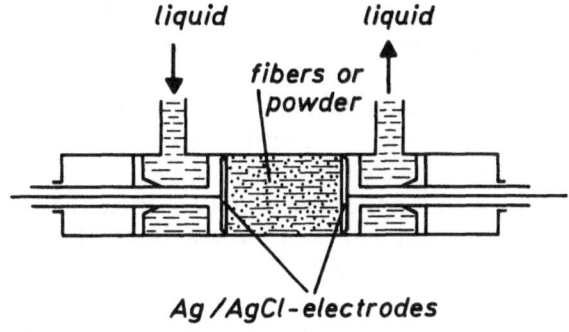

Fig. 8. Measuring cell for fibers, powders, etc.

only necessary to use measuring cells allowing a reproducible determination of the quotient L/Q, and to determine the quotients $U/\Delta p$ or $I/\Delta p$. This concept was realized in the construction of the electrokinetic measuring system EKM by the Institute of Polymer Technology of Dresden, the Institute of Physical Chemistry, Graz, and the Anton Paar KG Company, Graz. Figures 7 and 8 show measuring cells for fibers, powders, and similar materials, and for plates, foils, etc. The measurement of the quotients $U_s/\Delta p$ and $I_s/\Delta p$ is done in the following way. By means of a microprocessor-controlled geared or centrifugal pump a variable pressure difference is produced at the measuring cell (Fig. 9). The electrical parameters are determined by means of Ag-AgCl electrodes, the pressure difference by means of a piezoelectric pressure transducer. The quotient is calculated using a personal computer.

3. Relations between zetapotential and the physical and chemical structure of polymers

We will assume that for most real capillary systems (as plugs of powder or fibers, filters, interstices between foils and plates) the Helmholtz-Smoluchowsky relations (1)–(4) hold. Then the potential at the shearing plane or at least a potential related to it in a well-defined way can be determined with the help of the methods described above.

Next we encounter the questions about the relationship between the zetapotential and the chemical and

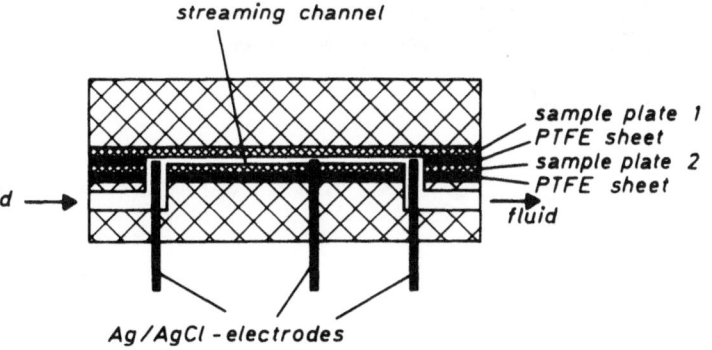

Fig. 7. Measuring cell for polymer plates

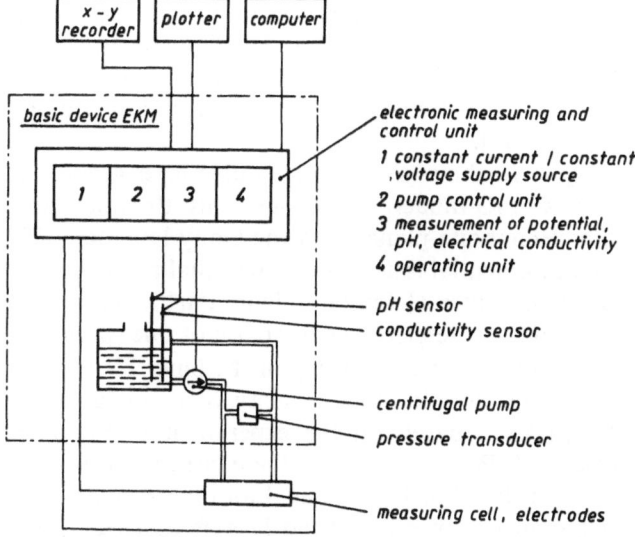

Fig. 9. Schematic of the electrokinetic meter EKM (A. Paar KG, Graz, Austria)

physical structure of polymers as well as about the correlation between zetapotential and interfacial chemical parameters.

In general, for the quantitative description of electrokinetic phenomena, the model of the electric double layer is used, as connected with the names Gouy, Chapman, Stern, and Grahame (GCSC model). For polymers in contact with aqueous solutions of electrolytes, the electric double layer can be formed by dissociation of suitable molecular groups or the preferential adsorption of one sort of ions or both. With regard to the use of zetapotential measurements for polymer characterization, we then have to consider how we can separately characterize dissociation and adsorption and if it is possible to determine, by means of zetapotential measurements, the parameters relevant for this phenomena, namely the pK-number of functional groups, adsorption free energy, surface coverage, and the number of adsorption sites.

Many papers have been published by such authors as Ottewil, Shaw, Rendall, and Smith [15–17] concerning the determination of acidic or basic functional groups by means of zetapotential measurements. The description of the relationship between adsorption and the zetapotential goes back to the paper published by Stern in 1924, and the extension of the Stern model by Esin and Markow [cf. 17]. For the description of real systems, one generally takes into account only one of these mechanisms, namely dissociation or adsorption, and neglects the other. For this limiting case, the dependence of zetapotential on the concentration of H^+ or OH^--ions or of adsorbing ions shown in Figs. 10 and 11 is valid, if negatively charged polymers are primarily considered. Börner in the Dresden group has attempted to model the dissociation and adsorption taking place at polymer surface simultaneously, and to calculate from the functions $\zeta = f(\text{pH})$ and $\zeta = f(\text{c})$, the parameters characteristic for dissociation and adsorption [18].

Figure 12 shows the electric double layer according to the CGSG model. It is divided into the inner and outer Helmholtz planes, and the Gouy layer [cf. 17]. The zetapotential is assumed to be equal to the potential at the outer Helmholtz plane. Furthermore it is assumed that preferential adsorption of one species of ions will take place primarily in the inner Helmholtz plane. The double layer capacity is assumed to be $< 25\ \mu\text{F/cm}^2$, based on the experimental results of Ferse and Paul [19]. With these assumptions, the equations listed in Table 3 and 4 for the calculation of the parameters of Figs. 10 and 11 were derived. These equations show that the dissociation constants of acidic or basic groups as well as the adsorption poten-

tials (nonelectrostatic part of the adsorption free energy) for anions and cations can be determined by means of zetapotential measurements.

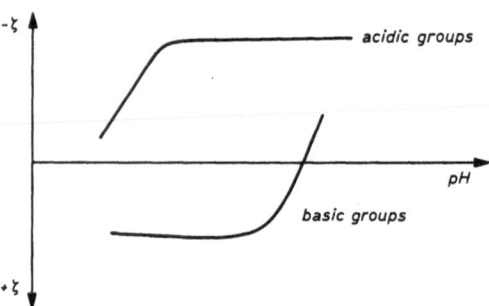

Fig. 10. Schematic of the dependence of zetapotential on pH for polymers with acidic or basic functional groups

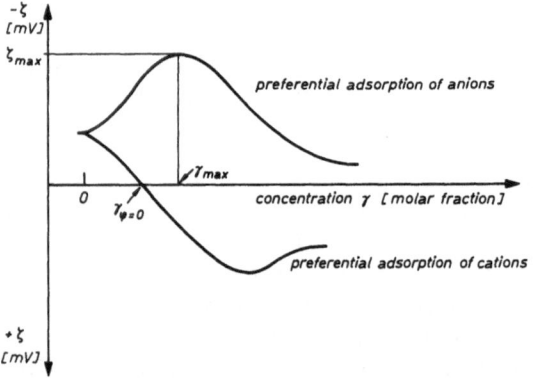

Fig. 11. Dependence of the zetapotential of negatively charged polymers on concentration of anionic or cationic surfactants in the measuring solution

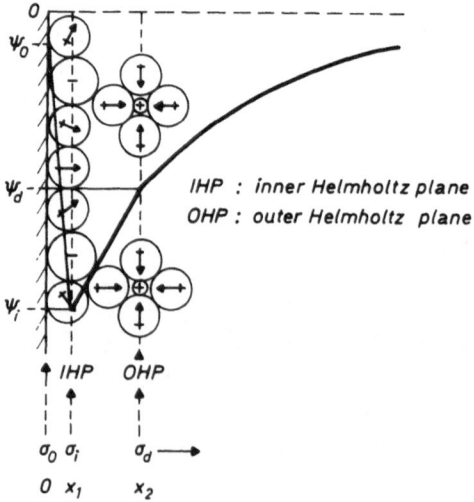

Fig. 12. Model of the electrical double layer according to Hunter (cf. [17])

Table 3. Determination of dissociation constants by zetapotential measurements

I. Acid groups

$$HA + H_2O \rightleftharpoons H_2O^+ + A^-$$ (10)

$$K_A = \frac{[H_3O^+]_s[A^-]}{[HA][H_2O]} \qquad s = \text{surface} \quad (11)$$

$$[H_3O]_s = [H_3O^+]_b \exp\left(-\frac{F\zeta}{RT}\right) \qquad b = \text{bulk} \quad (12)$$

$$\log K_A = pK_{(A)} = pH + 0.4343$$ (13)

$$\times \left[\frac{F\zeta}{RT} + \ln\left\{\frac{\sinh\left[-(F\zeta\text{plateau})/(2RT)\right] - 1}{\sinh\left[-(F\zeta)/(RT)\right]}\right\}\right]$$

II. Basic groups

$$B + H_2O \rightleftharpoons BH^+ + OH^-$$ (14)

$$K_B = \frac{[BH^+][OH^-]_s}{[B][H_2O]} = \frac{[BH^+]K_W}{[B][H^+]_s}$$ (15)

$$[OH^-]_s = [OH^-]_b \exp\left(\frac{F\zeta}{RT}\right)$$ (16)

$$\log K_B = pK_{(B)} = pK_w - pK_{(A)}$$ (17)

F = Faraday constant
R = Gas constant
T = Temperature
K_w = Ion product of water

Table 4. Determination of the adsorption free energy of ions by means of zetapotential measurements

I. $\quad \Delta G^0_{ads} = z \cdot F \cdot \zeta + \phi_\pm$ (18)

z valency
F Faraday's constant
$\phi_\pm$ nonelectrostatic term of adsorption free energy

II. 1-1-valent electrolyte on primarily negatively charged polymer surfaces (according to Stern)

$$\phi_- + \phi_+ = 2RT\ln c_{max}$$ (19)

$$\phi_- - \phi_+ = 2F \cdot \zeta_{max}$$ (20)

III. Preferential adsorption of cations at primarily negatively charged polymer surfaces

$$\phi_+ = RT\ln c_{\zeta=0} + \phi_{OH^-} - 2.3RT(pH\ 14)$$ (21)[a]

$$\phi_+ \approx RT\ln c_{\zeta=0}$$ (21 a)[b]

[a] The adsorption of OH^- ions in the Stern layer is taken into consideration.
[b] No difference is made between dissociation of acid groups and adsorbed OH^- ions.

The colloid chemist, interested in material or technological problems will, of course, question the material and technological relevance of the parameters obtained by zetapotential measurements. Therefore, in the following some typical examples are presented without claim to completeness.

Determination of the pK values of dissociating groups

Figures 13 and 14 show the dependence of the zetapotential of different polymers on the pH of the solution (10^{-3} N KCl). The strongly hydrophilic cellulose fibers show an extended plateau above a pH of about 4.5. A second plateau at lower pH is in-

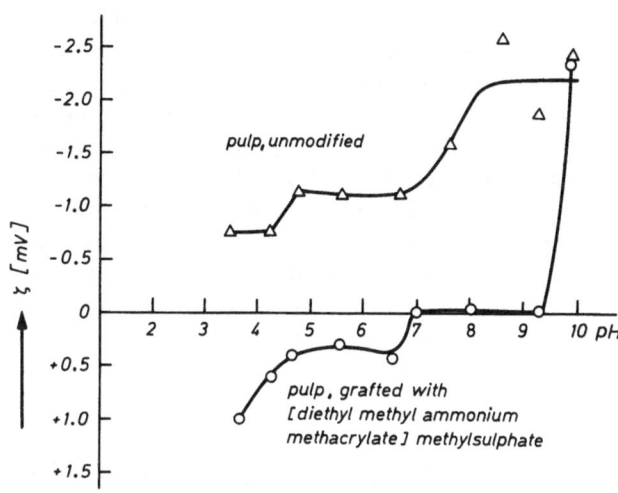

Fig. 13. pH dependence of zetapotential of unmodified and cationically modified cellulose pulp

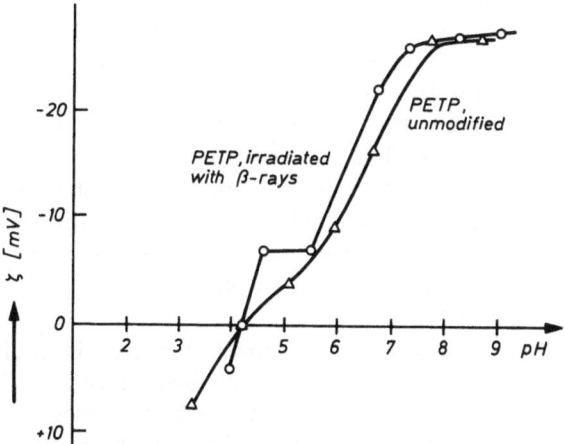

Fig. 14. pH dependence of zetapotential of unmodified and modified poly (ethylene terephthalate) sheets

dicated. The cationically grafted cellulose fibers exhibit a positive zetapotential in acidic and neutral medium. The hydrophobic polyethylene terephthalate foil shows a plateau in the alkaline range. A polyethylene terephthalate (PET) foil, irradiated for improved adhesion and wetting shows, in addition to the plateau in the alkaline range, a second plateau with a position similar to that found for cellulose fibers. According to Eq. (13) we calculate for the plateau of unmodified cellulose fibers and irradiated PET a $pK_{(A)}$ of 4.2. This value is characteristic of carboxyl groups. The second plateau indicated by the cellulose fibers would belong to more acidic groups. The plateau of the PET foils in the alkaline range is caused by the adsorption of OH^- ions.

A computer simulation of the adsorption-dissociation mechanism yields a practically identical ζ/pH function, if (in the case of PET) an adsorption free energy of $-50\,kJ/mol$ is assumed for the OH^- ions. For the cellulose fibers, an adsorption potential of $-20\,kJ/mol$ for the OH^- ions is obtained by this method [18]. Thus, the adsorption of OH^- ions will yield a ζ/pH function which simulates the dissociation of very weak acid groups only with hydrophobic polymers. We believe that the measuring of ζ/pH curves represents an excellent possibility for the characterization of acidic or basic surface groups of polymers, but the adsorption of OH^- ions must be taken into consideration.

Determination of the adsorption potential of anions

The nonelectrostatic part of the adsorption free energy of anions at primarily negatively charged polymer surfaces can be determined from the coordinates of the extreme value in the ζ/c curve (Eqs. 19, 20). This is true for such electrolytes as KCl and for anion-active surfactants. The value of the maximum zetapotential in the ζ/c curve has proved to be a characteristic figure for polymers which contain no dissociable groups. The value of ζ_{max} in KCl solutions often correlates with other parameters in a striking way. For instance, in Fig. 15, ζ_{max} in KCl solution is shown for several polymers as a function of their contact angle (or its cosine) with water [19]. An exactly linear relationship is found.

In former papers it was shown that ζ_{max} in KCl solutions correlates with the Hamaker constant of modified PAN fibers and with the adhesiveness of both synthetic and glass fibers [21]. Furthermore it was shown, that ζ_{max} can be lowered by competitive adsorption of, e.g., water or highly disperse oxides

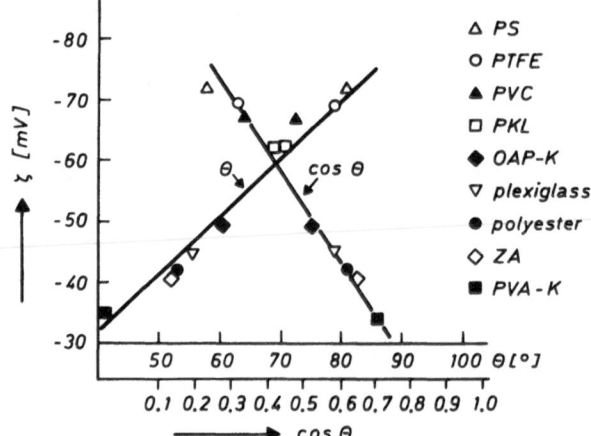

Fig. 15. Maximum zetapotential in KCl solutions of polymers as a function of the contact angle θ with water and $\cos \theta$ according to [19]

[21]. We finally arrive at the conclusion, that the maximum of the zetapotential in KCl solution is a measure of the dispersion forces exerted by the polymers.

Investigation of preferential adsorption of cations

The nonelectrostatic term of the adsorption free energy of preferentially adsorbed cations at primarily negatively charged polymers can be calculated according to Eq. (21) (in Table 4) from the concentration leading to charge reversal. The second and third term of Eq. (21) can be neglected for weak adsorption of OH^- ions and for absence of functional groups. The significance of the respective measurements can be demonstrated by the example of the zetapotential of primarily negatively charged polymers in solutions of cation active surfactants. The nonelectrostatic term of the adsorption free energy of tensides at solid surfaces is generally ascribed to the association energy of the surfactants caused by hydrophobic interactions. According to Schubert [22] we can write

$$\phi_{\pm} = \phi_{ass} = k \cdot S(n-1) \cdot \varphi \cdot N_L$$

$\phi_{\pm}$ molar free energy of adsorption
ϕ_{ass} molar free energy of association
k constant, taking into account deviations from ideal behavior
S degree of association
n number of CH_2 groups in the main chain
φ change of free energy per CH_2 group caused by complete association
N_L Avogadro number

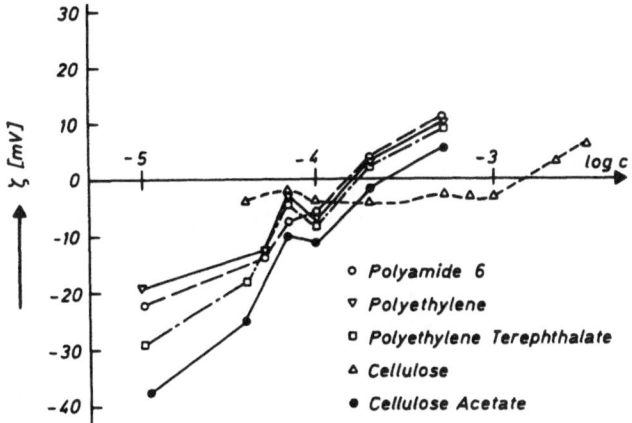

Fig. 16. Zetapotential of polymer sheets in solutions of dodecyl-pyridinium chloride

the number of functional groups on the surface arises. We think that here even the term "surface" is not quite precise, since with polymers we will not find a plane surface, but a more or less accessible surface region. The number of "surface groups" will, therefore, always depend on the method of determination. Their technological relevance, e.g., in adhesion problems, will depend on the mutual penetration of the adhering partners.

We can no longer expect to obtain the same number of functional groups by zetapotential measurements and by potentiometric titration. In the first case we will obtain, at best, the number of groups at the shearing plane. In the second case, we shall determine the number of ionogenic groups within a layer accessible for protons or hydroxide ions.

In Table 5 we show, as an example, the dependence of $c_{\zeta=0}$ of PET fibers in solution of alkylpyridinium salts on the length of the alkyl residue. We note that the charge reversal concentration decreases with increasing length of the alkyl chain, and that the value of the association energy given in the literature [17], 0.77 kT per CH_2 group, is approximately obtained. From Table 4 we calculate an average association energy of 0.81 kT per CH_2 group. This means that for a given substrate the zetapotential can be used for a characterization of the structure, but also for the quantitative analysis of adsorbed ionogenic surfactants. For a given surfactant the determination of the charge reversal concentration can be used for the characterization of solid polymers. Figure 16 shows the ζ/c-function for different polymer foils in solutions of dodecylpyridinium chloride. We note that the concentration of charge reversal increases with the increase in the hydrophilic character of the polymer. At present we are trying to derive quantitative relations between the concentration of charge reversal and the adsorption potential for the surfactant cation and the number of charges or the thickness of the swollen layer. Here the basic problem of the determinability of

Conclusions

Returning to the questions we started out with concerning measuring methods and relevance of zetapotential, we are of the opinion that with the measuring apparatus we developed and the described model of the electric double layer, it will be possible to determine the surface properties of polymers in a satisfactory way. We know that the models described are based on approximations which are not completely safe. In addition, any apparently solved problem gives rise to new questions. These latter refer in particular to the effect of electric double layer forces in comparison with other components of interparticular interactions. Also, they concern the effect of electric double layers on the structure and properties of the solid material. Presumably it may even be wrong to speak about double layers on the surface of solid materials, rather we may expect the double layer to be formed within a surface region, especially for polymers. The exact determination of the extension of the double layer and its effect upon structure and properties of solids will be subject of future work.

Table 5. Zero point of charge and adsorption free energies of polyester-fibers in solutions of alkylpyridinium salts of different chain lengths at varied KCl content and pH

Electrolyte	C_{12}		C_{14}		C_{16}		$G^0_{el.stat.}$ (kJ/mole)
	$c_{\zeta=0}$ (mole/l)	ϕ_+ (kJ/mole)	$c_{\zeta=0}$ (mole/l)	ϕ_+ (kJ/mole)	$c_{\zeta=0}$ (mole/l)	ϕ_+ (kJ/mole)	
H_2O, pH = 6.7	$7 \cdot 10^{-5}$	-33.9	$3.8 \cdot 10^{-5}$	-35.6	$7 \cdot 10^{-6}$	-39.7	0.8
10^{-3} N KCl, pH = 6.2	$5.5 \cdot 10^{-4}$	-28.9	$5.0 \cdot 10^{-5}$	-34.8	$1.04 \cdot 10^{-5}$	-38.7	1.0
10^{-3} N KCl, pH = 9.2	$6.5 \cdot 10^{-4}$	-28.4	$1.14 \cdot 10^{-4}$	-32.7	$2.4 \cdot 10^{-5}$	-36.6	4.0

References

1. Schurz J, Erk G (1985) Progr Coll & Polym Sci 71:44−48
2. Schausberger A, Schurz J (1979) Angew Makrom Chem 80:1
3. Bauböck G (1983) Diss Univ Graz
4. Jacobasch HJ, Bauböck G, Schurz J (1985) Coll & Polymer Sci 263:3
5. Jacobasch H-J, Bauböck G, Schurz J (1986) Monatsh Chem 117:1133
6. Jacobasch H-J, Börner M (1983) Acta Polymerica 34:374
7. Happel J (1959) AIChE J 5:174
8. Ciriacks JA (1967) Ph D Thesis, Lawrence Univ
9. Biefer G, Mason S (1959) Trans Faraday Soc 55:1234
10. Chang M, Robertson A (1959) Can J Chem Engng 45:66
11. Fairbrother F, Mastin H (1924) J Chem Soc 125:2319
12. Zorin ZM, Lashner VI, Sidorova MP, Sobelev VD, Churaev NV (1974) Kolloid Zhur 39:1154
13. Van der Put A (1980) Diss Agricultural University Wageningen
14. Saleh N (1975) Diss Techn Univ Dresden
15. Ottewill RH, Shaw JN (1967) Kolloid Z Z Polym 218:34
16. Rendall HM, Smith AL (1978) J Chem Soc Faraday I 74:1179
17. Hunter JR (1981) Zeta-Potential in Colloid Science, Academic Press. London, New York
18. Börner M (to be published)
19. Ferse A, Paul M (1973) Z Phys Chem (Leipzig) 252:198
20. Kühn N, Jacobasch H-J, Lunkenheimer K (1986) Acta Polymerica 37:394
21. Jacobasch H-J (1984) Oberflächenchemie faserbildender Polymerer. Akademie-Verlag, Berlin
22. Schubert H (1971) Archiwum Gornictwa 16:157

Received February 26, 1988;
accepted June 5, 1988

Authors' address:

H.-J. Jacobasch
Institut für Technologie der Polymere
Akademie der Wissenschaften der DDR
Dresden/DDR

Progress in Colloid & Polymer Science

Progr Colloid & Polymer Sci 77:49−54 (1988)

Measuring the zetapotential of fibers, films and granulates

V. Ribitsch, Ch. Jorde, J. Schurz, and H. J. Jacobasch[1]

Institut für Physikalische Chemie, Universität Graz, Austria, Institut für Technologie der Polymere der Akademie der Wissenschaften der DDR, Dresden, GDR[1])

Abstract: An instrument to measure the zetapotential of solids as fibers, films, plates, and granulates with the streaming current potential method is described. The quantities U and I are continuously monitored as a function of the linearly increasing driving pressure. Specially designed cells guarantee the exact determination of the geometric parameters of the probes under investigation. The data recording and evaluation is performed by a personal computer.

Key words: Zetapotential, streaming potential, films, fibers, plates, granulates

1. Introduction

Electrokinetic surface properties of materials are important for a number of industrial processes as well as for biological and medical tasks, i.e., the performance of an artificial vessel (Table 1). These properties are generated by the electrochemical double layer which exists at phase boundaries between solids and electrolyte solutions. Results correlating with important material properties, shown in Table 1, were obtained with laboratory-built equipment. But they are often not comparable due for different assumptions.

We therefore developed an instrument to measure the streaming potential, the streaming current, and to determine the zetapotential of fibers, granulates, and films (flat surfaces) according to two accepted methods (Fairbrother and Mastin; Chang and Robertson). The reproducibility and comparability of the results is ensured due to the cell geometries chosen which satisfy the above mentioned equations.

2. The electrochemical double layer

Electrical charges at the boundary between the surface of solids and electrolyte solutions are caused by dissociation of suitable chemical groups or by adsorption of ions. The distribution of electrical charges at this boundary is different from that in the bulk phase. According to the Stern model [1], the charges at the surface of the solid are compensated by counter ions forming two different layers. The layer close to the surface is immobile while the other one allows diffusion of counter ions due to the thermal motion (Fig. 1). A number of papers describe the details of the electrochemical double layer [2−5].

Unfortunately the basic values, interface potential ψ_0, and interface charge density σ_0 cannot be measured. However, the electrokinetic or zetapotential ζ can be determined by electrokinetic measurement. The quantity ζ is defined as the potential at the slipping plane where the counter ions are removed from the

Table 1. Characterization of materials by electrokinetic measurements

Field of application	Object to be investigated
Characterization of wood	Wood origin, pulping process, bleaching
Paper making industry	Relation between zetapotential and flocculation, retention, additive uptake, sizing, stability
Fiber processing	Characterization of antistatic agents, lubrication, uptake of processing additives
Textile finishing	Uptake of additives
Dyeing of textiles	Adsorption of dyestuff and additives
Polymer films	Characterization of chemical groups
Filter industry	Performance of different filters
Offset printing	Characterization of printing plates
Medicine	Investigation of artificial implantation material

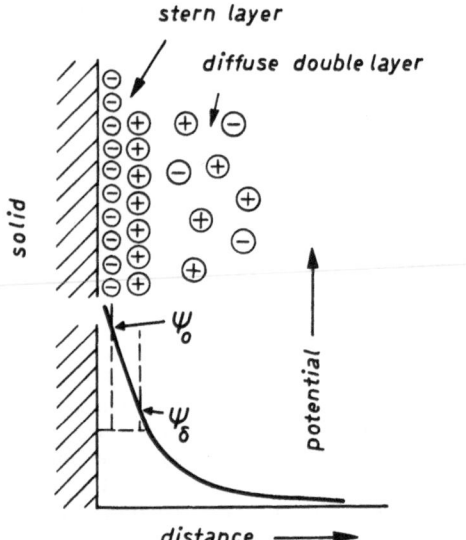

stern layer

diffuse double layer

solid

potential

distance ⟶

ψ_0

ψ_δ

Fig. 1. The electrochemical double layer according to the Stern model

surface by streaming forces. This is caused by the relative movement of solid and liquid toward each other shearing off a part of the counter ions causing merely a partial compensation of surface charges.

The model of the electrochemical double layer is generally accepted. The zetapotential itself is not well defined and is frequently assumed to be identical with the potential at the Stern layer ψ_δ. The results obtained for the zetapotential seem to depend on the method used for its determination. However this problem may be solved by using specially designed measuring cells as discussed in this paper.

3. Methods to determine the zetapotential

The zetapotential can be determined using any one of the electrokinetic effects, i.e., streaming current/potential, electro-osmosis, electrophoresis, sedimentation potential. We focus on the streaming potential/current method where an electrolyte solution is forced by external pressure through a bundle of capillaries (plug) or a small channel built of two flat surfaces. A current or a potential resulting from the motion of ions in the diffuse layer can be measured. The equations relating mechanical and electrical forces were derived by Helmholtz [6] and Smoluchowski [7]. After extension from a single capillary to a bundle of capillaries, the streaming current-potential can be calculated by using

$$\zeta = \frac{I_s \eta L_D}{\Delta p \, \varepsilon \, \varepsilon_0 \, Q_D} \tag{1}$$

$$\zeta = \frac{U_s \eta L_D}{\Delta p \, \varepsilon \, \varepsilon_0 \, R \, Q_D} \tag{1a}$$

I_s = streaming current
U_s = streaming potential
ζ = zetapotential
ε_0 = influence constant
ε = relative dielectric constant
η = viscosity
Δp = pressure difference
R = electrical resistance
L_D = length of the fiber bundle
Q_D = cross-sectional area of the bundle

The first term of Eq. (1) or (1a), $I/\Delta p$ or $U/\Delta p$ is determined by the experiment. We have shown in earlier papers [8], that U and I react immediately to an alteration of the pressure difference. A continuous increase of p causes a continuous increase in U or I and the quotient of $U/\Delta p$ or $I/\Delta p$ can be detected as the slope of a linear relation. The accuracy of the detection is remarkably increased if the proposed method is used instead of the I_s or U_s values obtained at a constant driving pressure. The second term of Eq. (3), $\eta/\varepsilon\varepsilon_0$ contains well-known material parameters. The third term L_D/Q_D in Eqs. (1) and (1a) describes geometrical parameters. The parameters L_D as well as Q_D cannot be measured directly, but the quotient L_D/Q_D can be determined by the different methods discussed in the following. It was therefore sought to develop measuring cells enabling an exact definition of L_D/Q_D.

3.1. Method of Fairbrother and Mastin

The ratio of L_D/Q_D is determined by measuring the electrical resistance R of the fiber plug [9]. Under the assumption that the fibers do not conduct current, electrical conductance can only occur in the electrolyte solution between the fibers. The specific electrical conductance $\varkappa$ of an electrolyte solution is given by

$$\varkappa = \frac{1}{R} \frac{L_M}{Q_M} \tag{2}$$

L_M = distance of electrodes
Q_M = cross-sectional area of the electrodes

The ratio L_M/Q_M is called the cell constant (resistance capacity) C

$$C = \frac{L_M}{Q_M} \, . \tag{3}$$

The quotient Q_D/L_D in Eq. (1) can be substituted by $\varkappa R$ using

$$\zeta = \frac{I_s \eta \varkappa R}{\Delta p \varepsilon \varepsilon_0} \, . \tag{4}$$

In general the surface conductance of the fibers must be taken into consideration. But it is assumed that the surface conductance can be neglected if the capillaries are filled with concentrated electrolyte. Accordingly the resistance of the plug has to be measured in 0.1 N KCl to avoid the surface conductance and to obtain a correct value of L_D/Q_D according to

$$\frac{L_D}{Q_L} = C_D = R_{0.1\,N\,KCl}\,\varkappa_{0.1\,N\,KCl} \, . \tag{5}$$

Practically, the electrolyte solution is exchanged for a 0.1 N KCl solution and the resistance of the plug $R_{0.1\,N\,KCl}$ is measured [10]. Finally the following are used to determine the zetapotential of fiber plugs

$$\zeta = \frac{I_s \eta R_{0.1\,N\,KCl}\,\varkappa_{0.1\,N\,KCl}}{\Delta p \varepsilon \varepsilon_0} \tag{6}$$

$$\zeta = \frac{U_s \eta R_{0.1\,N\,KCl}\,\varkappa_{0.1\,N\,KCl}}{\Delta p \varepsilon \varepsilon_0 R} \tag{6a}$$

3.2. Capillary bundle model by Chang and Robertson

The cell constant C_D can also be obtained using the geometrical considerations of Goring and Mason [11]. They introduced a porosity factor depending on the packing density

$$C_D = \frac{L_D}{Q_L}\,\frac{1}{(1-ad)\cos^2 \vartheta} \tag{7}$$

$$\pi = (1-ad)\cos^2 \vartheta \tag{7a}$$

ϑ = average angle
d = packing density
a = specific volume of swollen fiber
π = porosity factor

and the zetapotential can be expressed as

$$\zeta = \frac{I_s \eta L_D}{\Delta p \varepsilon \varepsilon_0 Q_D}\,\pi \, . \tag{8}$$

Chang and Robertson [12] developed and confirmed experimentally an exponential relation between the packing density and the porosity factor

$$\pi = e^{Bd} \tag{9}$$

B = constant = 2.5.

This approach proved to be more accurate even for swelling fibers. The zetapotential can be written

$$\ln \zeta - Bd = \ln \frac{I_s \eta L_D}{\Delta p \varepsilon \varepsilon_0 Q_D} \tag{10}$$

$$\ln \zeta - Bd = \ln \frac{U_s \eta L_D}{\Delta p \varepsilon \varepsilon_0 R Q_D} \, . \tag{10a}$$

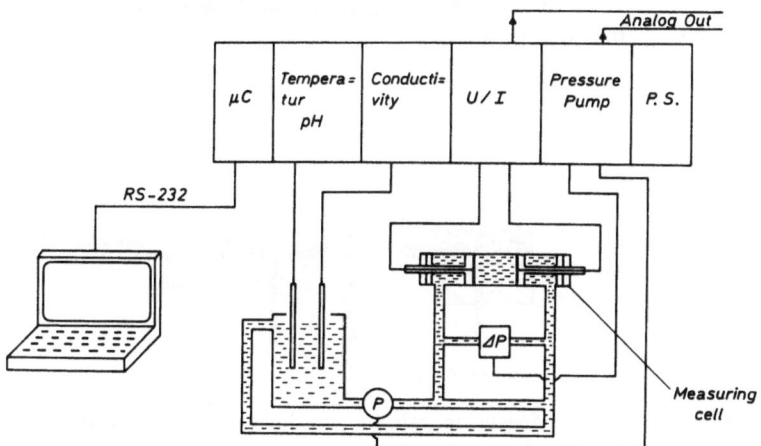

Fig. 2. Schematic of the electrokinetic analyser

The evaluation of this exponential relation is obtained by altering the packing density of the plug; ζ is calculated from the intercept of Eq. (10) or (10a).

4. Electrokinetic analyser

Based on the earlier discussed requirements a modular system was developed to determine the zetapotential of fibers, films, plates, and granulates according to Eqs. (6) and (10).

The electrokinetic analyser (EKA) is shown schematically in Fig. 2. It consists of a measuring cell, electrolyte solution pump to create the flow, pressure difference measuring system, U/I detector system, conductivity meter, pH measuring system, and a temperature sensor. The mechanical unit consists of a device to produce and measure the driving pressure and to attach different kinds of cells. Temperature, pH, and conductivity sensors are also included. The electronic unit contains the boards to measure streaming potential/current, conductivity, pH, and temperature.

The plug resistance included in Eq. (9) can be measured using an AC bridge at three different frequencies (500, 1000, 2000 Hz) or by a DC method described by Schausberger and Schurz [8]. They measured the DC resistance by means of the quotient of streaming potential over streaming current

$$R = \frac{U_s/\Delta p}{I_s/\Delta p} \; . \tag{11}$$

The instrument is under the control of a personal computer. The linear increase of the driving pressure as well as the data evaluation is performed by the computer program. The experimentator can choose between the different methods to measure the resistance as well as the two methods of data evaluation (Fairbrother and Mastin or Chang and Robertson).

4.1. Measuring cells

4.1.1. Fiber cell

The fiber cell (Fig. 3) consists of two perforated sliding electrodes. Ag/AgCl electrodes meet the requirement of low asymmetry potential best. The fibers are placed between these electrodes and the electrolyte solution is forced through the plug. The pressure difference is increased continuously and the quotient $U/\Delta p$ or $I/\Delta p$ is permanently recorded

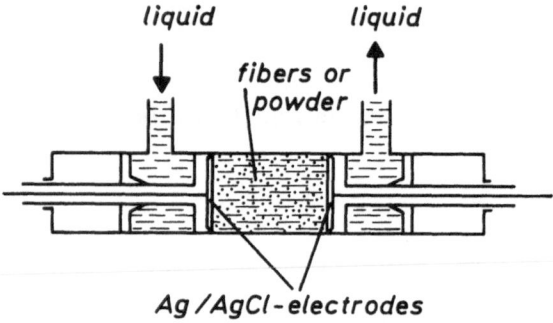

Fig. 3. Measuring cell for the investigation of fibers and granulates

The data evaluation can be done using either the approach of Fairbrother and Mastin, (Eq. (6) or (6a)) or using different packing densities as shown by Chang and Robertson. Using this method the distance between the electrodes must be decreased in order to squeeze the plug. The slope of $U/\Delta p$ or $I/\Delta p$ at each packing density is recorded. The zetapotential is obtained from Eq. (10) or (10a).

Granulates can also be packed into special adapters of the fiber cell to detect the zetapotential according to Eq. (6).

4.1.2. Cells for films and plates

Parallel plates and films can be investigated in a cell (Fig. 4) creating a channel of well defined distance between two sample surfaces. The distance is adjusted with the help of Teflon films. The zetapotential is calculated according to Eq. (6) or (6a). The surface conductivity is determined using different electrode distances.

Irregularly formed probes can be placed in sealing material (i.e., wax). The distance between both surfaces is also adjusted with teflon films. Biological material such as skin and vessel walls were measured by this technique.

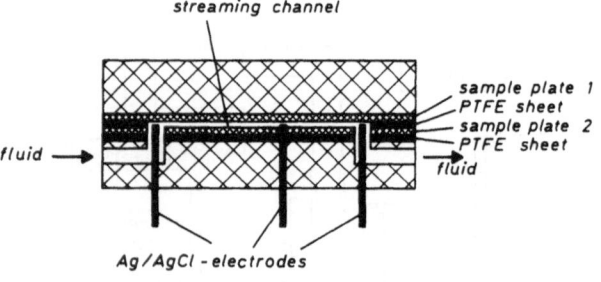

Fig. 4. Measuring cell for the investigation of films and plates

5. Results and discussion

5.1. Characterization of surfaces

If dissociable groups are present at the surface the following can be assumed: 1) alterations of ζ as a function of pH are due to dissociation of surface groups; and 2) the plateau of the ζ pH curve is due to complete dissociation of functional groups. According to this assumptions the *pK* of dissociable groups can be determined by investigations of pH dependence [13].

The ζ pH curve of cellulose fibers in 10^{-4} N KCl is shown as an example in Fig. 5. This figure also shows the influence of high energy plasma pulses on the surface characteristics of the fibers. The potential is increased due to the alteration of dissociable groups at the surface. The point apart from the curve shows the zetapotential obtained with "white water", that means the sieve water from a paper machine. This remarkable difference demonstrates the strong dependence of the zetapotential on the ionic strength and polyelectrolytes probably present in the white water.

Figure 6 demonstrates the pH dependence of the zetapotential of Al_2O_3 plates. The surface treatment caused two different qualities of the surface, a metallic one and a dim one. The zetapotential shows the expected behavior. Total dissociation is obtained at

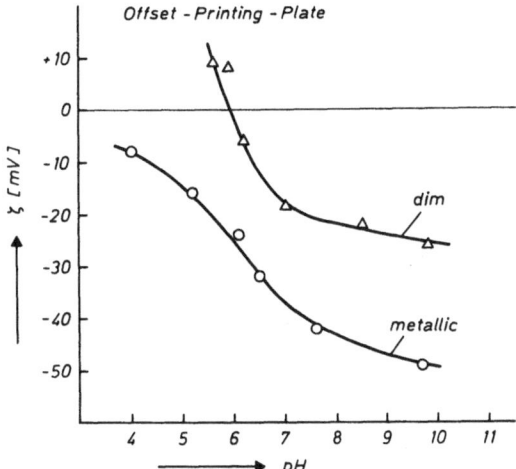

Fig. 6. pH dependence of Al_2O_3 plates

pH > 7 in the case of the dim surface, and at pH > 9 in the case of the metallic side.

5.2. Interaction between solid and solvent

The influence of interactions of the solid with components of the solvent (adsorption, desorption) depends on the sign of charge of both components. Surfactants of the same charge cause an increase of the zetapotential whereas surfactants of an opposite charge decrease the zetapotential, finally leading to a charge reversal at sufficient concentration. This phenomenon is shown by an example of cetidyl pyridinium chloride adsorbed on PAN fibers (Fig. 7). The increasing amount of surfactant decreases the charge and finally reverses the sign as discussed above.

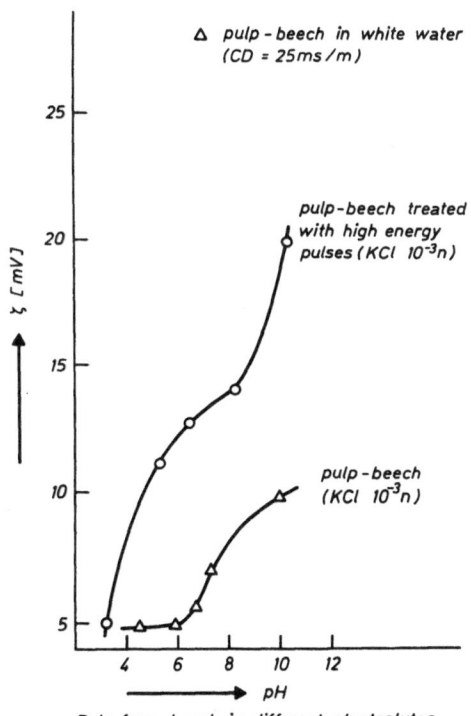

Fig. 5. pH dependence of cellulose fibers in 0.001 N KCl, untreated and treated with high energy pulses

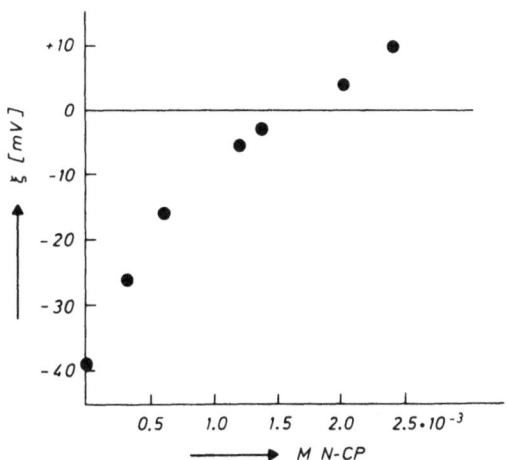

Fig. 7. Zetapotential of polyacrylonitrile fibers as a function of the adsorption of cetydil pyridinium chloride

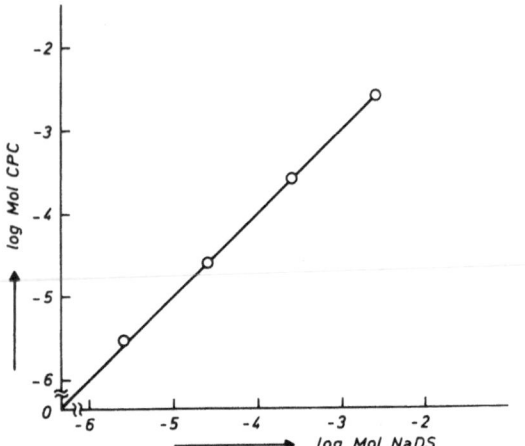

Fig. 8. Titration of sodium dodecylsulfate with cetydil pyridinium chloride using the zetapotential detected with the cell for films as indicator

The amount of charged groups at the surface can be calculated from the sign reversal concentration of the surfactant.

5.3. Polyelectrolyte titration

The zetapotential may also be used as an indicator for a polyelectrolyte titration. It is well known that anionic surfactants form neutral complexes with cationic surfactants. A small surplus of one of the surfactants causes a charge reversal. The error is therefore very small if this method is used as a detector. A polyelectrolyte titration using the EKA and the cell geometry for films can, for example, be performed with a hydrophobic polymer as indicator (Fig. 8).

Acknowledgements

This work was supported by the Austrian FFF, grant 6/427, which is gratefully acknowledged.

References

1. Stern O (1924) Z Elektrochemie 30:508
2. Sennet P, Olivier JP (1967) in: Chemistry and Physics of Interfaces. Amer Chem Soc Publ, New York, Vol 7
3. Duchin SS (1974) In: Matijevic E (ed) Surface and Colloid Science. Academic Press, New York
4. Hunter RJ (1981) Zeta Potential in Colloid Science. Academic Press, New York
5. Bockris JOM, Reddy AKN (1977) Modern Electrochemistry. Plenum Press, New York, Vol 2
6. v Helmholtz H (1879) Wied Ann 47:46
7. v Smoluchowski M (1903) Bull Intern acad Sci Cracovie 184
8. Schausberger A, Schurz J (1979) Angew Makromol Chem 80:1
9. Fairbrother F, Mastin H (1924) J Chem Soc 75:2318
10. v Stackelberg M, Kling W, Benzel W, Wilke F (1954) Kolloid Z 135:67
11. Goring D, Mason S (1950) Can J Res, Sect B 28:307
12. Chang M, Robertson A (1967) Can J Chem Eng 45:66
13. Börner M, Jacobasch H-J (1985) Elektrokinetische Erscheinungen '85, Tagungsband des Inst. f. Technologie der Polymere, Dresden

Received April 19, 1988;
accepted June 8, 1988

Authors' address:

Dr. Volker Ribitsch
Inst. f. Physikalische Chemie
Universität Graz
Heinrichstraße 28
A-8010 Graz

Progress in Colloid & Polymer Science   Progr Colloid & Polymer Sci 77:55–61 (1988)

Surfactant Aggregation

Relationship between solubility and micellization of surfactants: The temperature range of micellization

Y. Moroi

Department of Chemistry, Faculty of Science, Kyushu University, Fukuoka 812, Japan

Abstract: Many technical terms relating the solubility of surfactants with their aggregation as micelles are reviewed in order to derive a consistent concept of this relation, and the reason why the micelle temperature range (MTR) can clarify the relation between solubility and micellization of surfactants better than the Krafft point is given. It is also made clear that the temperature range for the MTR is based on the concentration range of cmc, which in turn results from the polydispersity of micelles. From the concept of MTR, there exist in principle two ways to decrease the MTR: one is to increase monomeric solubility of surfactants while leaving cmc unchanged and the other is to decrease cmc values of surfactants while leaving monomeric solubility unchanged. These are examined by the Gibbs' phase rule for surfactant solutions and by using many experimental results.

Key words: Dissolution of surfactants, micellization, micelle temperature range (MTR), Krafft point, cmc, phase rule of surfactant solution

Introduction

Detergent action, colloid formation, and surface activity are different manifestations of the same characteristics of surfactant solutions. The detergency is the most important and conventional function of soaps and is closely connected with their solubility. Usual surfactants, including soaps of course, are composed of a hydrophilic group and a bulky hydrophobic group in the same molecule and, therefore, their aqueous solubility is not expected to be high. This is the case where temperatures are below a certain temperature called the Krafft point. However, surfactant solubility increases very steeply above some narrow temperature range. This is because aggregation of monomeric surfactant molecules takes place.

In the previous paper [1], dissolution and micellization of surfactants in aqueous solutions were discussed from the viewpoint of degrees of freedom based on the phase rule, and the conclusion was reached that the mass action model for micelle formation can elucidate the above phenomena of surfactant solutions better than the phase separation model. Similarly, two concepts on the Krafft point have been reported, one a phase transition at the Krafft point and the other a solubility increase up to cmc for micellization

at the Krafft point. The recent example of the former concept is a melting-point model of a hydrated surfactant solid [2]. As for the latter, the most direct approach to the Krafft point rests absolutely on the solubility and cmc measurements of surfactants with temperature. From these measurements the concept of Krafft point can be made clear. In this paper then, the technical terms and concepts which have been presented in order to relate dissolution of surfactants with their micellization are reviewed. In addition, it is also shown that the term "Micelle Temperature Range" (MTR) can elucidate various phenomena concerning dissolution and micellization of surfactants, not only from a view point of the phase rule but also experimentally.

Krafft point and relational technical terms

The paper entitled, "On the Behavior of Fatty Acid Alkalies and Soaps in the Presence of Water", by Krafft and Wiglow [3] appeared in 1895. The authors worked with soap solutions and used the German term "Ausscheidungstemperatur" for the temperature at which a new phase separates from the soap solutions upon cooling. They found that the temperature was lower than the melting point of the corresponding fatty acid and that the difference between the separation temperature and the melting point increased with decreasing alkyl chain length of the fatty acids for the six fatty acids examined. The term "Krafft point" was not used. In 1926, McBain and Elford studied the potassium oleate-water system and drew a phase diagram where they recorded the minimum temperatures at which heterogeneous systems became homogeneous isotropic liquids [4]. It was found that these homogeneous solutions separated at the same temperature upon cooling. The term "Krafft point" first appeared in the paper by Lawrence in 1935, in which he interpreted the Krafft point as a phase transition [5]. He wrote: "The Krafft point is then due to loosening of the attractive forces between hydrocarbon chains throughout the micelle. This is a phase change of their adhesion; from solid to liquid. The degree of dispersion is thereby greatly increased".

In the same year Murray and Hartley described a rapid solubility increase over a narrow temperature range by using the mass action equation [6]. And was not long before another new concept of Krafft point emerged. In 1951, Eggenberger and Harwood performed conductometric studies on solubility and micelle formation of dodecylammonium chloride and found a point on the solubility curve corresponding to a sharp break in the solubility vs temperature curve which could be attributed to the "Krafft effect" or solubilization of the undissociated molecule by the micelle. Significantly, the cmc vs temperature curve intersected the precipitation curve at this point [7]. The very sharp break in the precipitation curve at the point was interpreted as indicating no appreciable formation of micelles below cmc. Thereafter the view of Krafft point was established as in the paper by Phillips, where Krafft point was defined as the temperature at which the cmc is equal to the saturation solubility [8]. In "Colloid Science" by Alexander and Johnson [9], the "Krafft phenomenon" was interpreted as an unusual property where solubility of soaps increased enormously over a small temperature range (Krafft point); they said, "from this point onwards the transfer (of soap molecules) is effectively from the solid phase to the micelle, since the concentration of single molecules only increases quite slowly once micelles are present". This concept of the Krafft point is totally correct.

To the contrary, Shinoda and Hutchinson presented a new interpretation that the Krafft point represented the freezing of the micelle or the melting of hydrated solid of surfactants [2]: "the Krafft point now can be interpreted as a point at which solid hydrated agent and micelles are in equilibrium with monomers: in terms of phase with two components the equilibrium hydrated solid ⇌ monomers ⇌ micelles is univariant, so that at a given pressure the point is fixed." Unfortunately this concept is incorrect as is made clear in the following sections. Another technical term "critical solution temperature" (cst) appeared to designate the inflection temperature from which the solubility of nonionic surfactants in organic solvents increased markedly showing the inflection in a solubility curve [10]. Furthermore, Mazer et al. used the term "critical micellar temperature" (cmt) to refer to the boundary between the hydrated solid phase and micellar phase [11]. The cmt values were then determined as the midpoint of a narrow temperature range over which the hydrated solid phase clarified after slow warming and with continuous vigorous shaking.

As was seen in the preceding review, the concept of the Krafft point can be divided into two parts, a phase transition of the solid surfactant and a solubility increase up to cmc for micellization. The Krafft point becomes a definite temperature for the former concept and a small temperature range for the latter. The existence of two different views of the same phenomena is quite inconsistent. Therefore, the present conflict will be resolved in the following section.

Physico-chemical meaning of MTR

The Krafft point has been defined as the temperature at which the solubility vs. temperature curve intersects the cmc vs. temperature curve. Let us think about what the above definition means by using Fig. 1 [1]. If the micelle is regarded as a phase, three phases (intermicellar bulk phase, surfactant solid phase, and micellar phase) coexist at the Krafft point, and the Gibbs' phase rule ($f = C - P + 2$) gives only one degree of freedom since the number of components is two (water and surfactant), where f, C, and P are the number of degrees of freedom, component, and phase. Thus, when a pressure is specified the Krafft point T_k is inevitably determined. From the figure this fact may seem to apply well and to be consistent with the phase rule, but the same three phases still coexist at every point − point P for example of the solubility curve above the Krafft point. In other words, there exist innumerable temperatures on the curve at which the three phases are in equilibrium. This is quite contrary to the conclusion derived from the phase-separation model: at one atm pressure, many equilibrium temperatures exist instead of one temperature; so we can conclude that the phase separation model cannot be applied to usual micelles whose aggregation number is

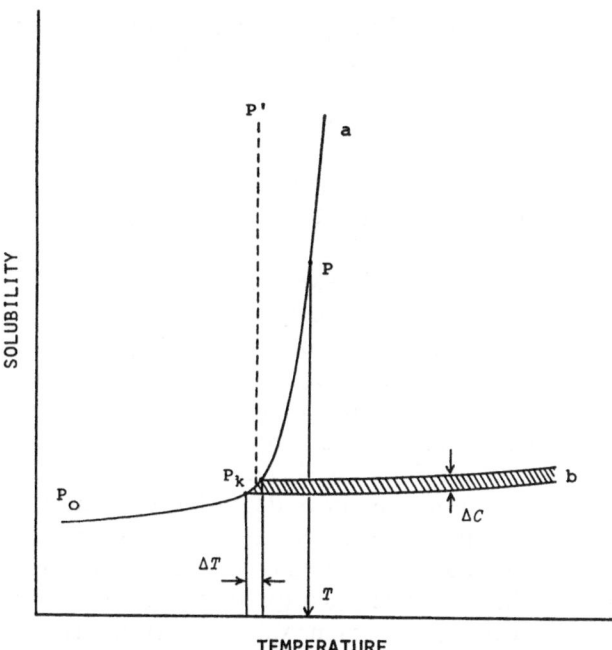

Fig. 1. Changes of solubility and cmc with temperature: (a) solubility curve, (b) cmc curve, ΔC narrow concentration range of cmc, ΔT micelle temperature range

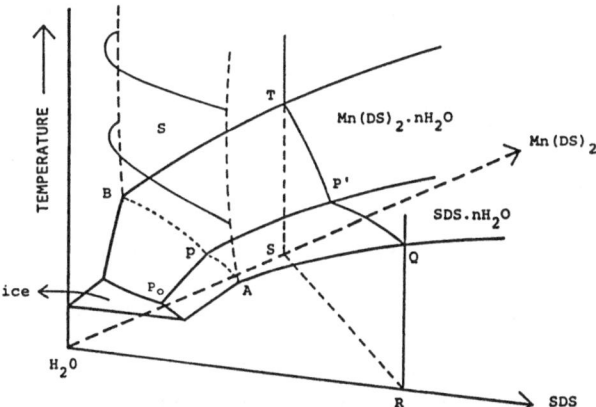

Fig. 2. Phase diagram of water-sodium dodecylsulfate-manganese (II) dodecylsulfate system: S mixed cmc surface

less than or near a few hundreds. However, if the aggregation number were infinite, a phase separation would take place and the solubility curve would approach the path $P_o \rightarrow P_k \rightarrow P'$ with increasing temperature, and the monomer concentration remains constant at S_{T_k}. In this case, one temperature T_k is specified at one atm pressure.

But when the mass action model is applied to the micellization, the following association equilibrium between surfactant monomers (S) and micelles (M) can be considered:

$$K_n = [M]/[S]^n \qquad (1)$$

where K_n is the equilibrium constant of the micellization and n is the aggregation number of the micelles. In this case, the number of degrees of freedom $(f = C - P + 2 - r)$ is two because of the two phases (surfactant solution phase and surfactant solid phase), three components (water, monomeric surfactant, and micelle), and one equilibrium equation where r is the number of equilibrium equations. Then, two intensive thermodynamic variables, temperature and pressure for example, can specify the system of the solution. This fact agrees totally with the experimental evidence that the solubility is determined only by temperature at atmospheric pressure. This simplified expression,

Eq. (1), may not be adequate to express the general micelle formation of ionic surfactants because the counterions and the size distribution of the micelles are not taken into account. However, Eq. (1) is satisfactory for the degrees of freedom, as was discussed previously [1].

Another consideration is the melting point model of hydrated solid surfactant [2, 12–15]. As for the model, the solubility curve below the Krafft point is for the hydrated surfactant solid and that above it is for the melted surfactant phase. If the micelle is regarded as a phase, the system is invariant $(f = 0)$ at the Krafft point because of four coexisting phases: intermicellar bulk, hydrated surfactant solid, melted surfactant, and micellar for two components. This contradicts the experimental evidence that cmc changes with pressure [16, 17]. Even if the micelle is regarded as a chemical species, the melting point model is not correct. In this case the system is monovariant $(f = 1)$ and the Krafft point is determined automatically at one atm pressure, which seems reasonable for a single surfactant solution. However, when the model is applied to mixed surfactant solution, it turns out to be incorrect.

The phase diagram of the water-sodium dodecylsulfate (SDS)-manganese(II) dodecylsulfate (Mn(DS)$_2$) system is shown in Fig. 2 with temperature as the ordinate [18]. The cmc of the surfactant mixture gives a curved surface between the cmc-temperature curves of each component. We can conclude from the three-component phase diagram in analogy with the two-component phase diagram that the rational Krafft point for a binary surfactant mixture is determined by the intersection of mixed cmc surface and both hump surfaces of the surfactant solids, A$\cdots$P$\cdots$B.

Now let us consider point P in Fig. 2. If the melting point model is applied to the phase diagram, there exist at the point five phases: micellar solution, SDS solid, melted SDS, $Mn(DS)_2$ solid, melted $Mn(DS)_2$. The number of degrees of freedom becomes zero, which means that such a phase diagram as Fig. 2 can be drawn only for one atm pressure. It is inconceivable, judging from many experimental facts that solubility and cmc of surfactants have been measured at pressure from one up to several thousand atmospheres [17, 19]. The shape $QRSTP'$ which is made by the intersection between both hump surfaces of two surfactant solids and a plane of constant total concentration far above cmc is very similar to a phase diagram of two components with a eutectic point. The shape is similar to that of melting two hydrated surfactant solids, and also to a thermodynamic analysis of the Krafft point of a binary surfactant mixture whose view is based on a freezing point depression of a binary mixture [20]. However, the model gives an absurd conclusion, neglecting the presence of the third component: water, which is an essential component in the discussion of micelle formation. In addition, if a micelle is regarded as a phase, the number of degrees of freedom becomes -1, which is also absurd. On the other hand, if a micelle is regarded as a chemical species and no melting of the surfactant solids is assumed, two degrees of freedom still remain along the line of $P_o \rightarrow P \rightarrow P'$ due to four components, three phases, and one equilibrium equation. This means that temperature specifies the binary surfactant system at a definite pressure as far as two surfactant solid phases coexist in the system, which is consistent with experimental evidence. Bivalent metal dodecylsulfates or sulfonates serve as typical examples of hydrated surfactant solids, $Cu(DS)_2 \cdot 4H_2O$ and $Cu(DSo)_2 \cdot 2H_2O$, for example. The Krafft points are $19.0 °C$ for the former [18] and $53.5 °C$ for the latter [21], although their phase transition temperatures are 44 and $66 °C$, respectively [12]. These two examples show the difference in temperature between the Krafft point and phase transition and are enough to indicate the misconception of the melting point model, although there is the case where the phase transition temperature of a hydrated surfactant solid happens to be very near to the Krafft point of the surfactant [22, 23]. Accordingly, it can be concluded that the Krafft point is the temperature at which the solubility of surfactants as monomers becomes high enough for the monomers to start to aggregate or micellize. An important point is that cmc depends on the method used for cmc determination and the cmc value should be defined as a narrow temperature range [24], although

the solubility is definitely determined only by a temperature under a constant pressure. This fact results from the polydispersity of micelles [25]. As a result, the MTR must be defined as the narrow temperature range where a solubility vs. temperature curve intersects a cmc range vs. temperature curve, although one temperature for MTR can be obtained from one cmc determination. In other words, the MTR is determined by the balance between cmc and solubility and by their dependence on temperature [21, 25].

Now that the concept of the micelle temperature range has been made clear, it is appropriate to make a few remarks about how to shift the MTR to lower temperatures because this is of considerable practical importance for ionic surfactants. It can be easily understood from the above definition that there are two ways in principle to decrease the MTR: one is to increase monomeric solubility of surfactants while leaving the cmc unchanged (process a in Fig. 3). The other way is to decrease cmc value of surfactants while leaving monomeric solubility unchanged (process b in Fig. 3) [1]. The former, based on the solubility increase, is closely related to the crystalline state of surfactant. The less energetically stable the solid surfactant, the higher its solubility becomes. The diminished stability could be brought about by dispersion of the electrical charge of the counterion, volume increase of the counterion, an increase in water content of crystallization, the introduction of a branched chain into a hydrophobic surfactant chain, etc. On the other hand, the latter method (due to cmc decrease) might be ex-

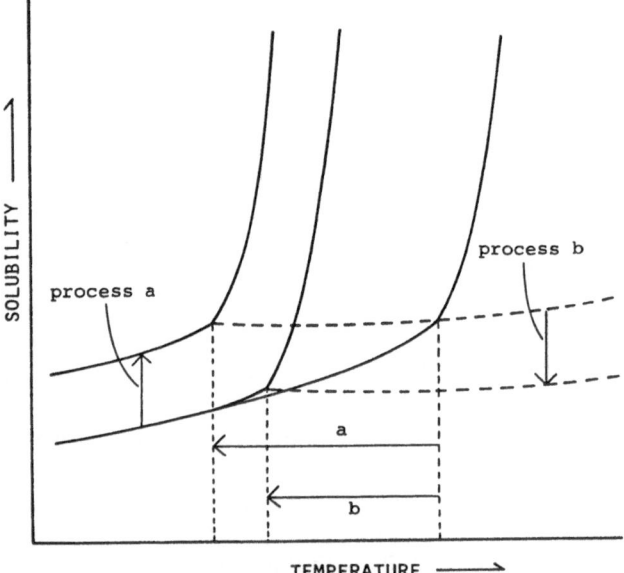

Fig. 3. Schematic diagram to shift MTR: process a increase of monomer solubility, process b decrease of cmc value

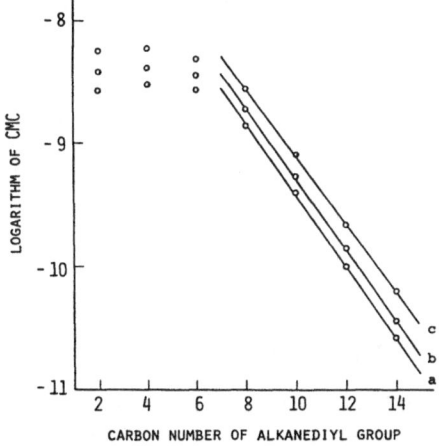

Fig. 4. Logarithm of cmc plotted against alkanediyl chain length at (a) 35 °C, (b) 45 °C and (c) 55 °C

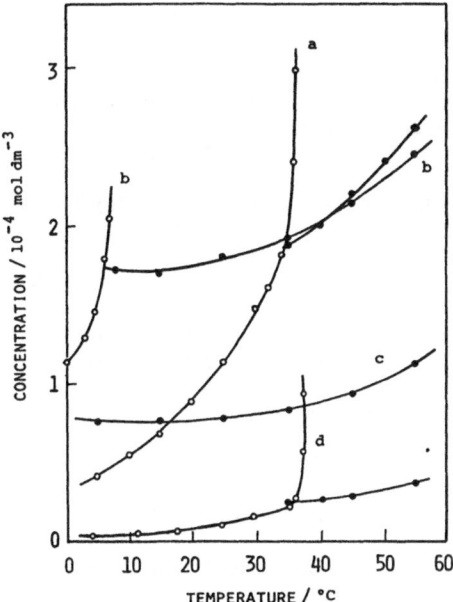

Fig. 5. Changes in solubility ($\bigcirc$) and cmc ($\bullet$) with temperature (a) $C_2BP(C_{14})_2$, (b) $C_6BP(C_{14})_2$, (c) $C_{10}BP(C_{14})_2$, and (d) $C_{14}BP(C_{14})_2$

Table 1. Micelle temperature range (MTR) and heat of dissolution (Δh^0) of 1,1′-[1,ω-alkanediyl]bispyridinium tetradecane-1-sulfonates

Surfactants	MTR (°C)	Δh^0 (kJ mole⁻¹)
$C_2BP(C_{14})_2$	34.6	106
$C_4BP(C_{14})_2 \cdot 2H_2O$	11.4	98
$C_6BP(C_{14})_2$	6.0	80
$C_8BP(C_{14})_2$	<0	–
$C_{10}BP(C_{14})_2$	<0	–
$C_{12}BP(C_{14})_2$	2.1~2.3	–
$C_{14}BP(C_{14})_2$	35.7	133

pected from an increase in the association of counterion to micelle, the increase in hydrophobicity of counterion and surfactant ion, etc. If the above discussion is taken into account the MTR change obtained from the aqueous solubility and cmc of tetradecane-1-sulfonates ($C_nBP(C_{14})$)$_2$) whose cationic counterion is a divalent one with separated electric charges, 1,1′-(1,ω-alkanediyl)bispyridinium ion,

$$\bigcirc N^+ - (CH_2)_n - {}^+N\bigcirc \; ; \quad n = 2, 4, 6, 8, 10, 12, 14$$

is easily understood (Figs. 4, 5 and Table 1) [26, 27]. The decrease in the MTR from $C_2BP(C_{14})_2$ to $C_4BP(C_{14})_2$ and $C_6BP(C_{14})_2$ evidently results from process *a* because their solubilities increase stepwise while their cmc values remain almost constant. Further decreases in the MTR of $C_8BP(C_{14})_2$ and $C_{10}BP(C_{14})_2$ are due to both solubility increases and cmc decreases (processes *a* and *b*). The increase of the MTR through the minimum for $C_{12}BP(C_{14})_2$ and $C_{14}BP(C_{14})_2$ clearly results from a pronounced stability of the solid surfactants, i.e., their low solubility. The decrease in the cmc of the two surfactants contributes to the decrease in the MTR, of course. However, the effect of a solubility decrease on the MTR increase is found to be much larger than that of a cmc decrease on the MTR decrease in this case, judging from the fact that the MTR depends absolutely on the balance between the monomeric solubility and the cmc of the surfactants. The following general conclusions can be drawn from the above cmc, solubility, MTR, and from Δh^0 values: (1) As long as their charge separation is small, the divalent counterions with separated electric charges easily move about the charged micellar surface, just like divalent concentrated (Cu^{2+}) or diffused charge (methylviologen, MV^{2+}) counterions [21, 28]. The divalent charges whose separation is more than six methylene groups become anchored to the micellar surface, and this leads to a decrease in the cmc. And, (2) the surfactant solids with a divalent counterion of separated electric charges become less stable, and their solubility increases as the charge separation increases. Beyond a certain charge separation, the surfactants become more stable and solubility decreases.

Pressure is another thermodynamic parameter that affects the MTR because the monomeric solubility and the cmc are both dependent on a pressure. Nevertheless, there are only a few reports on the pressure effect [29–31]. A cmc change with pressure takes a slightly elevated maximum and goes down very slowly, where the change is less pressure-dependent compared with a solubility change with pressure. The solubility,

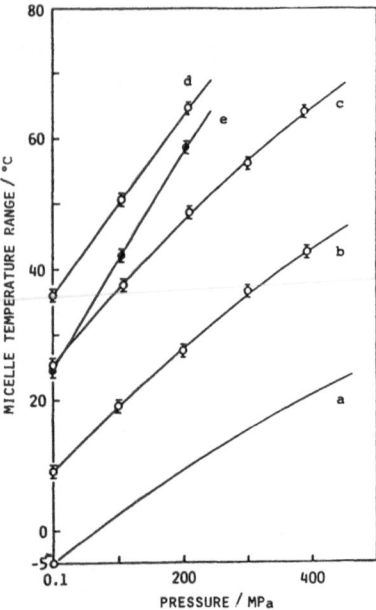

Fig. 6. MTR changes with pressure for typical ionic surfactants [30]: (a) sodium decylsulfate, (b) sodium dodecylsulfate, (c) sodium tetradecylsulfate, (d) sodium hexadecylsulfate, (e) hexadecyltrimethylammonium bromide

which is more pressure dependent, decreases more rapidly with increasing pressure due to a positive volume change on dissolution. Hence, the difference between the cmc and the solubility at a definite temperature increases with pressure. This leads to a higher MTR at high pressure so that the solubility increases up to the cmc. This is why MTR increases with pressure (Fig. 6).

Much practical effort to depress the MTR of ionic surfactants has been made. Examples are the introduction of oxyethylene groups [14, 32], or oxypropylene groups [33] between the hydrocarbon chain and the ionic head group. The decrease in the MTRs is brought about by the cmc decrease rather than by the solubility increase of the surfactants. Another interest has been the MTR of fluorinated surfactants. Their MTRs are much higher than those of the corresponding hydrocarbon surfactants. This is due largely to their low aqueous solubilities [13, 34]. The solvent effects on the solubility and the cmc is the other area of interest. When formamide is used as a substitute for water, the cmc of SDS and hexadecyltrimethylammonium bromide increase greatly, resulting in a higher MTR than in water: 55 °C compared with 43 °C [35]. The review article on the Krafft point by Sowada is very informative [36].

The conclusions reached from the above discussions are: (1) the micelle temperature range (MTR) is a bet-

ter technical term than the Krafft point to express the relation between solubility and micelle formation; (2) MTR is determined only by the balance between the cmc and the solubility and their dependence on the temperature; and (3) these can be explained perfectly by the mass action model of micelle formation [25].

References

1. Moroi Y, Sugii R, Matuura R (1984) J Coll Interf Sci 98:184
2. Shinoda K, Hutchinson E (1962) J Phys Chem 66:577
3. Krafft F, Wiglow H (1895) Ber Dtsch Chem Ges 28:2566
4. McBain JM, Elford WJ (1926) J Chem Soc 129:421
5. Lawrence ASC (1935) Trans Faraday Soc 31:206
6. Murray RC, Hartley GS (1935) Trans Faraday Soc 31:183
7. Eggenberger DN, Harwood HJ (1951) J Am Chem Soc 73:3353
8. Phillips JN (1955) Trans Faraday Soc 51:561
9. Alexander AE, Johnson P (1949) in: Colloid Science, pp 683–685, Oxford University Press, London
10. Kon-no K, Jin-no T, Kitahara A (1974) J Coll Interf Sci 49:383
11. Mazer AM, Benedek GB, Carey MC (1976) J Phys Chem 80:1075
12. Shinoda K (1981) J Phys Chem 85:3311
13. Nakayama H, Shinoda K, Hutchinson E (1966) J Phys Chem 70:3502
14. Hato M, Shinoda K (1973) J Phys Chem 77:378
15. Tsujii K, Saito N, Takeuchi T (1980) J Phys Chem 84:2287
16. Tuddenham RF, Alexander AE (1962) J Phys Chem 66:1839
17. Kaneshina S, Tanaka M, Tomida T, Matuura R (1974) J Coll Interf Sci 48:450
18. Moroi Y, Oyama T, Matuura R (1977) J Coll Interf Sci 60:103
19. Tanaka M, Kaneshina S, Kuramoto S, Matuura R (1975) Bull Chem Soc Japan 48:432
20. Tsujii K, Mino J (1978) J Phys Chem 82:1610
21. Moroi Y, Sugii R, Akine C, Matuura R (1985) J Coll Interf Sci 108:180
22. Kodama M, Seki S (1983) Progr Coll Polym Sci 68:158
23. Kodama M, Seki S (1984) Netsu Sokutei 11:104
24. Preston WC (1948) J Phys Chem 52:84
25. Moroi Y, Matuura R (1988) Bull Chem Soc Japan 61:333
26. Moroi Y, Matuura R, Kuwamura T, Inokuma S (1986) J Coll Interf Sci 113:225
27. Matuura R, Moroi Y, Ikeda N (1986) in: Mittal KL, Bothorel P (eds) Surfactants in Solution 4. Plenum Press, New York, pp 289–298
28. Moroi Y, Ikeda N, Matuura R (1984) J Coll Interf Sci 101:285
29. Tanaka M, Kaneshina S, Tomida T, Noda K, Aoki K (1973) J Coll Interf Sci 44:525
30. Nishikido N, Kobayashi H, Tanaka M (1982) J Phys Chem 86:3170
31. Offen HW, Turley WD (1983) J Coll Interf Sci 92:575

32. Hato M, Tahara M, Suda Y (1979) J Coll Interf Sci 72:458
33. Shinoda K, Maekawa M, Shibata Y (1986) J Phys Chem 90:1228
34. Shinoda K, Hato M, Hayashi T (1972) J Phys Chem 76:909
35. Rico I, Lattes A (1986) J Phys Chem 90:5870
36. Sowada R (1985) Chem Techn 37 Jg 11:470

Received February 2, 1988;
accepted June 6, 1988

Author's address:

Yoshikiyo Moroi
Department of Chemistry, Faculty of Science
Kyushu University
Higashi-ku, Fukuoka 812
Japan

Progress in Colloid & Polymer Science   Progr Colloid & Polymer Sci 77:62–66 (1988)

Critical micelle concentration
of some homogeneously ethoxylated nonylphenols

D. F. Anghel and M. Balcan

Central Institute of Chemical Research, Department of Physical Chemistry, Bucarest, Romania

Abstract: The surface tension and spectral change of the aromatic chromophore as functions of the concentration have been determined for homogeneously ethoxylated nonylphenols (NPE_8, NPE_{10-15}). Areas per molecule, surface tensions at the critical micelle concentration, and critical micelle concentrations are presented as functions of the ethylene oxide chain length. The areas per molecule increased with the degree of ethoxylation. The CMC values of homogeneously ethoxylated nonylphenols were lower than those of the octylphenol-based counterparts, higher than those of the ethoxylated nonylphenols with a natural distribution of the polyoxyethylene chains, and closely related to those of the homogeneously ethoxylated dodecylpolyoxyethylenemonoglycolethers.

Key words: Nonylphenols ethoxylated, critical micelle concentration, surface tension, areas per molecule

Introduction

Nonionic surfactants manufactured by the base-catalyzed ethoxylation of alkylphenols contain a large number of ethoxymers that have Poisson distribution [1–3]. The ethoxylate chain confers the hydrophilic character and the ratio of nonylphenol to ethylene oxide determines the surfactant properties. By varying the length of the ethoxylate chain, the performance can be suited to a full range of surfactant applications.

In contradistinction to polydisperse ethoxylates, homogeneously ethoxylated surfactants have a fixed polyoxyethylene chain. Their synthesis is much more laborious and yields are low [4]. A convenient alternative for obtaining these compounds is by means of preparative scale high-performance liquid chromatography (HPLC). In this respect, nonylphenol polyethoxylate oligomers (NPE's) with discrete lengths of the polyethoxy chains of 10–15 ethoxy units were obtained. Homogeneously ethoxylated nonylphenol with eight ethoxy units was prepared by direct synthesis.

The present study has been undertaken to establish the critical micelle concentration (CMC) of the above mentioned NPEs. The data were obtained by the surface tension and spectral change techniques.

Experimental

Materials

Homogeneously ethoxylated surfactants were obtained by normal-phase HPLC from a commercial product (i.e., nonylphenol ethoxylated with 15 moles of ethylene oxide). Fractions containing the ethoxymers with polyethoxy chains of 10–15 were collected and evaporated almost to dryness in a rotary evaporator under reduced pressure. The solvent was completely evaporated under a stream of nitrogen. A Hewlett-Packard model 1084B liquid chromatograph fitted with a DuPont Zorbax SIL (250×9.4 mm) semi-preparative column and UV detector set at 280 nm were used. The separations were done with the aid of the following elution system: (A) hexane-isopropanol (40:60 vol.), (B) ethanol-water (80:20 vol.). A flow rate of 3 ml/min and a linear gradient going from 10% (B) to 95% (B) in 45 min were used. The main advantage gained by using normal-phase chromatography instead of reversed-phase chromatography was that the alkyl chain isomers of the parent nonylphenol would have minimal effect on the separation of ethylene oxide units.

Nonylphenol octaoxyethyleneglycolmonoether was synthesized by the direct condensation of *p*-alkylphenol (the ortho isomer was carefully removed) with the respective monodisperse polyethyleneglycol in the presence of dicyclohexylcarbodiimide [5]. The dicyclohexylurea by-product was removed by filtration. The surfactant was purified by molecular distillation.

The compounds obtained were checked by analytical HPLC in order to determine their purity. Irrespective of the ethoxylation degree, only one peak was recorded. Confirmation of the identity was done by IR and ¹H NMR.

The solutions used in micellization studies were prepared with doubly distilled water.

Methods

The surface tension method is based on the fact that above the CMC the surface activity is almost constant and increases sharply below it. By plotting the data against the logarithm of the concentration the CMC is found at the intercept of the descending line with another one close to horizontal. The drop-volume method and a previously described apparatus [6] were used. Determinations were carried out at 25 °C in a thermostatic system. Lando and Oakley [7] correction factors were used.

In the spectral change method, micelle formation is associated with a change in the ultraviolet absorption spectrum of the aromatic chromophore [8, 9]. The CMC was determined by a modified Gratzer and Beaven method [8]. Solutions of known surfactant concentration were prepared and their UV spectra recorded at 25 °C, using a Specord M 40 spectrophotometer fitted with 1 cm cells. In the pre-micellar region, the maximum absorption was situated in the 273.8 − 275.3 nm range. The absorbances were below 0.16. After the CMC, a bathochromic effect of the chromophore associated with a considerable enhancement of the absorbance was noted. The plot of maximum absorption wavelength against concentration showed curvature at higher concentrations and gave erroneous CMC values. When absorbances at maximum of absorption were plotted versus the concentration (logarithmic scale), two distinctive straight lines joining the points below and above the CMC were obtained. The CMC was determined by their intersection.

Results

Surface tension data as a function of concentration for various NPE surfactants are presented in Fig. 1. Spectrophotometric determination of the CMC for NPE_8, NPE_{10}, NPE_{13}, and NPE_{14} is shown in Fig. 2. Figure 3 shows the surface tension at the critical micelle concentration (γ_{CMC}) and area per molecule against ethylene oxide chain length. Areas per molecule were computed from the slope of the surface tension isotherms below to the CMC. Figure 4 points out the critical micelle concentrations as a function of ethylene oxide chain length. Literature data concerning homogeneously ethoxylated octylphenols [10, 11] and dodecyl alcohols [12 − 14] and polydisperse alkylphenols with natural [10, 15, 16] and reduced distributions [17] of oxyethylene head groups were also included.

The results obtained in CMC determination are summarized in Table 1. The spectral change method gives the CMC values in good agreement with the sur-

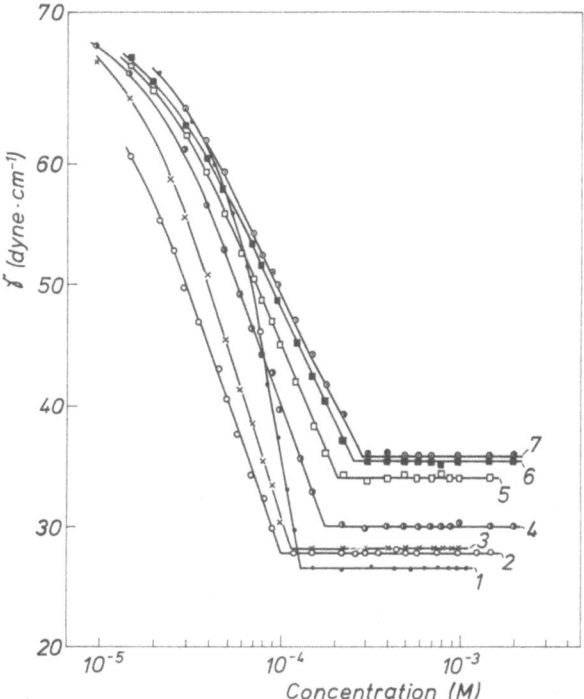

Fig. 1. Surface tension vs concentration of homogeneously ethoxylated nonylphenols (NPEs): 1) NPE_8, 2) NPE_{10}, 3) NPE_{11}, 4) NPE_{12}, 5) NPE_{13}, 6) NPE_{14}, and 7) NPE_{15}

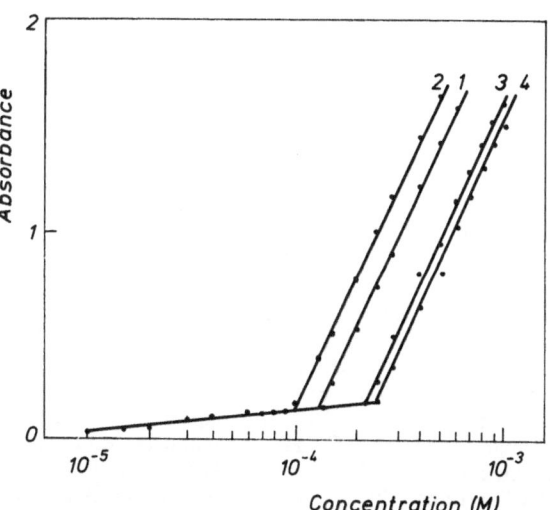

Fig. 2. The CMC determination by spectral change method. Nonionic surfactants: 1) NPE_8, 2) NPE_{10}, 3) NPE_{13}, and 4) NPE_{14}

face tension data. The method is simple and efficient. It is not time consuming because it does not require long time periods to reach equilibrium in each measurement as does surface tension.

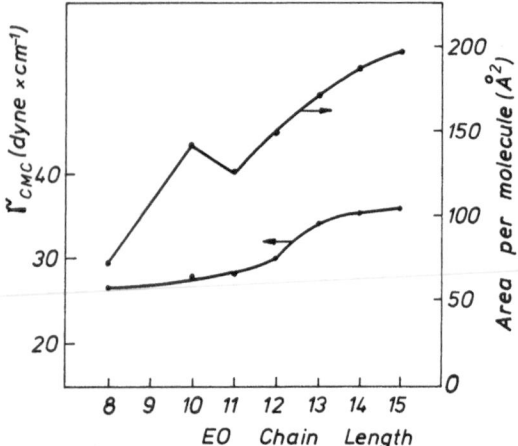

Fig. 3. Dependence of surface tension at CMC and area per molecule on ethoxylation degree of NPE surfactants

Table 1. Critical micelle concentrations of homogeneously ethoxylated nonylphenols

Compound	Critical micelle concentration (M × 10⁴)	
	Surface tension	Spectral change
NPE₈	1.32	1.26
NPE₁₀	1.00	0.98
NPE₁₁	1.15	–
NPE₁₂	1.77	–
NPE₁₃	2.04	2.18
NPE₁₄	2.57	2.40
NPE₁₅	2.82	–

The area per molecule at the air-water interface increases with the degree of ethoxylation. The starting compound of our series (i.e., NPE_8) has an area per molecule of 72 Å². This is higher than the reported values of 22 Å² per molecule for an aliphatic hydrocarbon [18] and 25 Å² per molecule for a benzene group [19], but close enough to the about 62 Å² per molecule previously found for homogeneously ethoxylated octylphenol (OPE_8) [10]. The packing of NPE molecules at the air-water interface is determined by the hydrated ethylene oxide chain and not by the hydrophobic alkylphenyl group.

The results concerning the CMC dependence on the ethoxylation degree of the homogeneously ethoxylated nonylphenols fit the frame of the literature data well (see Fig. 4). They are higher than the values for nonylphenols with natural [15, 16] or narrow [17] distributions of the polyoxyethylene chains, and very close to the homogeneously ethoxylated dodecyl alcohol [12–14]. They are also lower than the CMC values of the ethoxylated octylphenols. The higher CMC values of the homogeneously ethoxylated alkylphenols confronted by polydisperse counterparts are related to the presence of shorter ethoxymers in the normal distribution surfactants. For example, it was proved by HPLC that a commercial nonylphenol ethoxylated with 9 EO contains as many as 20 ethoxymers; one-third of them being compounds with ethoxy units shorter than nine [20]. These compounds are inherently more hydrophobic than the longer ethylene oxide chain members. Such inherently higher hydrophobia leads to the formation of micelles at lower concentrations than in the case of simple species materials where only one type of molecule is available for micellization.

The CMC data for NPEs 10–15 fit the equation of Hsiao et al. [16] very well:

$$\ln \text{CMC} = A + Bn$$

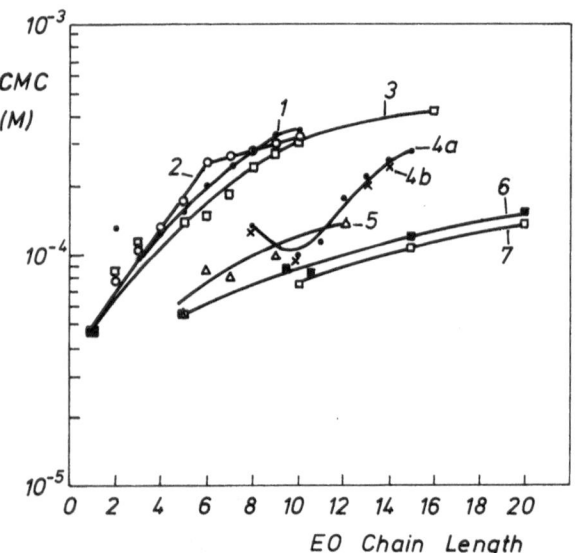

Fig. 4. CMC as a function of ethylene oxide chain length. Nonionic surfactants by reference: 1) Homogeneous OPEs [10], 2) Homogeneous OPEs [11], 3) Polydisperse OPEs [10], 4a) Homogeneous NPEs (this paper, data from γ vs lg c), 4b) Homogeneous NPEs (this paper, data from spectral change method), 5) Homogeneously ethoxylated dodecyl alcohol [12–14], 16) NPEs with reduced distribution of EO groups [17], and 7) Polydisperse NPEs [15, 16]

Discussion

The results presented above showed NPE_8 as the most surface active member of the series (see Figs. 1, 3). The surface activity increases gradually up to NPE_{12}. A greater increase in γ_{CMC} was recorded between NPE_{12} and NPE_{13}, followed by a leveling tendency (NPE_{14} and NPE_{15}).

where n is the ethylene oxide chain length of a given nonionic surfactant. The values found for A and B were -11.2308 and 0.2082, respectively. The coefficient of correlation was 0.9783. When n was substituted for R_0 (ethylene oxide mole ratio) the values of A and B changed to -16.2287 and 10.6953, respectively. The coefficient of correlation was only 0.8326, denoting a worse fit.

A striking position among the CMC values of homogeneously ethoxylated nonylphenols is hold by NPE_8. The value is higher than for NPE_{10} and NPE_{11}, denoting a more hydrophilic surfactant. The phenomenon may be assigned to the change of the polyoxyethylene chain conformation.

In aqueous solution the polyoxyethylenes are typical random-coil polymers [21, 22], and one would therefore expect that polyoxyethylene chains extending from the micelle core to be likewise randomly coiled. However, the parallel alignment of the chains in the micelle mantle may effect the preferred conformation, and three possible rigid conformations have been considered in the literature: the fully extended (zig-zag) conformation [23] with a length of 3.5 Å per monomer; the helical conformation seen in polyoxyethylene crystals [24, 25], with a length of 2.76 Å per monomer; and the meander conformation [23] with a length of $1.8-2$ Å per monomer. The presence of the helical structure has also been proven in some micelles formed by surfactants with long polyoxyethylene chains [26]. The polyoxyethylene chain in bulk is transformed from an extended to a meander conformation at a degree of polymerization of $20-40$, whereas in solution the transition occurs at a degree of polymerization of nine. The physical reason for the transformation is that with increasing length the polyoxyethylene chain acquires a large electrical moment in the coaxial direction, which contracts the main chain and transforms it into the meander conformation. However, no direct data are available on micellar systems. The existing data on the aggregation number and the hydrodynamic radius of nonionic micelles strongly suggest that even short polyoxyethylene chains do not take the fully extended conformation [23]. The interpretation of the result is not simple because it involves also a discussion of the shape of the micelle. For instance, in the case of Triton X-100 (a commercial ethoxylated octylphenol having an average of 9.5 oxyethylene units per molecule) Robson and Dennis [27] have shown that the spherical micelle is possible only if several oxyethylene groups are embedded in the hydrophobic core. Otherwise, an oblate ellipsoid of revolution is the most consistent candidate with intrinsic viscosity measurements and volume calculations.

On the other hand, unlike octylphenol, the nonylphenol based nonionic surfactants contain a considerable number of isomers derived from the parent hydrophobe. *p*-Octylphenol has only one major component (1,1,3,3-tetramethylbutylphenol) because it is synthesized from dimerized isobutylene. *p*-Nonylphenol is produced from trimerized propylene and phenol and therefore consists of a complex mixture of isomers with differently branched nonyl substituents [28], that may influence the micellar behavior of the resulting surfactants. Thus, the study of the micellar properties of the ethoxylated nonylphenols is much more complex than that of alkylpolyoxyethyleneglycolmonoethers or octylphenol-based nonionic surfactants.

To conclude, the homogeneously ethoxylated nonylphenols have lower CMC values than the octylphenol-based counterparts, but higher than the ethoxylated nonylphenols with natural or reduced distributions of the polyoxyethylene chains. The dependence of the CMC on the degree of ethoxylation has a discontinuity between the values for NPE_8 and the series of $NPE_{10}-NPE_{15}$. It was assigned to the change of the polyoxyethylene chain conformation. However, additional data are needed to support this and the work is now in progress.

References

1. Huber JKF, Kolder FFM, Miller JK (1972) Anal Chem 44:105
2. Rothman AM (1982) J Chromatogr 253:283
3. Anghel DF, Balcan M, Voicu A, Elian M (1987) Rev Chim (Bucharest) 38:148
4. Mulley BA (1967) In: Schick MJ (ed) Nonionic Surfactants. Dekker, New York, chap 13
5. Elian M, Voicu A (unpublished data)
6. Mândru I, Ceacăreanu DM (1976) "Chimia coloizilor şi suprafeţelor. Metode experimentale", Editura Tehnică, Bucureşti, p 265
7. Lando JL, Oakley HT (1967) J Colloid Interface Sci 25:526
8. Gratzer WB, Beaven GH (1969) J Phys Chem 73:2270
9. Ray A, Némethy G (1971) J Phys Chem 75:804
10. Crrok EH, Fordyce DB, Trebbi GF (1963) J Phys Chem 67:1987
11. Crook EH, Trebbi GF, Fordyce DB (1964) J Phys Chem 68:3592
12. Lange H (1961) Proc Intern Congr Surface Activity 3rd Cologne 1:279
13. Corkill JM, Goodman JF, Harold SP (1964) Trans Faraday Soc 60:202
14. Corkill JM, Goodman JF, Ottewill RH (1961) Trans Faraday Soc. 57:1627
15. Ginn MG, Kinney FB, Harris JC (1960) J Am Oil Chemists' Soc 37:183
16. Hsiao L, Dunning HN, Lorenz PB (1956) J Phys Chem 60:657
17. Schick MJ (1962) J Colloid Sci 17:801

18. Harkins WD, Florence RT (1938) J Phys Chem 6:847
19. Schick MJ, Beyer EA (1963) J Am Oil Chemists' Soc 40:66
20. Vonk HJ, Van Wely AJ, Van der Ven LGJ, De Breet AJJ, Biemond MEF, Van der Maeden FPB, Venema A, Huysmans WGB (1976) Proc 7th Intern Congr Surface Active Subst, Moscow, Publ 1977, vol 1, p 435
21. Flory PJ (1969) Statistical Mechanics of Chain Molecules. Wiley, New York, Chap V
22. Bailey Jr. FE, Koleske JV (1967) In: Schick MJ (ed) Nonionic Surfactants. Dekker, New York, Chap 23
23. Rösch M (1967) In: Schick MJ (ed) Nonionic Surfactants. Dekker, New York, Chap 22
24. Tadokovo H, Chatani Y, Yoshihara T, Tahara S, Murahashi S (1964) Makromol Chem 73:109
25. Koenig JL, Angood AC (1970) J Polym Sci, A 2, 8:1787
26. Kalyanasundaram K, Thomas JK (1976) J Phys Chem 80:1462
27. Robson RJ, Dennis EA (1977) J Phys Chem 81:1075
28. Empart CR (1967) In: Schick MJ (ed) Nonionic Surfactants. Dekker, New York, p 44

Received September 16, 1987;
accepted June 1, 1988

Authors' address:

D. F. Anghel
Central Institute of Chemical Research
Dept. of Physical Chemistry
Spl. Independenţei 202
79611 Bucarest, Romania

Progress in Colloid & Polymer Science   Progr Colloid & Polymer Sci 77:67–71 (1988)

Studies on the conformation of polypeptides in reverse micelles

G. Ebert, M. Plachky, M. Senō*), and H. Noritomi**)

Fachbereich Physikalische Chemie – Polymere – der Philipps-Universität Marburg, Marburg (Lahn), FRG
 *) Institute of Industrial Science, University of Tokyo, Minato-ku, Tokyo, Japan
**) Institute of Materials Science, University of Tsukuba, Sakuramura, Ibaraki, Japan

Abstract: It could be shown that the c.d. spectra of basic poly-a-amino-acids (BPA) solubilized in reverse micelles of the system AOT/H_2O/iso-octane depend not only on w_0 and the basic side-chain but also on the concentration of the BPA. Moreover the purification procedure applied to the AOT seems to influence the conformation of the BPA in the reverse micelles remarkably. For discussing the effect of AOT reverse micelles on the conformation of the solubilized BPA one has to take the ratio r' of AOT-molecules to the number of basic peptide residues and therefore the concentration of empty reverse micelles into consideration.

Key words: Reverse micelles, aqueous solutions, Bis(2-ethylhexyl)-sodiumsulfo-succinate), polypeptides, conformation, c.d. spectra

Introduction

Homo and copolymers of a-amino acids are useful in studying the interaction of polypeptides and proteins with the medium in which they are dissolved. It is well known that the solvation of the amino acid side chains strongly influences the conformation of these macromolecules. Because of the high degree of hydrophilic hydration poly-(a-amino acids) with ionic side-chains have been studies extensively in this regard. The solubilization of such molecules in reverse micelles, e.g., of the system AOT*)-H_2O-isooctane (i-C_8) allows variation in the number of water molecules available for the hydration of the amino acid side chains by varying the molar ration w_0 of water to the surfactant AOT. Senō et al. [1–3] have studied the conformation of various kinds of ionic and polar homopolypeptides and of some copolymers of amino acids with ionic and with apolar side chains in reverse micelles and in aqueous solution.

The conformations of the polypeptides used were changed completely by solvation in reverse micelles compared with normal aqueous solutions. Basic homopolypeptides like poly-(L-lysine) (Lys)$_n$, poly-(L-ornithine) (Orn)$_n$, poly-(L-arginine) (Arg)$_n$ in an aqueous solution of AOT showed c.d. spectra which could be interpreted as the superposition of a-helix and I-β-structures. In reverse micelles, however, the

c.d. spectra down to 210 nm seemed to indicate the presence of the I-β-structure only. The conformation of the polypeptides studied depended remarkably on the concentration of the components of the system used and on their pretreatment.

Experimental

Substances

Poly-(L-lysine) was prepared by the polymerization of 0.1 mol N-carboxyanhydride (NCA) of N,ε-car-bobenzoxy-L-lysine in 200 ml anhydrous tetrahydro-furan with triethylamine (anhydride/initiator ratio 180/1). After 48 h the highly viscous solution was diluted with twice its volume of chloroform and treated with HCl for 2 h and with HBr for 4 h. The precipitated polymer was isolated, washed with ether, dissolved in water, dialysed against water of pH 3, and freeze dried.

Poly-(L-arginine) was obtained by guadinization of poly-(L-ornithine) with 3,5-dimethylpyrazolyl-for-mamidiniumnitrate of pH 9.3 at 37 °C for 48 h, according to Ariely et al. [4, 5].

The copolymers of L-Lys and L-Leu were prepared by copolymerizing the corresponding NCA in a 1:1 molar ratio with triethylamine.

To determine the molecular mass, the viscosity of the Cbo-blocked polymers was determined in dimethylformamide. The mol-mass was evaluated us-

---

*) Bis(2-ethylhexyl)-sodium sulfosuccinate

ing the constants $a = 1.26$ and $K = 2.24 \cdot 10^{-7}$ of the Kuhn-Mark-Houwink-Sakurada equation obtained by Daniel and Katchalski for PBLG in acetic acid [6]. Friehmelt has found a good agreement between the DP of $(Cbo-Lys)_n$ obtained in this way and that obtained by ultracentrifuge measurements [5].

The molar mass of random $(L-Lys^{0.5}, L-Leu^{0.5})$ was determined to be 116 000 $(\overline{DP} = 832)$ and those of the block-copolymer $(L-Lys^{0.5}, L-Leu^{0.5})$ to be 150 000 $(\overline{DP} = 1020)$. For the homopolymers it was determined to be 100 000 $(\overline{DP} = 600)$ for $(Lys)_n$ and 445 000 $(\overline{DP} = 2320)$ [5] by ultracentrifuge measurements [5, 7], for $(Arg)_n$.

Preparation of reverse micelle solutions

The finely powdered polymers were added to a reverse micellar system consisting of isooctane, AOT, and water and stirred as long as necessary to obtain a clear solution. Compared with the injection method, this kind of preparation takes more time, but the micellar solutions obtained in this way are very stable for several weeks.

All measurements were carried out at 20°C. Usually an AOT-concentration of 0.1 mol/l was used.

Because of the water solubilizing capacity of AOT and probably also that of the polypeptides, further u.v. absorptions are seriously affected by impurities such as salts, the surfactant was carefully purified according to the method described by Martin and Magid [8].

Results

c.d. spectra of aqueous solutions

In rather dilute solutions $(2.5 \cdot 10^{-3} \text{ mol/l})$ compared with those reported earlier [1], $(Arg)_n$ as well as $(Lys)_n$ show only a weakly positive c.d. absorption band with a maximum in the vicinity of 217 nm, and

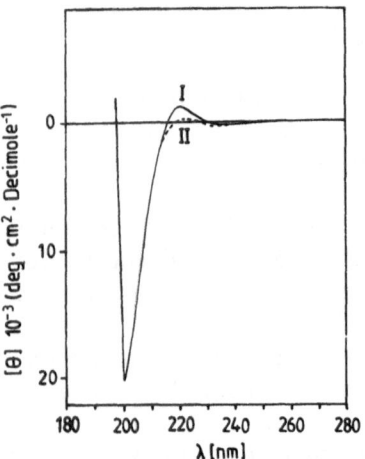

Fig. 1. The c.d. spectra of $(L-Lys)_n$ in water $(2 \cdot 10^{-3} \text{ mol/l})$ (I) and in a $0.1 \cdot 10^{-3}$ mol/l AOT solution, (II) $T = 20°C$

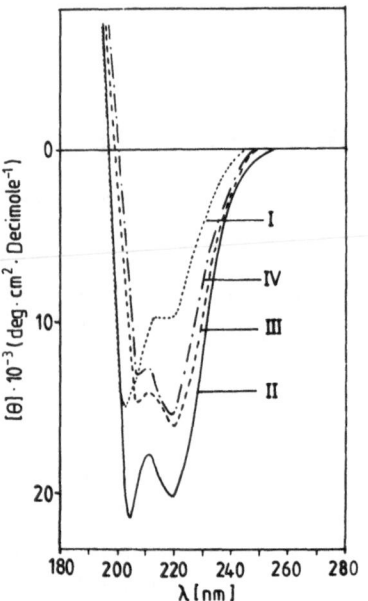

Fig. 2. The c.d. spectra of random copoly-$(L-Lys^{0.5}, L-Leu^{0.5})$ $(2.5 \cdot 10^{-3} \text{ mol/l})$ in water and AOT solutions, $T = 20°C$. I: in water; II: $c_{AOT} = 1.2 \cdot 10^{-3}$ mol/l; III: $c_{AOT} = 2.4 \cdot 10^{-3}$ mol/l; IV: $c_{AOT} = 4.4 \cdot 10^{-3}$ mol/l

a strongly negative narrow c.d. band at 203 nm. The addition of AOT diminishes the absorption at 217 nm. In the case of aqueous solutions of a random copoly-$(L-Lys^{0.5}, L-Leu^{0.5})$ not only does the c.d. spectrum of the pure aqueous solution differ remarkably from those of the basic homopolypeptide, but also from those of the AOT containing solutions. As one can see from Fig. 2, two troughs, one at 220 nm and the other near 203 nm are observed corresponding to the α-helical conformation of the polypeptide. Interestingly, the amount of the specific ellipticity $|\theta|$ decreases with increasing AOT concentration from $1.2 \cdot 10^{-3}$ to $4.4 \cdot 10^{-3}$ mol/l as does the ratio $|\theta|_{205}/|\theta|_{220}$.

c.d. spectra in reverse micelles

According to Fig. 3, the c.d. spectra of $(L-Lys)_n$ solubilized in $AOT-H_2O-i-C_8$ reverse micelles are completely different form those obtained from the aqueous solutions at similar polymer and AOT concentrations. They depend strongly on the water/AOT ratio w_0. It becomes obvious from these c.d. spectra that the polypeptide conformation changes from the α-helix at $w_0 = 6$ to a mixture of α-helix and β-structure with increasing w_0. The specific ellipticity at 208 nm amounts to $\approx 37 \cdot 10^3$ and at 220 nm to $31 \cdot 10^{-3}$ degcm²/dmol which corresponds approximately to the values found in aqueous solutions of α-helical polypeptides.

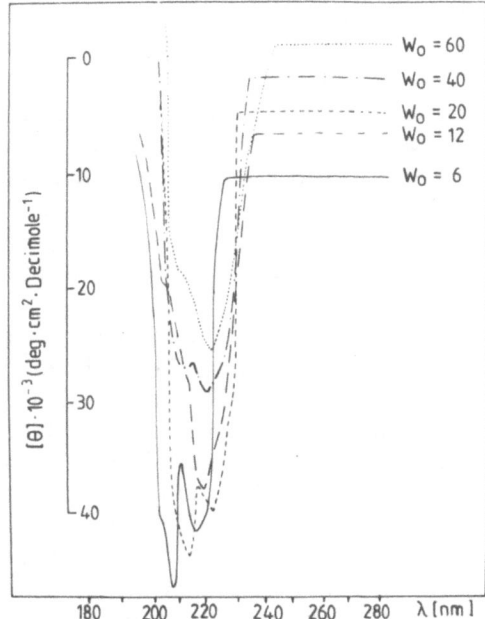

Fig. 3. The c.d. spectra of (L-Lys)$_n$ solubilized in AOT–H$_2$O–i-C$_8$ reverse micelles at different w_0 values. Polymer concentration: $1 \cdot 10^{-3}$ mol/l; AOT concentration: $100 \cdot 10^{-3}$ mol/l; $T = 20\,°$C

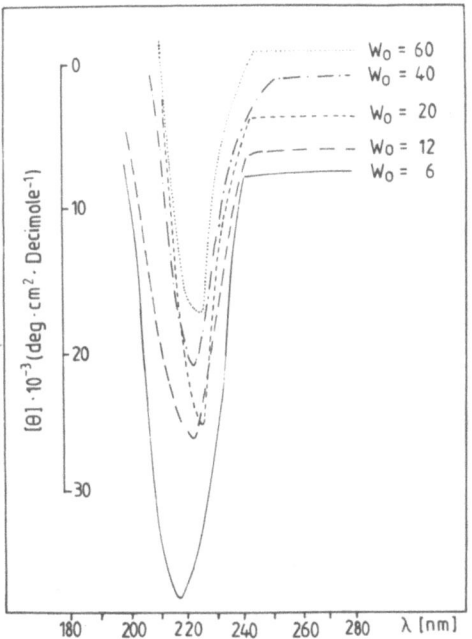

Fig. 5. The c.d. spectra of a block copoly-(L-Lys$^{0.5}$, L-Leu$^{0.5}$) solubilized in AOT–H$_2$O–i-C$_8$ reverse micelles at different w_0 values, Polymer concentration: $1 \cdot 10^{-3}$ mol/l; AOT concentration: $100 \cdot 10^{-3}$ mol/l; $T = 20\,°$C

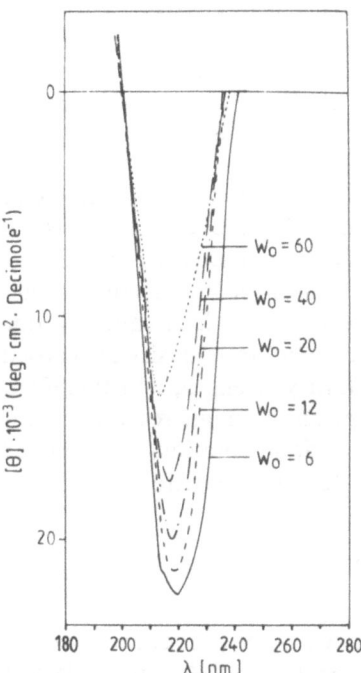

Fig. 4. The c.d. spectra of (L-Arg)$_n$ solubilized in AOT–H$_2$O–i-C$_8$ reverse micelles at different w_0 values. Polymer concentration: $0.1 \cdot 10^{-3}$ mol/l; AOT concentration: $100 \cdot 10^{-3}$ mol/l; $T = 20\,°$C

With increasing w_0, not only does the shape of the c.d. spectra and concomitantly the type of ordered periodical conformation change, but also the specific ellipticity $|\theta|$ changes to less negative values, especially above $w_0 = 20$.

In the case of (Arg)$_n$ (Fig. 4), the c.d. spectra show only one trough with a minimum near 220 nm which is shifted to lower wavelengths with increasing w_0. At $w_0 = 60$ it is situated at 213 nm and the c.d. absorption curve is broadened at higher wavelengths. As one can see from Fig. 4, $|\theta|_{220}$ is significantly lower than in the case of (Lys)$_n$. The intensity of the c.d. absorption decreases with increasing w_0 in an analogous manner.

By copolymerizing L-lysine with L-leucine the c.d. spectra, and therefore the conformation of the resulting copolymer in reverse micelles, is changed to a great extent compared with the basic homopolypeptide. This is shown in Fig. 5 for a block-copoly-(L-Lys$^{0.5}$, L-Leu$^{0.5}$). At $w_0 = 6$ a strong absorption at 218 nm with a specific ellipticity $|\theta|_{218} = 38 \cdot 10^3$ degree cm^2/dmol occurs. In this case, in contrast to the homopolymers already between $w_0 = 6$ and $w_0 = 12$, a pronounced decrease in the absorption intensity accompanied by a shift of the minimum to longer wavelengths takes place.

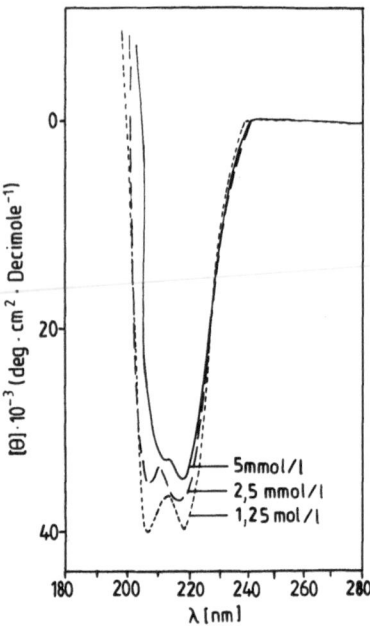

Fig. 6. The c.d. spectra of $(L-Lys)_n$ solubilized in AOT$-H_2O-i-C_8$ reverse micelles at $w_0 = 6$ but different polymer concentrations, AOT concentration: $100 \cdot 10^{-3}$ mol/l, Polymer concentration: $\cdots 1{,}25 \cdot 10^{-3}$ mol/l, $---$ $2{,}5 \cdot 10^{-3}$ mol/l, $\text{---} 5{,}0 \cdot 10^{-3}$ mol/l; $T = 20\,°C$

Polypeptide concentration dependence of the c.d. spectra

Figure 6 shows the c.d. spectra of AOT$-H_2O-i-C_8$ reverse micellar systems with different amounts of $(Lys)_n$ solubilized at $w_0 = 6$. As one can see, between a polypeptide concentration of 1.25 and $5.0 \cdot 10^{-3}$ mol/l not only does the ratio between the specific ellipticity at 205 and 221 nm typical for the α-helix change, but the absolute value of $|\theta|$ also decreases.

Discussion

Conformation in aqueous solutions

The ratio of AOT molecules to the number of basic peptides residues r is useful for discussing the effect of AOT on the formation of ordered, periodical structures of basic homopolypeptides, as shown by Noritomi [3]. In very dilute solutions such as $0.17 \cdot 10^{-3}$ mol/l of basic peptide residues, ordered structures are observed above $r = 0.25$. This means that more than 25 AOT molecules per 100 peptide residues are necessary for significant conformation changes. At $r = 1.2$, c.d. spectra indicating a mixture of α-helix and $I\beta$-structure have been detected [1, 3].

With increasing r the random coil content increases, and at $r = 2.5$ a ratio of α-helix : β-structure : random coil = 0.37 : 0.13 : 0.50 was found [3, 9].

In studying the conformation changes at higher polypeptide concentrations, $2.5 \cdot 10^{-3}$ mol/l solutions were used. However, in this case precipitation occurred above an AOT concentration of $0.1 \cdot 10^{-3}$ mol/l ($r = 0.04$). According to the previous results [1, 3] this r value is too low to change the polypeptide conformation.

On the other hand, a solution containing $2.5 \cdot 10^{-3}$ mol/l of random copoly(L-Lys$^{0.5}$, L-leu$^{0.5}$) does not precipitate when the AOT concentration is brought up to $4.4 \cdot 10^{-3}$ mol/l, which corresponds to $r = 0.96$ to 3.52, if L-Lys is considered singly. This is probably due to the amphiphilic pattern of apolar isobutyl side chains of L-leucine and the ionic ammonium side groups of L-lysine along the polypeptide chain. Therefore, AOT molecules interact by their apolar chains with the leucine side groups directed with their hydrophilic ionic groups into the water phase, preventing an aggregation of the AOT-covered polypeptide molecules, as is the case for the basic homopolypeptides.

For this reason it is reasonable to refer r not only on the L-Lys residues but on the whole molecule, i.e., L-Lys and L-Leu.

As Fig. 2 shows, $r = 0.5$ is most effective in forming an α-helix. At higher r values, $|\theta|_{205}$ becomes lower than $|\theta|_{221}$, probably indicating a change in the solvation of the polypeptide molecules, and moreover the $|\theta|$ values of both are shifted to less negative values.

From the shape of the c.d. spectra at higher AOT concentrations, one can conclude that there is no overlapping with other conformations, e.g., increasing random coil formation leading to the decreasing absolute values of $|\theta|_{205}$ and $|\theta|_{221}$. Consequently, one can suppose that these changes in intensity are caused by a decrease in the rotatory strength, and therefore, by a change in the transition moments responsible for the value of the mean residue rotation $|m|_\lambda$ linked with $|\theta|$ by the Kronig-Kramers transform.

Conformation in reverse micelles of $AOT-H_2O-i-C_8$

In more dilute systems with polymer concentrations between 0.16 and $0.28 \cdot 10^{-3}$ mol/l and 0.05 mol AOT/l an ordered periodical conformation characterized by a sharp minimum in the c.d. spectra near $215-217$ nm occurs [1]. In order to discuss these different conformation changing effects, one must consider that in the case of polymers with a high $\overline{DP}$, especially at low w_0 values, a rearrangement of the empty reverse micelles is necessary to form filled reverse micelles, i.e., micelles containing solubilized polypeptide molecules. This was shown by ultracentrifuge experiments [7, 10] and had been postulated earlier [1–3]. Therefore the number of AOT molecules, or in other words the concentration of empty reverse micelles, may be of importance for the formation of ordered structures. In the systems used previously [1, 3] the ratio r' between AOT and L-Lys residues was 178:312, whereas in this study the value was 100. Figure 6 shows that by further decreasing the ratio r' to 80, 40, and 20, corresponding $(L\text{-}Lys)_n$ concentrations of 1.25, 2.5, $5.0 \cdot 10^{-3}$ mol/l. The c.d. spectra indeed depend on this ratio r'. Because of the constant w_0 value used, the concentration of empty reverse micelles has a substantial effect on the conformation of the solubilized $(Lys)_n$. Considering the fact that the lifetime of one micelle is very short and that a continual rearrangement of empty and filled reverse micelles takes place accompanied by the formation of intermediate complexes containing the solubilized polymer, in the c.d. spectra an average conformation is observed which probably depends not only on the number of water molecules and therefore on w_0, but also on the ratio r', and therefore on the concentration of empty reverse micelles. Further studies are necessary in this regard.

Acknowledgement

The authors are grateful to "Deutsche Forschungsgemeinschaft" for financial support.

References

1. Senō M, Noritomi H, Kuroyanagi Y, Iwamoto K, Ebert G (1984) Colloid & Polymer Sci 262:727
2. Senō M, Noritomi H, Kuroyanagi Y, Iwamoto K, Ebert G (1984) Colloid & Polymer Sci 262:897
3. Noritomi H (1986) Thesis, University of Tokyo
4. Ariely S, Wilchek M, Patchornik H (1966) Biopolymers 4:91
5. Friehmelt V (1981) Thesis, Marburg (L)
6. Daniel E, Katchalski E (1962) Polyamino Acids, Polypeptides and Proteins, Stahmann MA (ed). University of Wisconsin Press, Madison
7. Plachky M (1987) Thesis, Marburg (L)
8. Martin CA, Magid LJ (1981) J Phys Chem 85:3938
9. Townend R, Kumosinski TF, Timasheff SN, Fasman GD, Davidson B (1966) Biochem Biophys Res Commun 23:163, p 183
10. Ebert G, Plachky M, Senō M, Shoji A (1988) Progr in Colloid & Polymer Sci 76

Received February 24, 1988;
accepted June 1, 1988

Authors' address:

G. Ebert
Fachbereich Physikalische Chemie – Polymere
Philipps-Universität Marburg
Hans-Meerwein-Straße
3550 Marburg (Lahn)

Progress in Colloid & Polymer Science

Progr Colloid & Polymer Sci 77:72–76 (1988)

Phase transitions in phospholipid monolayers and rheological properties of the corresponding membrane model

G. Kretzschmar

Academy of Sciences of the GDR, Central Institute of Organic Chemistry

Abstract: Insoluble spread monolayers of Dipalmitoyllecithin and Dipalmitoyl-cephalin are investigated by π/A isotherms and periodic compression and dilatation to determine their elasticity and the phase angles between strain and stress.

Depending upon the temperature, the phase angle becomes noticeable in the film pressure region in which the phase transition takes place.

From the rheological point of view, we conclude that the relaxation processes in the two-phase film state make the main contribution to the phase angle.

Key words: Phospholipid monolayers, phase transition, pressure-area isotherms, surface rheology, marangoni effect

1. Introduction

This paper deals with measurements of surface rheological properties in connection with the behavior of phospholipid bilayers that consist of different states. Phospholipid bilayers are well-known examples of biological membranes. One of the main questions is the relevance of monolayer studies to the properties of bilayers. Bilayers hold forces in the normal direction of the separated monolayers. These interaction forces cannot be simulated by monolayer experiments. For membrane deformations produced at slow strain rates with no permanent material alteration due to the membrane forces, elastic effects are reasonable representations of the material character of the membrane. However, if the rate of deformations increases, thermodynamically irreversible processes which result from internal friction and heat dissipation within the membrane become evident. The time dependence of the force relaxation is determined by the rate of viscous dissipation and the lateral transport processes of matter. The laster is forced by a compression or dilatation of each monolayer. A simple experiment will show this situation. A thin bilayer film is fixed in a frame and one of the wires is forced to oscillate. The film will be stretched and compressed periodically. During this process the concentration of the film-forming molecules may increase and decrease more near the movable barrier than far from it. Therefore a surface-pressure gradient results. The compensation of such a gradient, named the Marangoni Effect, takes place by lateral transport of the film components.

This process is characterized by the relaxation time of the pressure gradient.

If the relaxation time becomes greater than the time for a period of compression and dilatation of the monolayer, a phase angle between strain and stress occurs. This relationship can be described by a complex modulus of the monolayer elasticity.

$$\varepsilon = \frac{-d\pi}{d\ln A} = \varepsilon' + i\varepsilon'' = |\varepsilon|\cos a + i\,|\varepsilon|\sin a \ , \qquad (1)$$

where A is the area of the film.

Experiments and mathematical analysis of the Marangoni Effect for phospholipid films were published by Dimitrov et al. [1]. The static, or better the quasistatic, π/A isotherms of spread and compressed phospholipid films at different temperatures offer some fundamental problems. Some of these are related to the phase transition regions and to the hysteresis effects.

Surveys of this problem are given by Nagle [2] and Dörfler [3].

According to Erbrich, Septinus, and Zimmermann, the phase transition is characterized by a two-phase region in which liquid expanded and condensed domains exist [4]. The equilibrium between these two film states depends on the pressure which acts on the film. This process induces a characteristic time. If this time becomes greater than the time for a period of compression and dilatation of the film, a relaxation appears which can be observed by the phase angle a between strain and stress.

The imaginary part of Eq. (1) leads to

$$\eta_d = \frac{|\varepsilon| \sin a}{\omega} , \quad \text{where} \quad \varepsilon = \frac{-d\pi}{dA} \cdot A . \qquad (2)$$

From the works of Van den Tempel [5], Lucassen [6], Lucassen-Reynders [7], Kretzschmar, Lunkenheimer, and Miller [8], it is well known that the contribution of mass transfer between interface and bulk is made by altering the area of interface continuously or periodically. This is due to the propagation of transverse and longitudinal waves or radial oscillating bubbles. Here the phase angle a between strain and stress is the consequence of the vertical diffusion exchange. Less work has been published about the lateral transport processes triggered by a compression or dilation of a monolyer. Vollhardt and Wüstneck investigated long-chain fatty acids [9] and the special case of phopholipids is shown in ref. [1]. The initial state for our problem is the observed hysteresis in π/A isotherms of spread phospholipids. Phospholipids show phase transitions between so-called liquid expanded and liquid condensed films that depend upon the chain lengths and the chemistry of their head groups. In the phase transition region the π/A curves show plateau sections and concave slopes. Erbrich et al. [4] show that the phase transition region is distinguished by an equilibrium between the two adjacent phases. In other words, a two-phase region which contains liquid expanded and liquid condensed domains exists. The conversion is a rate-determining step in the

registration of π/A isotherms and leads to a noticeable phase angle between strain and stress. It means a difference between the π/A curves in the region of phase transition exists that is obtained by compression and dilatation. The aim of this paper is to show the possibility of detecting the phase-transition region by the method of surface rheology. The principle complications in detecting the phase transition by static π/A isotherms have been shown by Nagle [2]. In [4] plot A vs. T for different values of π as a constant parameter is used. Generally the phase transition shifts to a higher film pressure if the temperature increases.

2. Experimental

For measuring the retardation of the equilibrium in film pressure along the x-axis we used a device which was described by Kretzschmar and König [10] and is similar to the equipment of Van Voorst Vader [11], Van den Tempel [12], and Wasan [13]. The principle of the apparatus is shown in Fig. 1.

In the first step of the experiments the chromatographically purified DL-β,γ-Dipalmitoyl-α-Lecithin

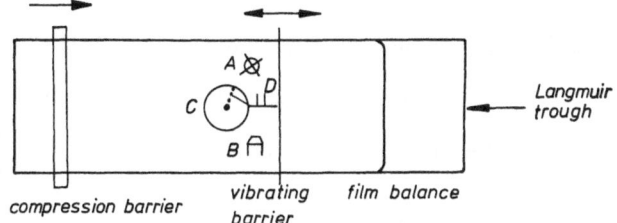

Fig. 1. Apparatus for determining the rheological parameters of monolayers; A, B, D Light barrier for the detecting of the phase angle a; C Controllable amplitude ΔL

Periodic compression dilatation of DL-β,γ-Dipalmitoyl-α-Cephalin

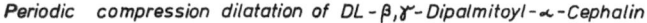

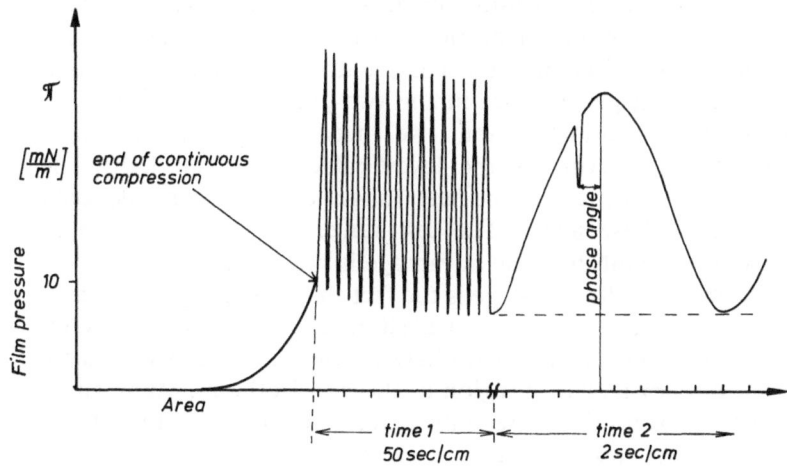

Fig. 2. Plot of $\Delta\pi$ for the periodic compression and dilatation of monolayers. The film elasticity ε is calculated using $\dfrac{-d\pi L}{dL} \sim \dfrac{-\Delta\pi L}{\Delta L}$. L is the maximum distance between the vibrating barrier and the film balance

(DPL) and DL-β,γ-Dipalmitoyl-α-Cephalin (DPC; Fluka) was dissolved in chloroform and methanol and spread on the cleaned electrolyte solution. After 20 min the film was compressed to a definite film pressure π. With the help of an electromagnetic device, a second barrier was placed at a distance of about 7 cm from the force balance. Moving this second barrier periodically with an amplitude of 1.5 cm and a frequency of 0.05 Hz produced the typical plot of Fig. 2. The phase angle a can be determined using the time difference between the end position of the oscillating barrier and the maximum of film pressure $\Delta\pi$.

This phase angle is due to the Marangoni Effect along the x-axis of the Langmuir trough. Additional evidence is provided by the experiment with talcum particles placed between the oscillating barrier and the force balance. The oscillation of the barrier results in a longitudinal surface wave that is propagated to the opposite wall and reflected there. The superposition of the single waves leads to local area changes. Lucassen and Van den Tempel [14] used the following formula to describe the particle movement Z

$$Z = Z_B \, e^{i\omega t} \, \frac{\sin k(L-x)}{\sin kL} \,, \tag{3}$$

wherein Z_B is the displacement of the barrier, k is the complex wave number, defined as $k = \dfrac{2\pi}{\lambda} - i\beta$, β is the damping coefficient, and L is the length of the space.

3. Discussion

The wavelength used for the low frequency is about 8 m. The damping coefficient becomes important only for short wavelengths. In these experiments the particle movement Z vs. x must be practically the same for the surface pure electrolyte solution and for a surface covered with phospholipid. This is illustrated in Fig. 3. The phase angle a increases. This means there is a gradient of surface pressure along the x-axis. The particle movement offers an unsymmetric amplitude because of the actual surface force. This leads to a deviation from the characteristic slope for a pure water surface. The film pressure gradient is a direct consequence of the significant relaxation process in the film transition region where the conversion between the two film states is a time dependent process in comparison with the low frequency of oscillation. Several papers deal with differences of π/A isotherms and the phase transition of phospholipids that consist of choline or

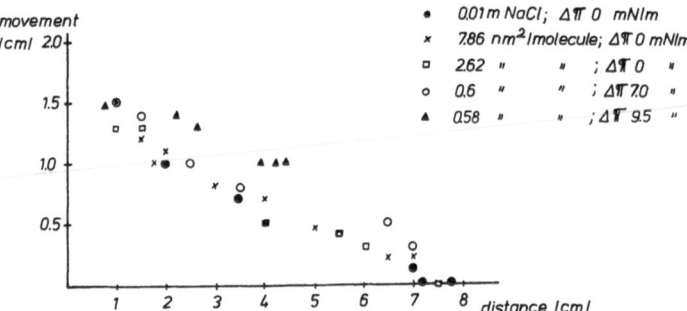

Fig. 3. Amplitude of talcum particles placed at the pure water surface and then covered by DPL. Deviation from Eq. (3) is due to the gradient of film pressure along the x-axis

ethanolamine head groups [15]. In the present paper we used Dipalmitoyllecithin and Dipalmitoylcephalin. For the structure of the latter, the intermolecular attractive force is the greatest. Comparing both types at the same temperature and pressure showed that the DPC film is more condensed than the other and the phase inversion temperature is higher. Interference with the motion of particles placed at the film covered surface offers a permanent gradient in the surface tension due to the relaxation processes in phospholipid monolayers and is one of the detectable effects. The other is related to measurements of viscoelastic properties using the device shown in Fig. 1. Characteristic values are obtained by calculating the complex modulus ε and by detecting the phase angle a. According to Lucassen-Reynders [7], for large amplitudes: η_d may be correctly expressed by K_d in the equation of Boussinesq [7]

$$\Delta\sigma = K_d \, \frac{dA}{dt} \cdot \frac{1}{A} \,. \tag{4}$$

This expression for the surface dilatational viscosity K_d is more realistic than the often used formula for small amplitudes. For the determination of rheological peculiarities in the two-phase state of the phase transition region, we need the quasistatic π/A isotherm for the investigated DPL and DPC, the elasticity in the test range, and the corresponding phase angle. The test range was limited to about 15 mN/m. The advantage is higher accuracy, whereas the disadvantage is the fact that in higher temperature regions the phase transition appears above 15 mN/m. In some cases the phase transition is simply detectable by a plateau region of the π/A isotherm. Many phase transitions are only detectable by the shape of the curves π vs. A or A vs. T.

4. Conclusion

The present investigations offer a new possibility for the determination of phase transitions by the described rheological method. Tables 1 and 2 represent the rheological results for DPL and DPC. By comparing the corresponding π/A isotherms, as shown in Figs. 4 and 5, we can conclude that surface rheology is usefully applied to the determination of phase tran-

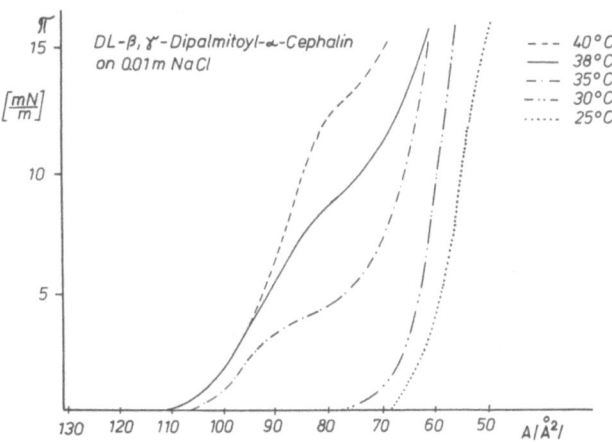

Fig. 4. π vs. A for DPL. The dimension is $|\text{Å}^2| = 10^2\,\text{nm}^2$

Table 1. Elasticity and phase angle for DL-β,γ-Dipalmitoyl-α-lecithin

Temperature [°C]	Film pressure π [mN/m]	Elasticity ε [mN/m]	Phase angle a [degree]
15	5.5	157	0
	10.0	170	18
	15.0	185	21
20	5.5	30	28
	10.3	65	28
	13.0	120	6
	15.2	120	6
25	5.5	28	26
	10.2	40	28
	13.0	45	28
	15.2	60	26
30	5.2	36	31
	10.3	45	27
	13.0	45	27
	15.2	45	27
40	5.0	59	3
	10.0	65	23
	15.0	65	23

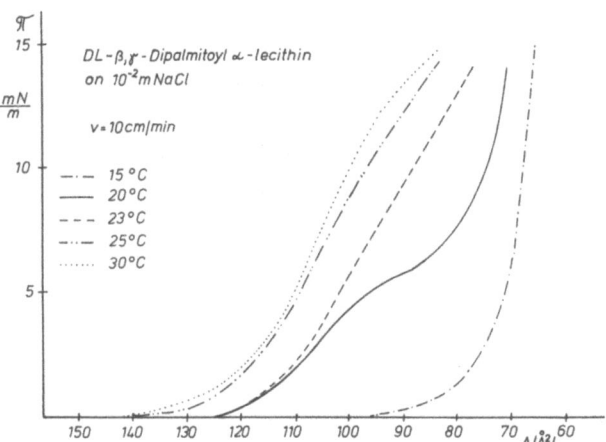

Fig. 5. π vs. A for DPC

Table 2. Elasticity and phase angle for DL-β,γ-Dipalmitoyl-α-cephalin

Temperature [°C]	Film pressure π [mN/m]	Elasticity ε [mN/m]	Phase angle a [degree]
20	5.0	90	5
	10.3	175	0
	15.0	250	0
25	5.0	45	0
	10.1	125	0
	15.1	160	0
30	5.7	80	0
	10.3	110	0
	12.7	150	5
	15.7	180	0
35	5.0	40	3
	10.0	62	11
	15.0	120	0
40	5.5	33	0
	9.8	50	0
	12.0	72	13
	15.0	95	6

sition regions in such π/A plots which do not show characteristic plateau regions. This observation may be understood as the time dependent conversion between the two film states in the phase transition region.

References

1. Dimitrov DS, Panajotov I, Richmond P, Ter-Minassian-Saraga L (1978) J Colloid Interface Sci 65:483–494
2. Nagle JF (1980) Ann Rev Phys Chem 31:157–195
3. Dörfler H-D, Koth C (1984) Wiss Z Univ Halle XXXIII:69–79
4. Erbrich U, Septinus M, Zimmermann (1982) Ber Bunsenges Phys Chem 86:724–728
5. Van den Tempel M (1977) J Non-Newt Fluid Mech 2:205–219
6. Lucassen J, Van den Tempel M (1972) J Colloid Interface Sci 41:441–498

7. Lucassen-Reynders E (1981) In: Surfactant Science Series Vol 11. Marcel Decker, New York, pp 173–216

8. Kretzschmar G, Lunkenheimer K, Miller R (1986) IXth European Chemistry of Interfaces, Zakopane. In: Material Science Forum. Trans Tech (1988) 25–26:211–222

9. Vollhardt D, Zastrow L, Wüstneck R (1978) Colloid & Polymer Sci 256:973–982

10. Kretzschmar G, König K (1981) J Signal AM 9:203–212

11. Van Voorst Vader F, Erkens TF, Van den Tempel M (1964) Trans Faraday Soc 60:1170–1177

12. Veer FA, Van den Tempel M (1973) J Colloid Interface Sci 42:418

13. Maru HC, Wasan DT (1979) Chem Eng Sci 34:1295–1307

14. Lucassen J, Van den Tempel M (1972) Chem Eng Sci 27:1283–1291

15. Helm CA, Laxhuber L, Mösche M, Möhwald H (1986) Colloid & Polymer Sci 264:46–55

Received February 19, 1988;
accepted June 7, 1988

Author's address:

Prof. Dr. Günter Kretzschmar,
GDR Berlin 1199
Rudower Chaussee 5

Structures in aqueous solutions of nonionic tensides

R. Heusch and F. Kopp*

Bayer AG, Leverkusen, F. R. G., *Diabetes Forschungsinstitut, Düsseldorf, F. R. G.

Abstract: Since Bordier succeeded in isolating solutions of hydrophobic proteins with the aid of aqueous Triton X-114 (reaction product of p-octylphenol with 7–8 mol ethylene oxide), more and more biochemists and biotechnologists have become interested in such systems. We have been able to demonstrate that lamellar structures formed in miscibility gaps of polyglycol ether/water systems are responsible for this isolation. We then succeeded in isolating and electron-micrographing the insulin receptor in one of these lamellar systems.

Structures in the isotropic solution in the region between the micelles and mesomorphous phases of the Triton X-114/water system have been detected with the electron microscope for the first time and are considered to be responsible for the formation of micro-emulsions.

The phase diagram of the reaction product of an alcohol mixture with 5 mol ethylene oxide and water is presented and interpreted. Polarization and electron micrographs are again used for this purpose.

The author explains why it has taken so long for the existence of defined tenside hydrates to receive recognition.

Key words: Polyglycol ether, liquid crystal hydration, polarization microscopy, electron microscopy

1. Introduction

Since the beginning of the 1970s, more and more biochemists and biotechnologists have begun to use nonionic tensides to isolate or purify their products. The interest in aqueous solutions of polyglycol ethers in biotechnology has constantly increased since Bordier's success in separating hydrophobic proteins from hydrophilic ones with the aid of aqueous Triton X-114 solutions [1].

In order to cast some light on the background of such working procedures, we have investigated structures in such solutions.

We could demonstrate that lamellae which deposit in the miscibility gap developed by aqueous Triton X-114 solutions at increasing temperatures are responsible for the isolation of hydrophobic substances [2–4]. We have succeeded in isolating the insulin receptor in such a lamellae system and depicting it using an electron microscope; the size and shape of this receptor thus became visible for the first time [3].

In the conclusion we will attempt to show why the existence of defined tenside hydrates is even now occasionally doubted.

2. Structures in Triton X-114-water mixtures

The structures which appear in the phase diagram of Triton X-114-water mixtures have meanwhile been published [4]. It is remarkable that at room temperature only a single mesomorphic phase exists, which we ascribe to a polydihydrate [5–7], a complex lamellar compound in which two water molecules are bonded to each oxygen atom of the polyether chain.

We employed the electron microscope to depict the structures present in the isotropic solution in the range between the mesomorphic phases and the zone of micelle formation [4]. Above 50% water, water-filled vesicles whose envelope consists of overlapping lamellae are formed. Figure 1 was taken from a 40% Triton X-114 sample, in which water-filled vesicles of 400–600 Å size can be recognized.

When further water is added the vesicles assume a drawn out form and finally begin to conflux, their aqueous contents and their lamellar exterior becoming continuous over a broad area [4].

In the range of isotropic solutions there exists the zone of anomalous viscosity. In this region, the viscosity rises with increasing temperature [6, 8]. In con-

Fig. 1. Electron micrograph of a 40% aqueous solution of Triton X-114® at 23 °C

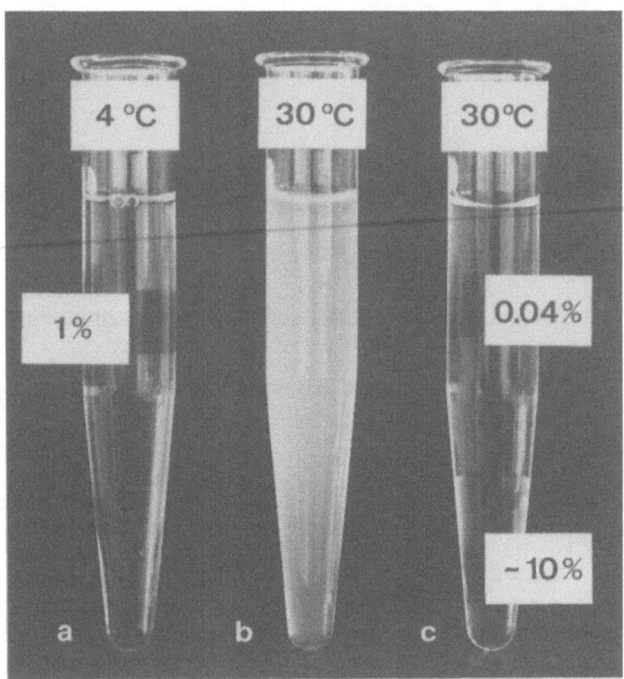

Fig. 2. Separation of a 1% Triton X-114 solution in the miscibility gap:
a) solution at 4 °C, b) after heating to 30 °C and, c) after standing at 30 °C

trast to our previous interpretation [2], the anomaly appears to be caused by the broadly contiguous lamellae, which mutually hinder one another by their confluction with increasing temperature.

The electron photomicrographs in the range of the isotropic solutions also throw a new light on the concept of "microemulsions". If a liquid phase can be finely divided in such a structured solution so that the isotropy of the system is retained, we are then confronted with a microemulsion. It should therefore be possible to replace at least a part of the water surrounded by the vesicular lamellae with other materials.

If we heat an isotropic solution above its cloud point [9], two phases are formed: an aqueous phase containing very little polyglycol ether, and a phase rich in polyglycol ether.

Figure 2 illustrates that a 1% solution of Triton X-114 is clear at 4 °C. If this solution is heated to 30 °C, it becomes cloudy. Two phases separate if it is allowed to stand; the polyglycol ether collects in the lower phase, and the upper aqueous phase contains only 0.04% of the Triton.

In the organic polyether phase electron microscopy shows that unbound water is surrounded by tightly-packed lamellae. Bordier isolated his hydrophobic proteins in such a lamellar system [1]. We could depict the

insulin receptor and erythrozytes band-3 protein in such lamellae [3, 4].

3. Structures in aqueous mixtures of the reaction product of Dobanol 91 with 5 moles of ethylene oxide

Dobanol 91 is a mixture of linear fatty alcohols made by Deutsche Shell Chemie with a composition of around 18% C_9, 50% C_{10} and 32% C_{11} alcohol. The abbreviation "Dob. 91/5 -EO" in the following text will stand for the reaction product of Dobanol 91 with 5 mol of ethylene oxide.

Figure 3 illustrates the phase diagram of aqueous mixtures of Dob 91/5 EO and the temperature is plotted against the weight percent of the polyglycol ether. To make this graph, the cloud points (+), thermoanalytical melting points (●), changes of optical polarization (○) and deviations from the normal slope of viscosity (▲) were determined in dependence on the polyglycol ether concentration. In addition, optical polarization and electron microscopic studies were carried out in order to determine and depict the structures which arise.

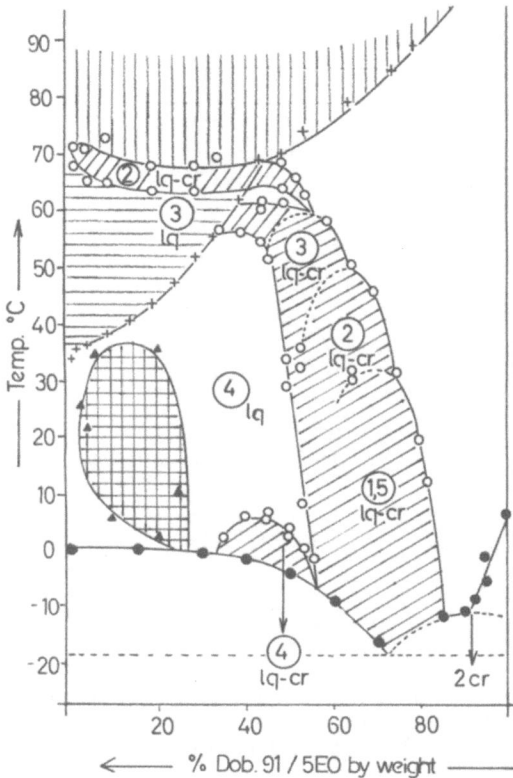

Fig. 3. Phase diagram of aqueous mixtures of the reaction product of Dobanol 91 with 5 mol of ethylene oxide. Signatures are explained in the text

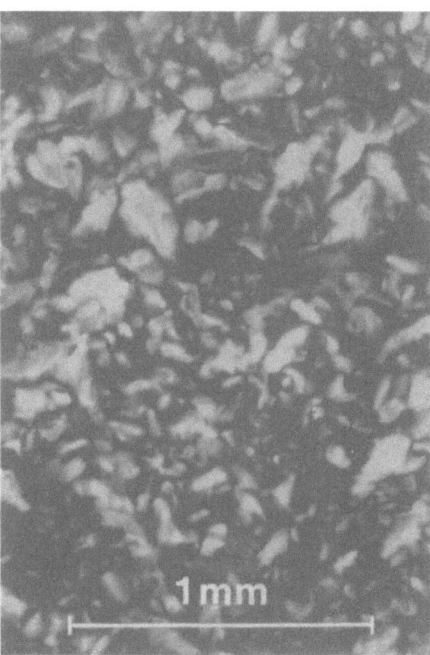

Fig. 4. Optical polarization structure of the polytetrahydrate of the reaction product of Dobanol 91 with 5 mol of ethylene oxide (45%) at 2,5 °C

At the lower part of the phase diagram in the crystalline range of this technically interesting product (which has a molecular weight of 380, low for polyethers, as well as a wide chain distribution on the hydrophilic and hydrophobic sides), we see a peritectic zone indicating the existence of a crystalline hydrate in which two water molecules are incorporated into the crystal structure of the anhydrous polyglycol ether.

Above the crystalline range two liquid crystalline phases — which we have marked with oblique lines — stand out from one another. We see a liquid crystalline zone between 33 and 55% Dob. 91/5 EO, whose maximum lies at 45%, corresponding to a polytetrahydrate whose melting point lies at 5.5 °C. A polytetrahydrate is a mesomorphic phase in which four water molecules are bound to every oxygen atom of the polyether chain.

The optical polarization structure of this polytetrahydrate can be seen in Fig. 4.

In contrast to the conical hexagonal structure of a polytetrahydrate (which is always sharply defined), this appearance, undoubtedly due to the short polyoxyethylene chains, is not especially characteristic. The photomicrograph was taken on a 0.2 mm thick

Fig. 5. Electron micrograph of a 45% aqueous solution of the reaction product of Dobanol 91 with 5 mol ethylene oxide and water at 0 °C

layer of a 45% sample at +2.5 °C. The corresponding electron micrograph taken of a 45% sample at 0 °C is shown in Fig. 5 where spheroidal structure is found.

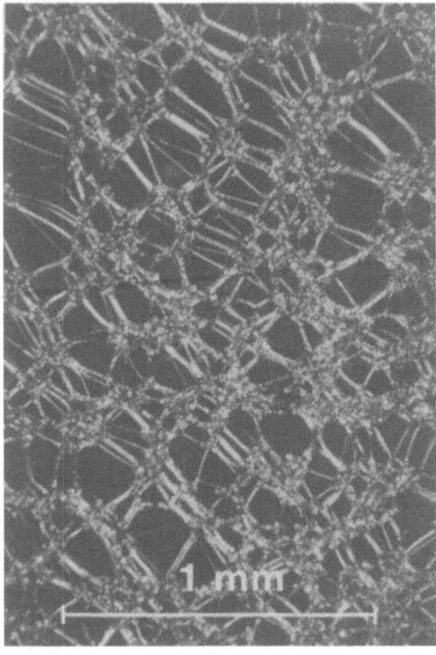

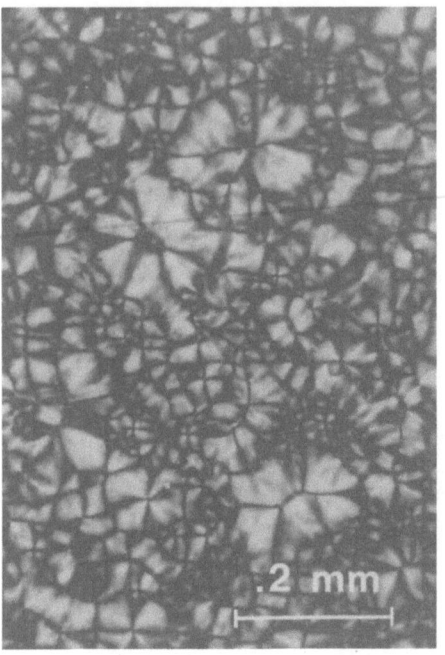

Fig. 6. Optical polarization structure of a 70% mixture of the reaction product of Dobanol 91 with 5 mol of ethylene oxide and water at 20°C in a thin layer

Fig. 7. Optical polarization structure of a 70% mixture of the reaction product of Dobanol 91 with 5 mol of ethylene oxide and water at 20°C in a 0.2 mm thick section

A liquid crystalline phase in which the individual hydrate stages intermingle is formed to the right next to the polytetrahydrate maximum between 55 and 85% Dob. 91/5 EO. In optical polarization we see a lamellar structure forming the familiar networks in a thin layer in this zone, as shown in Fig. 6. The photomicrograph was taken on a 70% sample at 20°C. At a layer thickness of 0.2 mm, the lamellar structure of this sample appears as an indistinctly defined spherolitical structure, as it is shown in Fig. 7.

Electron microscopy demonstrates that lamellae are actually present in this broad crystalline phase. In the freeze-fracture electron micrograph, the individual tenside layers of a 65% sample show irregular fracture (Fig. 8). Since the lamellae are undulated the surface of the fracture often fluctuates (jumps).

At higher magnification, the lamellae show fine ribbing of micellar size (Fig. 9), which we did not see with Triton X-114. The distance between the individual ribs lies between 50–60 Å.

Figure 10 establishes that no structural change takes place over the broad mesomorphic range between 55 and 85% Dob. 91/5 EO. At 40°C, the lamellar structure of an 0.2 mm thick section of a 55% sample was unchanged in comparison to that of a sample at 20°C. Electron microscopy confirms this.

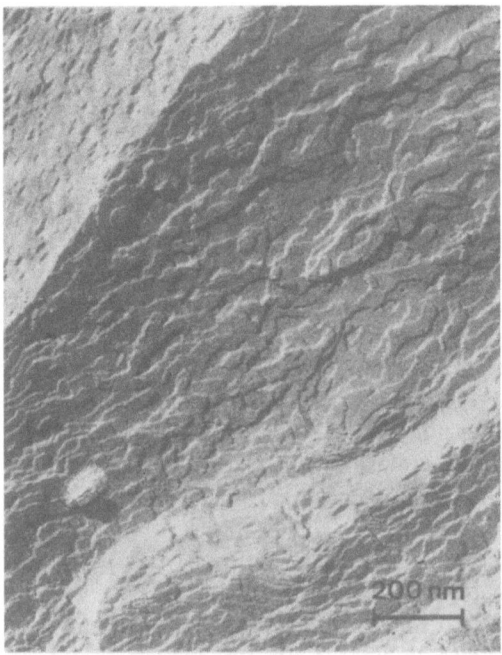

Fig. 8. Electron microscopic structure of a 65% mixture of the reaction product of Dobanol 91 with 5 mol of ethylene oxide and water following freeze-fracture processing

Fig. 9. Electron microscopic structure as shown in Fig. 8 in higher magnification

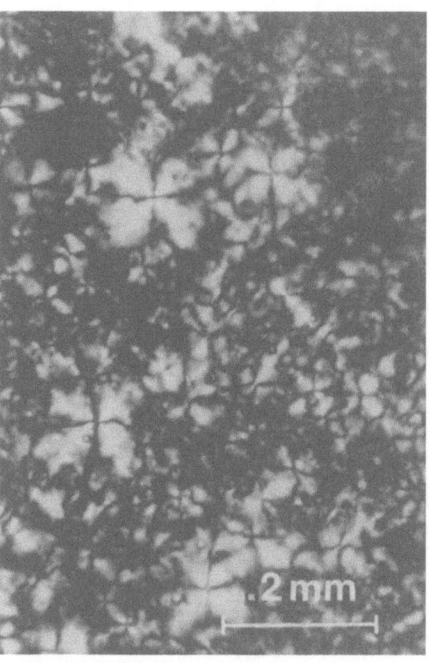

Fig. 10. Optical polarization structure of 55% mixture of the reaction product of Dobanol 91 with 5 mol of ethylene oxide and water at 40 °C in a 0.2 mm thick section

Our studies show that lamellae in which water molecules are stored exist in the mesomorphic range between 55 and 85% Dob. 91/5 EO. Mathematically speaking, three water molecules are available for each oxygen atom of the polyether chain at 54% polyether.

The stepwise incorporation of water molecules into the polyglycol ether lamellae is not clear at first sight in the present phase diagram (Fig. 3).

The individual hydration stages mingle and we have indicated them with dotted lines. However, there is also experimental evidence for the existence of the individual hydration stages. Water is successively cleaved off with increasing temperature, a process which can be followed by polarization microscopy. When a 54% sample is slowly heated, it can be seen to become clear between 31° and 35 °C; a 65% sample has a much shorter range of clarity, between 30° and 31 °C.

If we slowly heat a 50% sample the polytetrahydrate melts at 2−3 °C. Above 23 °C, a mesomorphic lamellar structure corresponding to Fig. 7 crystallizes out of the isotropic polytetrahydrate solution. This structure, which we term the polytrihydrate, melts at 61 °C. A liquid crystalline phase again appears in the polarization microscope at 64 °C, and melts at 68 °C. This structure, which we view as the polydihydrate, appears in the form of needles (bâtonnets) as illustrated in Fig. 11. The narrow zone in which the mesomorphous polydihydrate exists does not permit complete crystallization of the polydihydrate structure.

The 45% sample displays similar behavior; we must assume that the mesomorphous polytrihydrate forms the shaded tongue between 55 °C and 61 °C and that this breaks off at the miscibility gap.

A second tongue whose optical polarization structure can be recognized into far inside the miscibility gap is found between 64° and 68 °C. The polarization micrograph in Fig. 12 demonstrates the latter statement. In the miscibility gap we see a lamellar structure in an 0.2 mm-thick section of a 30% sample which does not differ from the mesomorphous lamellar structure, shown in Figs. 7 and 10.

The thermotropic behavior with isotropic intermediate zones of these aqueous polyglycol ether solutions confirms the existence of defined tenside hydrates which gradually give up their water of hydration with increasing temperature.

We again must rely on electron microscopy to understand the behavior of the ethoxylated Dob. 91/5 EO-water system in the miscibility gap. We assume a 20% solution in this case. Figure 13 shows the structure of the isotropic solution.

Water-filled tubes whose walls are composed of several tenside lamellae can be seen. The photomicrograph shows a side view of such a tubular system. In this case, the fracture predominately follows the tenside lamellae; the filling with water is invisible.

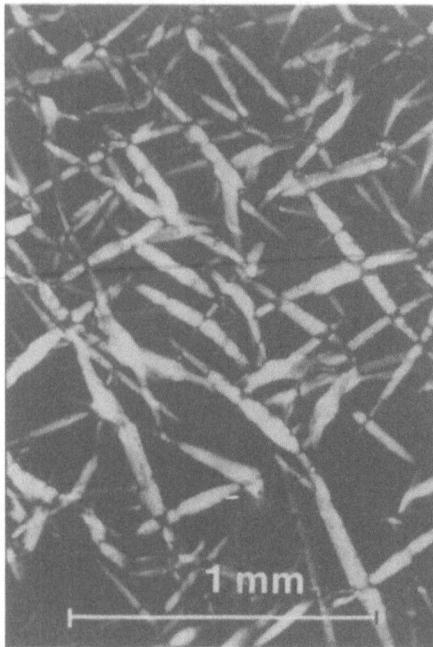

Fig. 11. Optical polarization structure of a 47% mixture of the reaction product of Dobanol 91 with 5 mol of ethylene oxide and water at 67°C in a 0.5 mm thick section

Fig. 12. Optical polarization structure of a 30% mixture of the reaction product of Dobanol 91 with 5 mol of ethylene oxide and water at 66°C in a 0.2 mm thick section

A cross fracture of the tubes is illustrated in Fig. 14. In addition to the detergent lamellae forming the tubular wall, the frozen free water is illustrated in the smooth diagonal fracture. Since both the tenside lamellae and the ice are continuous over a large area, the system may be termed bicontinuous. The last two micrographs were again taken using the freeze-fracture procedure, starting with a solution at 18°C.

If we now heat the 20% solution to 55°C, two phases appear. The relationship of densities is now such that the polyglycol ether ($d = 0.980$) collects in the organic upper phase. Very little polyether remains in the lower, aqueous phase ($d = 1.016$).

Figure 15 reproduces the electron microscopic structure if we freeze fracture the 55°C organic phase.

The water-filled tubes have disappeared, and stacks of the lamellae of a hydrated tenside remain. Whereas, an analogous elimination (such as in the case of Triton X-114), leads to a state in which lamellae and free water are present in a bicontinuous system. The elimination of the lamellae or the expulsion of free water, which is synonymous, is more complete in this case. Therefore, we see large stacks of lamellae in the tenside-rich phase. The photomicrograph shows the border between two liquid crystals.

If we heat the 20% solution to 75°C, the aqueous phase expands and the polyglycol ether layer shrinks.

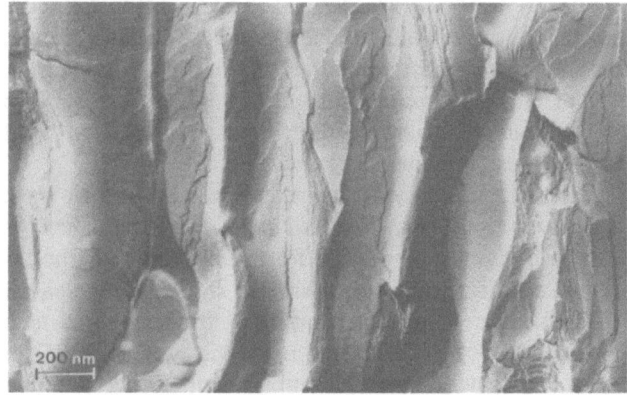

Fig. 13. Electron microscopic structure of a 20% isotropic solution of the reaction product of Dobanol 91 with 5 mol of ethylene oxide at 20°C (side view)

The expulsion of water in dependence on the temperature is illustrated in Fig. 16. The obvious plateaus between 57 and 66°C, and above 71°, indicate stepwise expulsion of water. We see a stepwise dehydration of the polyhydrates in the miscibility gap, which in return insures their existence.

The upper polyether phase now still contains about 40% water, a concentration which approaches that of the Dob. 91/5 EO polydihydrate.

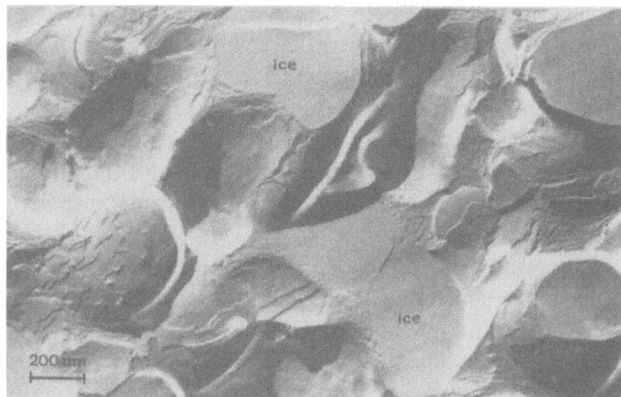

Fig. 14. Electron microscopic structure of a 20% isotropic solution of the reaction product of Dobanol 91 with 5 mol of ethylene oxide at 20 °C (cross fracture)

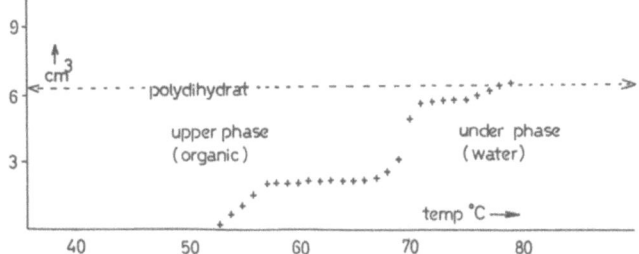

Fig. 16. Separation of water in the miscibility gap in dependence on temperature in a 20% solution of the reaction product of Dobanol 91 with 5 mol of ethylene oxide

Fig. 15. Electron microscopic structure of the organic phase separated in the lower part of the miscibility gap at 55 °C from a 20% isotropic solution of the reaction product of Dobanol 91 with 5 mol of ethylene oxide

Fig. 17. Electron microscopic structure of the organic phase separated in the upper part of the miscibility gap at 75 °C from a 20% isotropic solution of the reaction product of Dobanol 91 with 5 mol of ethylene oxide

Electron microscopy additionally provides Fig. 17, which was made using the freeze-fracture procedure after shock freezing the 75 °C solution.

Figure 17 shows a structure which is visually reminiscent of a turbulent flow. The lamellar stacks are permanently distorted. Packets of lamellae are only formed over relatively short ranges with diameters of about 0.1 – 0.2 µm. The phase is fluid and simultaneously regular.

We can assume that the polyglycol ether phase in the upper part of the miscibility gap consists of molten polydihydrate, which is liquid crystalline in the central part of the miscibility gap. A hydrate with a higher water level, probably a liquid polytrihydrate, exists in the lower part of the miscibility gap.

Below the miscibility gap the range of anomalous viscosities lies between 3 and about 25% Dob. 91/5 EO, and is designated in Fig. 3 by crosshatching. The viscosity again rises with increasing temperature in this range. Figure 18 reproduces our viscosity measurements.

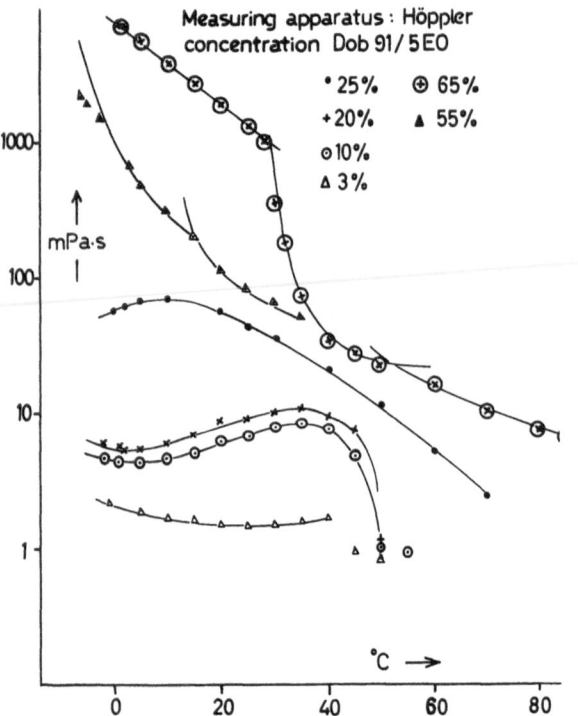

Fig. 18. Viscosity measurements on aqueous mixtures of the reaction product of Dobanol 91 with 5 mol of ethylene oxide

The reason for this increase in viscosity is the mutual hinderance of the multilamellar tubules (Fig. 13 and 14). Since the zone of anomalous viscosities continues up to concentrations of around 3% polyglycol ether, we must also accept the existence of lamellae in dilute tenside solutions.

In addition to the anomalous viscosities between 3 and 25% Dob. 91/5 EO, two viscosity-temperature curves of 55 and 65% Dob. 91/5 EO-water mixtures, whose viscosities lie in the range of the mesomorphic phases, are plotted in Fig. 18. These two viscosity curves for highly concentrated polyglycol ether solutions show that no viscosity maxima develop in this polyether. The polyether contains only short polyoxyethylene chains. The occurence of break points in the viscosity-temperature curves once more indicates phase transformation shown in Fig. 3.

In sum, we can make the following statements regarding the system Dobanol 91 with 5 moles ethylene oxide: the crystalline polyglycol ether can accept two water molecules into its crystal structure. At temperatures above the crystal phases we see two mesophases, a lamellar phase characterized by the fact that it can assimilate up to three water molecules per oxygen atom of the polyether chain, and a polytetrahydrate phase which does not display the usual sharply delineated structure under optical polarization.

Electron microscopy shows vesicles surrounded by hydrated tenside lamellae in the range of isotropic solutions. The vesicles expand with increasing water levels by storing water in their interior. Water bags are created, still surrounded by the tenside lamellae. When even more water is added, both, water and tenside lamellae are begin to fuse and bicontinnous system is formed.

The water molecules of the mesomorphic hydrate phases are successively cleaved off as temperature is increased. The consequence of this is that mesomorphic zones (which are illustrated as shaded zones in the phase diagram) are created in the optically isotropic range; of these, one phase can even by seen in the miscibility gap.

The phase diagram also shows the sharp separation between two mesomorphic zones. The clearly delineated minimum between the two phases indicates that the structures must be so different that no mixed crystal formation is possible. Even a slight increase in the water level causes this massive structural change.

4. On the existence of the polyhydrates

In conclusion, an attempt will be made to answer the question of why it took so long for the existence of defined hydrates in tensides to be accepted, although a large volume of experimental evidence independent of our results is presently available.

The phase diagram of the reaction product of Dubanol 91 with 5 mol of ethylene oxide provides an answer to this question. The individual hydrate stages are more mobile in this short-chain polyglycol ether with the low molecular weight of 380 than in polyethers with a longer ethylene oxide chain. One of us (R. H.) had already pointed out this situation in 1980 at the Polymer Colloquium in Hamburg.

In tensides, the formation of mesomorphic structures has a close connection to their surface-active properties [7].

Both the enrichment at an interface and the association in solution are closely related to the solubility of the tensides. If this solubility is low, the substance is present in the form of a precipitate and the few molecules in solution are insufficient for either adsorption at an interface or association in solution. If the solubility is too high, no reason for enrichment or association exists. However, at a critical solubility, the molecules attempt to remain in solution by forming certain structures at interfaces or in their solution. The formation of structures thus prevents precipitation of the tenside molecule from solution. Our assumption

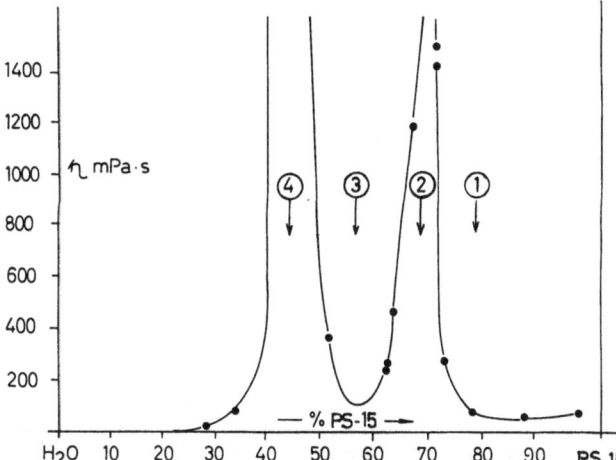

Fig. 19. Viscosity measurements on aqueous mixtures of the reaction product of 5 mol of ethylene oxide with the condensation product of phenol and styrene

that the not easily soluble portions of the molecule are turned toward the gas phase at the solution-air interface underscores this process.

When concentration is increased, aqueous zones — which are used as interfaces by the tenside molecules — are formed in the solution. Tenside lamellae then surround these aqueous zones, which shrink with increasing tenside concentration until a polytetrahydrate with a quasi-crystalline structure is reached. The question whether this liquid crystalline structure is always hexagonal still remains open.

Following the mesomorphic structures on the right side of the phase diagram in the direction of higher tenside concentrations, the water molecules are deposited in the crystal structure of the anhydrous polyglycol ether. Depending on the length of the polyglycol ether chain and/or (in ionic tensides) the strength of the electrical charge, the structures mentioned are more or less stable. In addition, intermediate stages can arise which can be broaden and complicate the appearance.

Figure 19 reproduces such a viscosity-concentration curve. At that time, we designated the polyglycol ether as "Emulsifier 2". It was obtained by reaction of the condensation product of phenol and styrene with 15 mol of ethylene oxide. It was also true that we studied a tenside with a sufficiently long ethylene oxide chain, thus, we could determine a defined polydi- and polytetrahydrate.

5. Concluding remarks

When polyglycol ether gels were studied in the 1950s for commercial reasons [10], we were fortunate — the particular emulsifiers studied displayed pronounced viscosity maxima, from whose composition we inferred the presence of polyhydrates.

References

1. Bordier C (1981) J Biol Chem 256:1604
2. Heusch R (1986) Biotech Forum 3:1
3. Kopp F, Meyer HE, Reinauer H (1985) Biol Chem Hoppe-Seyler 366:695
   Kopp F, Meyer HE, Reinauer H, Heusch R (1986) Tenside, Deterg 23:119
4. Heusch R, Kopp F (1987) Ber Bunsenges Phys Chem 91:806
5. Heusch R (1978) Ber Bunsenges Phys Chem 82:970
   Heusch R (1979) Ber Bunsenges Phys Chem 83:834
   Heusch R (1984) Ber Bunsenges Phys Chem 88:1083
6. Heusch R (1983) Tenside Deterg 20:1
   Heusch R (1984) Tenside Deterg 21:173, 298
7. Heusch R (1981) Makromol Chem 182:589
8. Becher P, Ara H (1968) J Coll Interf Sci 27:634
   Becher P, Ara H (1969) J Coll Interf Sci 31:583
9. DIN 53917 of January, 1981
10. Boehmke G, Heusch R (1960) Fette — Seifen — Anstrichm 62:87

Received November 6, 1987;
accepted June 7, 1988

Author's address:

R. Heusch
Bayer AG
5090 Leverkusen

Progress in Colloid & Polymer Science   Progr Colloid & Polymer Sci 77:86–93 (1988)

Phase equilibria and thermodynamics of liquid and solid crystal phases: formation of transition metal dodecyl benzenesulfonates

Đ. Težak, S. Heimer, S. Popović*), and B. Cerovec-Kostanić

Department of Physical Chemistry, Faculty of Science and *)Ruder Bošković Institute, University of Zagreb, Zagreb, Yugoslavia

Abstract: The precipitation solubility equilibria in detergent-metal ion aqueous solutions were characterized by the solubility constants (pK_{so}^0) of manganese, cobalt, nickel, and copper dodecyl benzenesulfonate as follows:

Mn (DBS)$_2$: 10.78±0.10
Co (DBS)$_2$: 11.17±0.19
Ni (DBS)$_2$: 11.01±0.12
Cu (DBS)$_2$: 11.33±0.29

The formation of solid and liquid crystal phases, as well as their mixtures due to the mutual nucleation of crystals (usually very fast) and the formation of smectic (slower processes) were characterized by reaction enthalpies.

The precipitation experiments were performed by the light scattering method (20°–90 °C); reaction calorimetry was used for the heat exchange measurements, and the characterization of phases was done by polarization microscopy and x-ray diffraction.

The precipitation enthalpies (ΔH_{prec}^0), as well as the reaction enthalpies (ΔH_{SC}), showed almost the same values for some of the examined dodecyl benzenesulfonates. The formation of liquid or solid crystal phases exhibited endothermic or exothermic behavior, respectively. Several periodicities exhibited in the smectics of the samples examined point to a different structural organization in smectic bilayers.

Key words: Liquid crystal, phase formation, solubility product

Introduction

Investigations of the aggregation processes in surfactant aqueous solutions [1–4], as well as their chemical reactions with metal ions, pose important problems from technological and ecological aspects [5, 6]. Dodecyl benzenesulfonic acid shows a lamellar periodicity in electrolytic solutions [7, 8]. For this reason it can be taken as a model system for the investigation of features related to the molecular structure of lyotropic liquid crystals as well as the interactions with bilayer phases [9] and other systems of scientific and technological importance. This work examined the solubility conditions and phase behavior of transition metal dodecyl benzenesulfonates. The solid-liquid equilibria were characterized by related constants; the formation of phases was characterized by thermodynamic parameters. These investigations follow the previous work presented in the papers [8, 10, 11] dealing with heterogeneous equilibria in detergent-metal ion precipitation systems.

Experimental

Materials

All aqueous solutions were prepared by dissolving the p.a. chemicals in double-distilled water as follows: Mn(II), Co(II), Ni(II) and Cu(II) nitrates from Kemika, Zagreb; dodecyl benzenesulfonic acid (HDBS) from Prva Iskra Barić, Belgrade. Metal nitrates were standardized complexometrically with EDTA [13]. HDBS is a mixture of 97–98% dodecyl benzenesulfonic acid, 1% sulfonic acid and 1.5–2% of a nonsulfonized part. The molar concentration of HDBS was calculated as a nominal value of a molecular mass of 326.5. It was used as supplied by the producer and standardized potentiometrically.

Techniques

The precipitation-solubility phenomena were followed by the light scattering method using a Virtis Brice Phoenix universal photometer, model DU2000, and a Zeiss tyndallometer connected to a Pulfrich photometer; the structures of crystalline forms were determined microscopically using a Leitz Wetzlar light microscope with polarizing equipment and by x-ray diffraction. The heat changes were measured using the reactions calorimeter described by Simeon [13]. The methods of sample preparation have been described earlier [8].

Treatment of data

For the solid-liquid equilibria calculations, a knowledge of ionic species found in aqueous solutions is necessary. It can be assumed that important ionic species appearing in aqueous solutions are nitrato and hydroxo complexes. The solubility products of metal(II) dodecyl benzenesulfonates were calculated using the linear regression in the following expression:

$$K_{so}^0 = a(M^{2+}) \cdot a^2(DBS^-) \ . \tag{1}$$

The mean activity coefficients were calculated using the Debye-Hückel law.

For calculations of enthalpy changes from calorimetric measurements the extent of the reactions was calculated applying the equilibrium constants obtained from precipitation experiments. Since the enthalpy changes for the formation of liquid crystals (ΔH_{LC}) and of solid crystals (ΔH_{SC}) were different (endothermic and exothermic, respectively), the enthalpy changes calculated from the heats of concurrent precipitation of liquid and solid crystals ($^*\Delta H_{LC+SC}$), could not be very reliable and the extent of each reaction could not be known. The heats of dilution of metal nitrates (from the concentrated stock solution to the working solution) were taken into account. They were considerable, especially for manganese nitrate.

The considerable heats of dilution, as well as the different reactions running in the experimental systems probably contributed to the errors in determination of the standard increase in the molar enthalpy. For this reason the values are considered as apparent.

The precision of the temperature jump determination was $5-10\%$ because of the simultaneous occurrence of two precipitation reactions: solid crystal nucleation and liquid crystals formation. The nucleation was a faster process. Therefore, the thermograms have the complex shape shown in Fig. 1. The temperature jump was determined in the usual way [14], although this might not be a reliable method in this special case and it will be necessary to pay special attention to it in future examinations.

Dilution of metal(II) nitrates in aqueous solutions

The appearance of the manganese nitrate ion pair $MnNO_3^+$ can be assumed according to our measurements in the concentration region $1 \ mol \ dm^{-3}$ to $2 \times 10^{-3} \ mol \ dm^{-3} \ Mn(NO_3)_2$. The dissociation reactions are supposed to be

$$Mn(NO_3)_2 \rightleftharpoons MnNO_3^+ + NO_3^- \tag{2}$$

$$MnNO_3^+ \rightleftharpoons Mn^{2+} + NO_3^- \tag{3}$$

There is some evidence of metal-nitrate ion pairing under specific conditions (a series of metallates in molten salts) [15, 16]. Therefore, there can be an indication of $MnNO_3^+$ ion pairing in aqueous solutions. Fedorov [17] et al. and Hoffman et al. [18] have found the formation of cobalt and nickel nitrato complexes under certain conditions.

In Fig. 2 the phenomenon is presented as a light scattering effect (the formation of the heterogeneous phase appeared after 10 min to several days. During the light scattering maximum pH gradually increased from 4 to 5.4. Ion pairing can be neglected below pH = 5.4. The manganese nitrate ion pair formation in aqueous solutions is followed by the fomation of $Mn(NO_3)_2 \cdot nH_2O$ associates. The evidence for this assumption was obtained by measuring the IR spectrum of the precipitate (KBr pellet) that showed the characteristic nitrate maximum, a very sharp absorption band at $1387 \ cm^{-1}$. This definitely indicates the presence of NO_3^- group, with no trace of an OH^- group. The heats of dilution of $Mn(NO_3)_2$ concentrated solutions (Q_{dil}), presented in Fig. 3, show the complete agreement with the light scattering and pH data, and it can be assumed that $MnNO_3^+$ forms as a transition state in the process of the total dissociation.

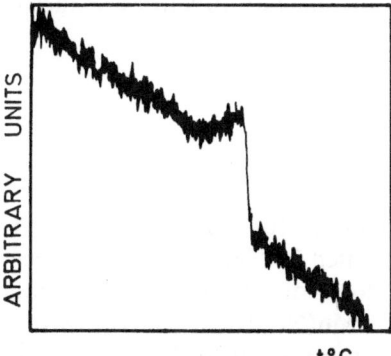

Fig. 1. Characteristic thermogram for the reaction system: $Cu(NO_3)_2 - HDBS - H_2O$

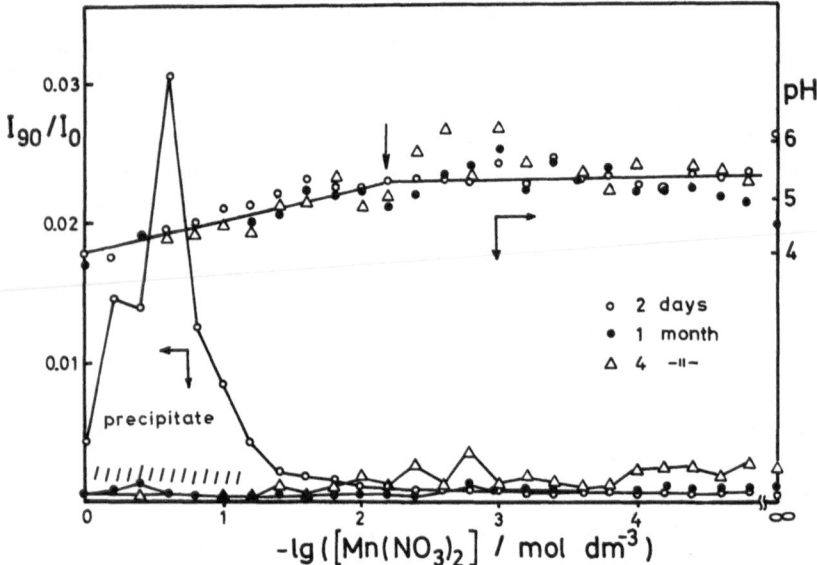

Fig. 2. Light scattering and pH plot versus Mn(NO₃)₂ concentration

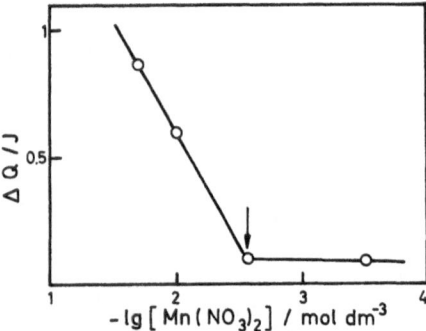

Fig. 3. Heat of dilution of Mn(NO₃)₂ versus concentration

The arrows in the Figs. 2 and 3 denote that below the $Mn(NO_3)_2$ concentration region 2×10^{-3} mol dm^{-3} there is no formation of a $MnNO_3^+$ ion pair in the aqueous solution.

The hydrolysis of the metal(II) ions

In general, the hydrolysis of manganese, cobalt, nickel, and copper nitrates in aqueous solutions does not occur below pH 5–6. Our experiments were performed below the above mentioned pH, so that according to the literature cited below, it was not necessary to take hydrolysis into account. In 1925, Britton [19] listed the solubility products $K_{so} = [M^{2+}][OH^-]^2$ for Mg, Mn, Co, and Ni, considering only pH values above 6.81.

The hydrolysis of the manganese(II) ion was studied by Perrin [20], assuming that the complex ion $MnOH^+$ is formed; a pH of 6.844 was the limiting value below which it was not possible to calculate an

equilibrium constant. Perrin also gave the data for nickel(II) hydrolysis, considering that the only important hydrolyzed species is $NiOH^+$ in equilibrium with hydrated Ni_{aq}^{2+} ion, but at a pH < 6.5 there is no appearance of $NiOH^+$ in an aqueous solution ($-\log K = 10.12$ for pH 7.276–8.472) [21].

Jena and Prasad [22] considered 1.8×10^{-16} the most probable value for solubility constant of $NiOH^+$ in a buffer solution with the resulting pH 6.8–7.11. The hydrolysis of copper(II) ion was also studied by Perrin [23]. The assumption that only the species $Cu_2(OH)_2^{2+}$ is formed for the pH range 6.01 and above, with $\beta_{22} = -10.95$. It is obvious that in experiments with pH less than 6 the hydrolysis of the above mentioned ions can be neglected.

Meanwhile, the considerable influence of hydrated and hydrolyzed species could be supposed in the processes of liquid crystal formation, because the very high concentrations of metal ions in water could be assumed to be parts of lamellar structures in mesophases.

Results and discussion

Solubility equilibria

The precipitation diagram for manganese dodecyl benzenesulfonate, presented in Fig. 4, is very characteristic. There are two slopes of the solubility limit line. In the higher excess of $Mn(NO_3)_2$ concentrations the relationship is 1 : 1; the presence of $MnNO_3^+$ has been assumed on the basis of IR investigations; the concentration solubility product for $MnNO_3DBS$ in

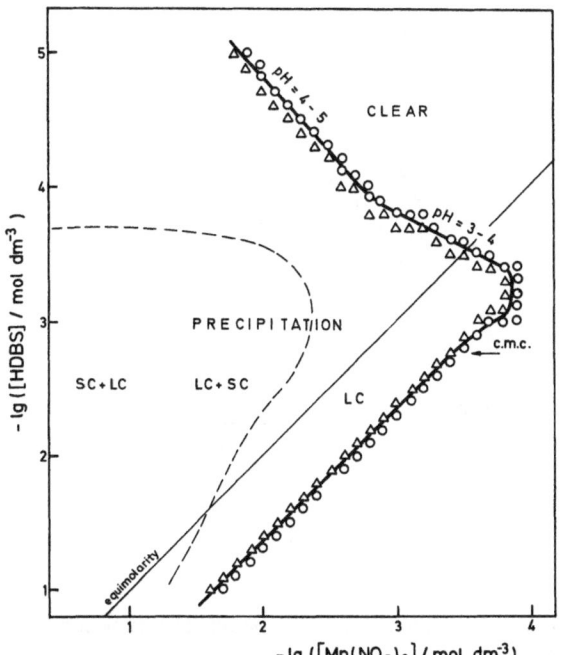

Fig. 4. Precipitation diagram of $Mn(NO_3)_2-HDBS-H_2O$ system at 20 °C

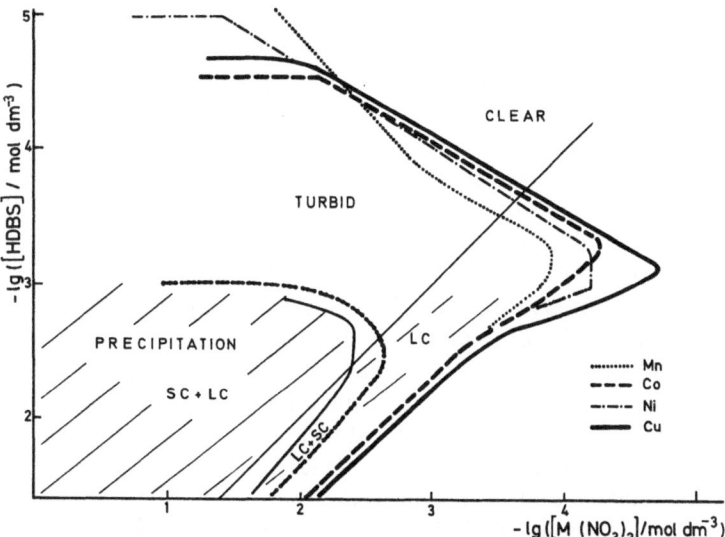

Fig. 5. The contours of solubility boundaries in the precipitation diagram of Mn, Co, Ni, Cu-DBS, and the precipitation regions for the larger concentrations of reacting components

$mol^2\,dm^{-6}$ was calculated to be: $pK_c = 6.82 \pm 0.06$. The equivalency line is intercepted by the linear part of the solubility limit, where the thermodynamic solubility product of $Mn(DBS)_2$ can be calculated because the stoichiometric relationship for the 2:1 electrolyte is exhibited. In the excess concentrations of HDBS the limit changes direction, i.e., complex ion formation occurs. These complexes can be characterized as micellar because the c.m.c. of HDBS is exceeded: it is the reason for the changed direction of the solubility limit. Inside the precipitation diagram the appearance of various liquid and solid crystal phases and their mixtures occurs, depending on the concentrations. The regions of predominancy of phases are shown in the precipitation diagrams (Figs. 4, 5). The characteristic lamellar periodicity (smectic textures) were found by microscopic observation under crossed polarizers for manganese, nickel, cobalt, and copper specimens. The contours of the solubility limits for all examined metal(II) dodecyl benzenesulfonates are characteristic and similar to the other electrolyte detergent systems [6, 7, 24] and are presented in Fig. 5. The limits between turbid and so-called clear systems were obtained by light scattering measurements. With higher concentrations of both precipitation components the formation of solid and liquid crystalline phases occurs, forming mainly their mixtures with solid crystals in excess of metal ions and liq-

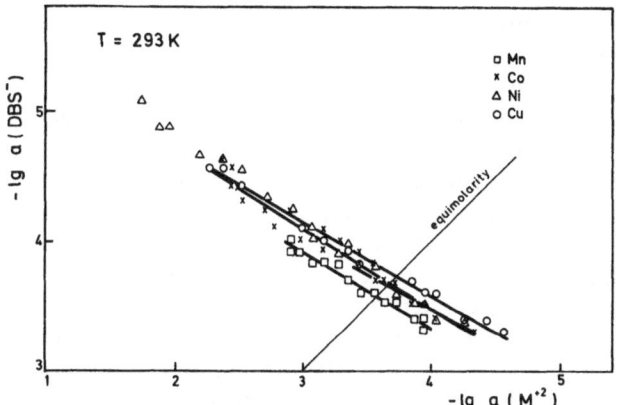

Fig. 6. Thermodynamic solubility limits of Mn, Co, Ni, Cu dodecyl benzenesulfonates

Table 1. The solubility products for $M(DBS)_2$ at 293 K

M^{+2}	pK_{so}^0	slope	Coefficient of correlation	c.c.
Mn	10.78 ± 0.10	-0.57	-0.98	$1.2 \cdot 10^{-4}$
Co	11.17 ± 0.19	-0.61	-0.98	$5 \cdot 10^{-5}$
Ni	11.01 ± 0.12	-0.58	-0.92	$6 \cdot 10^{-5}$
	*11.33 ± 0.29	-0.68	-0.99	$6 \cdot 10^{-5}$
Cu	11.26 ± 0.12	-0.56	-0.99	$2 \cdot 10^{-5}$

*pK_{so}^0 = apparent constant calculated from all experimental points.
c.c. = critical concentration = $[M^{+n}]$ in $mol\,dm^{-3}$ at which boundary changes slope.

uid crystals predominating in the negative equivalency region. The reason for this behavior lies in the faster nucleation of solid crystals in the large excess of electrolyte and, on the other side, in liquid crystal formation in a large excess of HDBS when the c.m.c. is exceeded, i.e., the equilibrium condition for micelle formation is fulfilled. The solubility limits of nickel, copper, and cobalt, show the intercept with the equimolarity line, mainly with 0.5 slope. The slopes from the calculations based on activities are listed in Table 1. The solubility products for all metal dodecyl benzenesulfonates were calculated only from experimental points lying on the straight lines in Fig. 6. The solubility product of nickel dodecyl benzenesulfonate calculated from all experimental points can be considered as an apparent value, $*K_{so}^0$, while the thermodynamic solubility product was calculated only from the points near the equivalency. It can be considered that under such limitations any possibility of nitrato and hydroxo complexes [15–23] can be neglected.

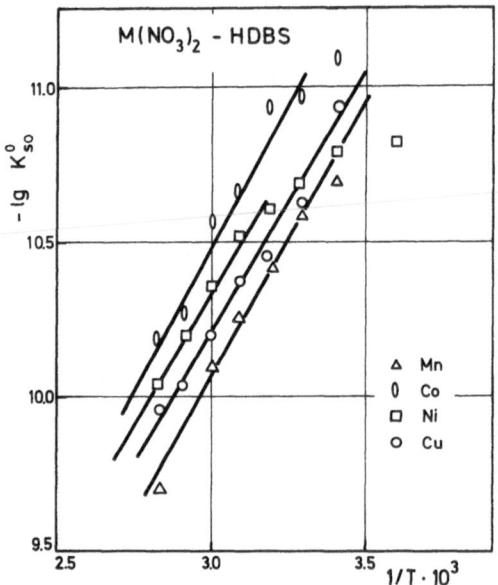

Fig. 7. The values of the solubility products as a function of temperature

Thermodynamics of phase formation

It is possible to calculate free energy and enthalpy changes from precipitation examinations. The solubility shift towards higher concentrations has been found by elevating the temperature. From the linear dependence of the logarithm of the solubility product and the reciprocal temperature, the reaction enthalpies (ΔH_{prec}^0) were calculated (Fig. 7). Almost the same slopes were obtained for all systems examined. The Gibbs free energy changes, as well as the reaction enthalpies and entropies are listed in Table 2. These values were calculated using solubility products (K_{so}^0 for 20 °C and K_{so} for 25–80 °C). The values show similar energy changes for the precipitation processes including all transition metal dodecyl benzenesulfonate solid crystal phases (ΔH_{prec}^0). It can be assumed that almost the same enthalpy changes can be expected by direct measurements in the concentration region of the precipitation diagram, where the solid crystals are microscopically detected. That is usually the region of a large excess of metal ions and the HDBS concentration under the c.m.c., and there is no possibility of liquid crystal phase formation.

In Fig. 8 the regions of calorimetric investigations are presented. The regions of monomeric form of HDBS, of micellar and liquid crystal forms characterized by the critical conditions for micelle (c.m.c.) as well as transition concentration for different micellar forms formation (c_t) [4] are denoted. The enthalpies listed in Table 2 present values obtained by calorimetry

and calculated from precipitation experiments. Even for iron(III) dodecyl benzenesulfonate the reaction enthalpy can be related to the enthalpies of the other transition metals(II). The entalpy of formation of the liquid crystal phase (ΔH_{LC}) obtained for iron(III) has shown a positive value. Meanwhile, calorimetric measurements of ΔH_{LC} and $*\Delta H_{SC+LC}$ (the mixture of solid and liquid crystal phases was detected microscopically) showed very small values. This is quite understandable taking into account that there are two mutual processes of phase formation: one process is isothermic and the other endothermic.

The values for the precipitation enthalpy of the solid crystal phase at the solubility boundary (ΔH_{prec}^0) and in the precipitation region of HDBS concentrations smaller than the c.m.c. (ΔH_{SC}) showed good agreement in the cases of Mn(II) and Cu(II) dodecyl benzenesulfonates, i.e., ΔH_{prec}^0 is identical to ΔH_{SC}. Such good results were not obtained in all systems. The reason of difference between ΔH_{SC} and ΔH_{prec}^0 in the cases of Co(DBS)$_2$ and Ni(DBS)$_2$ is probably due to difficulties in the temperature jump determination (see "Experimental"). Remarkable differences exist between the values obtained for ΔH_{SC} (below c.m.c.) and ΔH_{LC} (above the c.m.c.). The precipitation enthalpy must be equal to the reaction enthalpy below the c.m.c. of HDBS, because at these concentrations there is not liquid crystal formation. Meanwhile, $*\Delta H_{SC+LC}$ exhibits the values close to zero in all cases because of the mutual formation of both phases. The formation of liquid crystal phases usually exhibits a

Table 2. Thermodynamic parameters as calculated from: (i) precipitation method; (ii) calorimetric measurements

	By precipitation method			By calorimetry[a]		
	ΔG^0_{prec}	ΔH^0_{prec}	ΔS^0_{prec}	ΔH_{SC} [b]	*ΔH_{SC+LC}[c]	ΔH_{LC} [d]
	kJ mol^{-1}	kJ mol^{-1}	kJ mol^{-1}K^{-1}	kJ mol^{-1}	kJ mol^{-1}	kJ mol^{-1}
Mn(NO$_3$)$_2$	60.46	-31.51	-0.31	-28.5	-0.06	–
Co(NO$_3$)$_2$	62.66	-31.32	-0.32	-64.5	-0.8	–
Ni(NO$_3$)$_2$	61.76	-34.75	-0.33	-49.3	-1.47	–
Cu(NO$_3$)$_2$	63.17	-31.36	-0.32	-24.8	-0.4	–
[e] Fe(NO$_3$)$_3$	–	–	–	-33	–	23

$\Delta G^0_{prec} = \Delta H^0_{prec} - T\Delta S^0_{prec} = -RT\ln K^0_{SO}$
[a] Mean values from data obtained by reaction calorimetry for the systems chosen from the precipitation region below and above c.m.c. (HDBS)
[b] $c_{HDBS} <$ c.m.c.
[c] $c_{HDBS} >$ c.m.c.
[d] $c_{HDBS} \gg$ c.m.c.
[e] From reference [11]

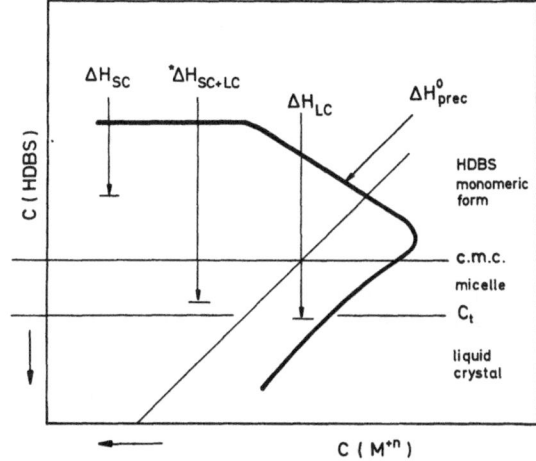

Fig. 8. The schematic presentation of the solubility boundary contour. ΔH_{SC}, *ΔH_{SC+LC}, and ΔH_{LC} are denoted as measured by calorimetry; ΔH^0_{prec} has been calculated from precipitation data

positive enthalpy of formation. The textures under crossed polarizers did not show so clear a "spherulite image" as did the other metal ions (Mg, Fe, Al) [8, 11]. The photomicrographs showed the textures of not only the unique form of liquid crystals (oily streaks or spherulites), i.e., the mixture of various textures belonging to different orderings in liquid crystals, but in most cases, the mixture of solid and liquid crystal phases, too.

X-ray investigations

The various phases have been detected by polarizing microscopy exhibiting the characteristic textures for

Table 3. Interplanar distances $D/\text{Å}$ calculated from X-ray maxima

M^{n+}	$D/\text{Å}$				
Mn^{2+}	34.6±0.6	33.1±0.1	29.50±0.03	21	18.2
			28		
Co^{2+}	35	33.5±0.5	30.0	21	
			30.5		
Ni^{2+}		33.5±0.3	24.5	20.0	
Cu^{2+}		33.7±0.5	27	21	
				22.5	
				20.5	18

the smectic phase or solid crystal. The interplanar distances between lamellar bilayers obtained from x-ray diffraction maxima are listed in Table 3. A characteristic reflection scan for manganese dodecyl benzenesulfonate in Fig. 9 shows the maxima for smectic phases. The liquid crystals have been formed in more than one periodicity in all experimental systems. The diffractogram is taken four days after preparing the system. The maxima are not very high because the structure is not very highly ordered; this fact can be related to the microscopic observations presented in the upper part of Fig. 9: in the section (a) the photograph of a mixture of birefrigent and dark crystals is taken one day after the preparation of the system. In this case the characteristic texture has not yet been observed. But some time later a completely birefringent smectic phase with the characteristic spherulites and the spherulite chains was found (presented in section (b)).

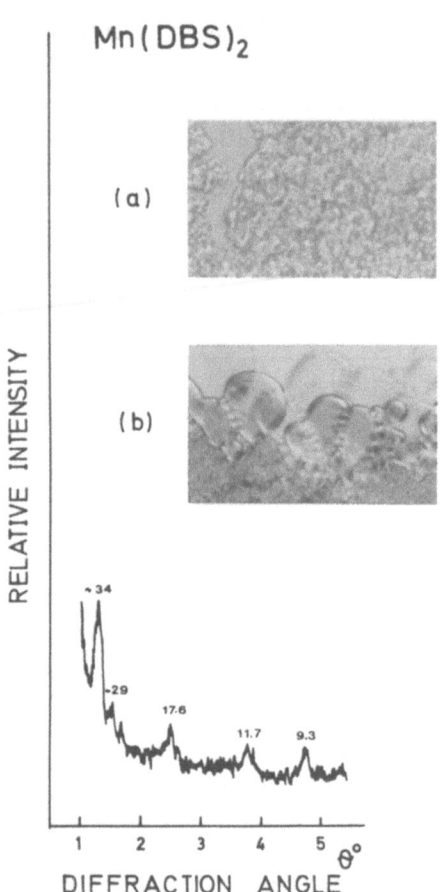

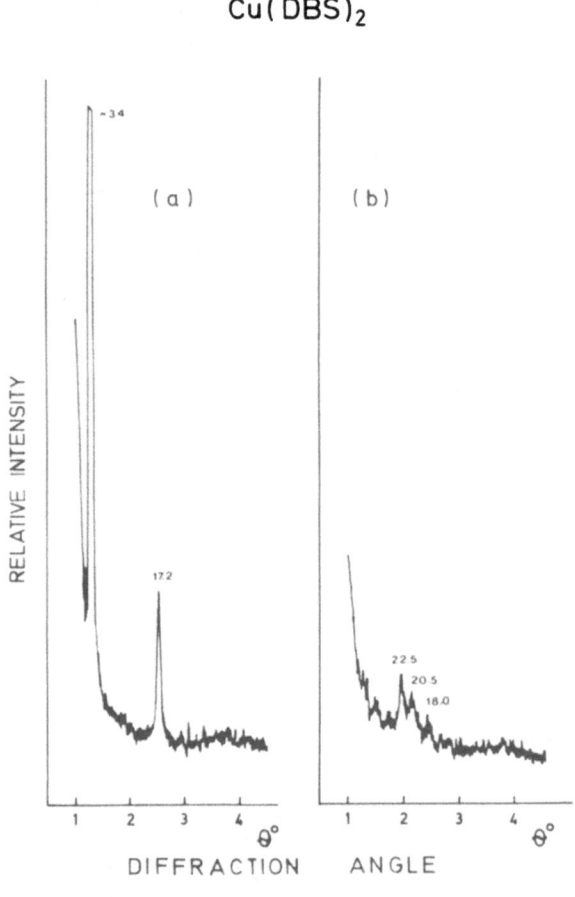

Fig. 9. The reflection scan of Mn(DBS)$_2$ four days after the preparation of the sample. Concentration in the sample are in mol dm^{-3}: [Mn(NO$_3$)$_2$] = 4×10^{-2}, [HDBS] = 4×10^{-2}. (*a*) photograph taken one day after preparation; (*b*) photograph taken several months after preparation

Fig. 10. The reflection scans of Cu(DBS)$_2$ samples exhibiting (*a*) liquid crystal formation; concentrations in mol dm^{-3}: [Cu(NO$_3$)$_2$] = 2.5×10^{-2}, [HDBS] = 5×10^{-2}; (*b*) mixture of solid and liquid crystal formation; concentrations in mol dm^{-3}: [Cu(NO$_3$)$_2$] = 6×10^{-3}, [HDBS] = 1×10^{-2}

Two reflection scans of copper dodecyl benzenesulfonate are presented in Fig. 10, presenting the highly ordered phase with the multilayer stack of lamellae in the part (a), (the sample taken from LC region), as well as the mixed SC+LC phase in the part (b) with not very well-defined structure. Three or even four different interplanar distances were found. For each periodicity, almost the same distance was displayed for the different cations. It can be assumed that the appearance of more than one interplanar distance has been caused by the possible different structuring in the lamellar layers, i.e., the depression of the water layer, the depression of the paraffin layer, or the tilt of the lamellar layers.

It can be also assumed that the hydration of the individual ions, along with the higher concentration of ions in the water layer of the lamellae, can influence the ordering within these layers.

Conclusion

In the concentrated and aged aqueous Mn(NO$_3$)$_2$ solutions, ion pairing is proposed on the basis of light scattering and pH measurements. It is confirmed by the very sharp adsorption band at the 1387 cm^{-1} in the IR spectrum, characteristic for the NO$_3^-$ group. Therefore, two dissociation equilibria can be assumed and described by relations (2) and (3). The existence of MnNO$_3^+$ and Mn^{2+} ion species in aqueous solutions was confirmed by the precipitation diagram in the system: Mn(NO$_3$)$_2$−HDBS−H$_2$O, where two equilibria with the related solubility constants were found:

$$pK_c = 6.82 \pm 0.06 \text{ (in mol}^2 \text{ dm}^{-6} \text{ for MnNO}_3\text{DBS)},$$

$$pK_{so}^0 = 10.78 \pm 0.10 \text{ (for Mn(DBS)}_2).$$

Nickel, copper, and cobalt dodecyl benzenesulfonate, exhibit solubility products of the same order of magnitude. The free energy of precipitation as well as the reaction enthalpies of precipitation (ΔH^0_{prec}) displayed fairly similar energetic changes (exothermic). The enthalpies of the liquid crystal formation present the opposite energy changes (endothermic). The apparent enthalpies in the cases of mutual nucleation of the crystal phase and the formation of mesophase ($*\Delta H_{SC+LC}$) are close to zero.

The reaction enthalpy can be used as an indication of the various precipitation processes, especially if the formation of different phases exhibits different behavior (endothermic or exothermic).

There are some differences in the interplanar distances of the smectic phases. These differences are caused by the structural organization of the smectic layers.

Acknowledgement

The authors are very grateful to Dr. Nevenka Brničević for the IR analysis, and to Dr. Zlatko Meić for the NMR characterization of dodecyl benzenesulfonic acid; both are with the "Rudjer Bošković" Institute, Zagreb, Yugoslavia.

References

1. Hoffmann H, Rehage H, Platz G, Schorr W, Thurm H, Ulbricht W (1982) Coll & Polym Sci 260:1042
2. van Os NM, Daane GJ, Bolsman TABM (1987) J Coll Interface Sci 115:402
3. Lianos P, Lang J (1983) J Coll Interface Sci 96:222
4. Težak Đ, Heimer S, Strajnar F, Popović S (1987) Farm vestn 38:223
5. Noïk C, Bavière M, Defives D (1987) J Coll Interface Sci 115:36
6. Peacock JM, Matijević E (1980) J Coll Interface Sci 77:548
7. Krishnamurti D, Somashekar R (1981) Mol Cryst Liq Cryst 65:3
8. Težak Đ, Strajnar F, Šarčević D, Milat O, Stubičar M (1984) Croat Chem Acta 57:93
9. Mann S, Hannington JP, Williams RJD (1986) Nature 324:565
10. Težak Đ, Strajnar F, Milat O, Stubičar M (1984) Progr Coll Polym Sci 69:100
11. Težak Đ, Čolić M, Hrust V, Popović S, Prgomet S, Strajnar F (1987) Progr Coll Polym Sci 74:87
12. Vogel A (1961) A Textbook of Quantitative Inorganic Analysis, Longman, London, Ed III
13. Simeon VL, Ivičić N, Tkalčec M (1972) Z Physik Chem, Neue Folge 78:1
14. Hemminger W, Höhne G (1984) Calorimetry, Verlag Chemie, Weinheim
15. Straub DK, Drago RS, Donoghue JT (1962) Inorg Chem 1:848
16. Liss CH, Hasson J, Smith GP (1968) Inorg Chem 7:2244
17. Fedorov VA, Šhmidko II, Robov AM, Simaeva LS, Šhchuhtina VA, Mironov VE (1973) Zhur Neorg Khim 18:1274 (E: 673)
18. Hoffman H, Janjić T, Sperati R (1974) Ber Bunsenges Phys Chem 78:223
19. Britton HTS (1925) J Chem Soc 127:2110
20. Perrin DD (1962) J Chem Soc: 2197
21. Perrin DD (1964) J Chem Soc: 644
22. Jena PK, Prasad B (1956) J Indian Chem Soc 33:122
23. Perrin DD (1960) J Chem Soc: 3189
24. Krznarić I, Božić J, Kallay N (1979) Croat Chem Acta 52:183

Received March 7, 1988;
accepted June 1, 1988

Authors' address:

D. Težak
Department of Physical Chemistry
Faculty of Science
University of Zagreb
Marulićev trg 19/II
P.O. Box 163
41001 Zagreb

Progress in Colloid & Polymer Science   Progr Colloid & Polymer Sci 77:94–99 (1988)

The influence of charged surfactants upon reverse osmosis

E. Hinke, D. Laslop and E. Staude

Institut für Technische Chemie, Universität Essen, Essen, F. R. G.

Abstract: Cationic as well as anionic surfactants influence the transport properties of cellulose acetate membranes. The rejection of the surfactants is decreased when electrolytes are added to the solution and can reach negative values at low pressures. Likewise, the volume flow is lower when the solution contains electrolytes and surfactants. Adsorption and the gel layer are responsible for the effects which can be better demonstrated using membranes that are not highly desalting.

Key words: Reverse osmosis, ionic surfactants, CA membranes, adsorption, gel layer

Introduction

The increasing production and use of surfactants in all areas of human life result in increasing releases of these substances into the environment. Consequently, surfactants are ubiquitous. This would not create severe problems if these surfactants could, like sodium dodecylsulfate, be metabolized by bacteria because of their chemical structure. However, especially for technical processes, e.g., in the plating industry highly resistant surfactants are necessary. In this case the carbon chain of the molecule is perfluorinated. Once released into the environment these substances resist bacterial attack longer. They can be adsorbed on surfaces and cell membranes. This may influence the transport behavior of the biologic membranes. Also, the synthetic membranes used for water recovery processes are exposed to interaction with surfactants when the raw water contains surface active agents. The characteristic properties of a membrane, volume flow, and selectivity can be changed by these interactions. This can be demonstrated when a reverse osmosis experiment is performed in the presence of polyvinyl methylether using an electrolyte solution. The salt rejection is enhanced whereas the volume flow through the cellulose acetate membranes used is slightly decreased [1]. In the beginning, reverse osmosis experiments were carried out using highly desalting membranes which reject very well, e.g., nonionic polyoxyethylated alkylphenols [2]. For earlier investigations, the change of membrane properties was attributed to a mechanical blocking of the pores by the surfactants used [1]. However, the conception of Markly [3] and Kesting [4] that a liquid membrane is built up adjacent to the membrane surface has prevailed. This liquid membrane can be called a secondary membrane made by the surface active agents exerting an additional transport resistance. Hence, two membrane resistances are put in series. Below the critical micell concentration (CMC), the liquid membrane is only partly developed upon the membrane, but increasing surfactant concentration reduces the mass transport through the membrane by increasing the liquid membrane area and its thickness. Once the CMC is reached, the liquid membrane is completely erected, and an additional transport resistance is now caused by the increasing density of this layer. The permselectivity of this membrane system depends on the hydrophilicity of the detergent. When the hydrophilicity of the detergents increases so does the permselectivity. This behavior is related to the ability of the surface active substances to bind water.

Another effect of the CMC upon the transport through differently annealed CA membranes was observed using charged detergents [5]. Neither the transport of cationic nor that of anionic detergents is influenced when the corresponding CMC is reached. Conversely, above the CMC of nonionic detergents their permeability is lower than below the CMC.

Further, the permeability of detergents through membranes can be affected by substrate-membrane interactions. When cation-exchange membranes are used the electrostatic repulsion hinders the adsorption of anionic surfactants, and a volume flow reduction does not occur below CMC. Above the CMC, the decreasing volume flow through the membrane is explained by the micell formation and by the gel polarization model, respectively [6]. Decreasing volume flow can also result from ionic bonding of cationic surfactants at cation-exchange membranes [7]. Generally, the influence of charged ion-exchange membranes is the consequence of the kind of counterion [8]. However, surfactant solutions and dense nonionic membranes are used in experimentation, interactions between the membrane polymer and the surfactant cannot be observed because rejection is very high. Reverse osmosis investigations using an anionic surfactant and membranes which possess less rejection properties recently demonstrated that under these conditions the volume flow through the membranes decreases remarkably [9]. Also, the rejection is strongly affected. This work has been continued in order to find out if there is an interaction between the membrane and the surfactant. The role of electrolytes added to the surfactant solution should also be investigated. Some of these results are presented in this paper.

Experimental

The surfactants used in the experiments described below were tetraethylammoniumperfluorooctane sulfonate, trade name FT 248, purchased from M & T GmbH, Stuttgart, F. R. G., and benzalkonium chloride, trade name BAC, purchased from Fluka GmbH, Neu Ulm, F. R. G.; other chemicals used were of analytical grade.

Experiments

The experiments were carried out in a closed-loop reverse osmosis plant on a laboratory scale that was fit with six cells connected in parallel [10]. Pressure of up to 40 bars was supplied by a membrane piston pump; the feed flow rate was 900 mL/min. The temperature was maintained at 20 ± 0.3 °C. Three types of cellulose acetate membranes were used for the experiments. They were kindly supplied by Hoechst AG, Werk Kalle, Wiesbaden, F. R. G. One type is usually employed for ultrafiltration, whereas the other is annealed at 75 °C for 8 min. By this procedure it becomes a reverse osmosis membrane of moderate salt rejection. A third membrane is annealed at 90 °C for 5 min. It was used only in the investigations for hydronium ion rejection (see last section). The characteristic values of these membranes are shown in Table 1.

Analysis: The volume flow J_v was measured gravimetrically. The chloride concentration was measured by potentiometry, and the magnesium concentration was measured by complexometry using EDTA. For the surfactants, photometry was employed, and FT 248 was dyed by methylene blue. BAC absorbs at 260 nm, therefore it could be measured directly.

Table 1. Membrane characteristics. 40 bars, pure water, and 0.085 m NaCl

	J_{H_2O} L/m²/d	J_v L/m²/d	R %
CA-10	5100	4500	21.5
CA-75	1700	1400	80.5
CA-90	1000	920	90.0

Results and discussion

Cationic surfactant: The surfactant BAC is a substituted quaternized ammonium base. One of the four substituents is an alkyl group with a carbon chain length from 12 to 16. The second is a phenylmethyl group and the other two substitutents are methyl groups. However, this bulky molecule does not remarkably influence the volume flow when the experiments are performed with the pure surfactant solution using both membrane types. This was measured over the whole range of applied pressures. Here, applied pressure means the effective pressure difference according to the volume flow equation: $J_v = L_p \cdot (\Delta P - \Delta \pi)$. Here, J_v is the volume flow (L/m²/d), L_p is the hydraulic permeability [L/(m² s bar)], ΔP is the transmembrane pressure difference, and $\Delta \pi$ the transmembrane osmotic pressure difference. Since the concentration of this surfactant solution was 0.11 mmol, the influence of the osmotic pressure difference could be neglected.

When electrolytes are added to the surfactant solution an effect on the reverse osmosis behavior can be observed which, in addition, depends on the solute properties. Thus, at low NaCl concentrations (5 mmol) the volume flow is reduced by about 20%, whereas the same concentration of MgCl₂ causes only a 10% decrease. The results for this experiment, as well as for those at higher electrolyte concentrations performed with CA-10 membranes, are shown in Table 2. Since these membranes are mostly used for ultrafiltration purposes, an effect of osmotic pressure on the volume flow need not be considered. In this table the

Table 2. Volume flow through a CA-10 membrane using a $1.11 \cdot 10^{-4}$ mol/L BAC solution and added electrolytes

ΔP bar	0.005 m NaCl	$J_{v,rel.}$ 0.005 m MgCl₂	0.1 m NaCl	0.1 m MgCl₂
10	0.84	0.92	0.73	0.76
20	0.84	0.91	0.69	0.76
40	0.80	0.90	0.68	0.74

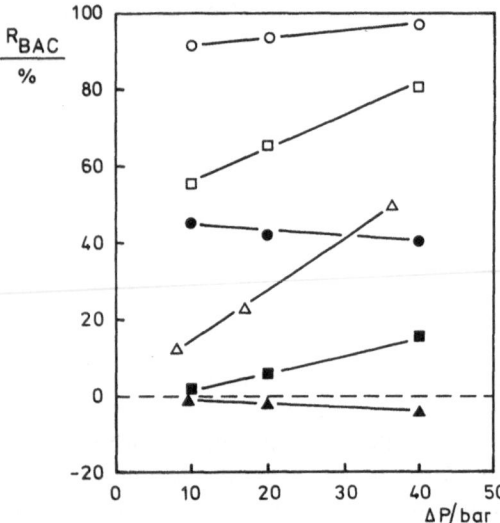

Fig. 1. Rejection of cationic surfactant R_{BAC} as function of effective pressure difference ΔP. Membranes: CA-10 filled symbols, CA-75 open symbols. Solutions BAC $1.11 \cdot 10^{-4}$ mol/L $\circ$, with 0.005 mol/L NaCl $\square$, with 0.1 mol/L NaCl $\triangle$. Temp. 20 °C

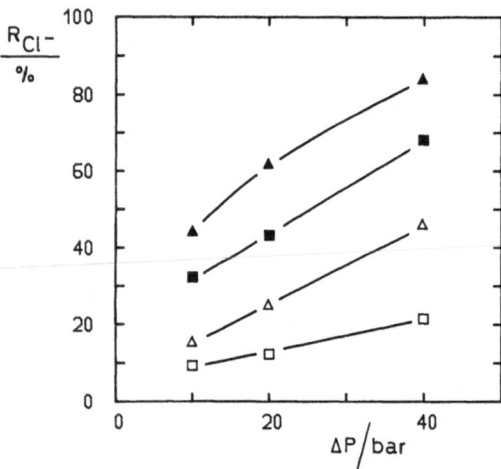

Fig. 2. Rejection of chloride R_{Cl^-} as function of effective pressure difference ΔP. Membrane: CA-10. Solutions: 0.005 mol/L NaCl $\square$, 0.005 mol/L MgCl$_2$ $\triangle$, open symbols without BAC, filled symbols with $1.11 \cdot 10^{-4}$ mol/L BAC, 20 °C

relative volume flow $J_{v,rel.}$ is the ratio of the electrolyte volume flow with surfactant to that without surfactant. Increasing electrolyte concentration results in a stronger volume flow decline which is caused by the electrolyte-surfactant interaction. In addition, the transport resistance through the membrane rises because of the preferred adsorption of the surfactant onto the membrane surface which follows from the diminution of the repulsion between the charged groups in the electrostatic double layer [11].

Besides the volume flow, the other characteristic quantity of reverse osmosis is the rejection of the substrate. This is likewise affected by adding electrolytes to a surfactant solution. In Fig. 1 the surfactant rejection is shown. In any case the rejection of the BAC is lessened and tends to negative values at higher NaCl concentrations when low-rejecting membranes are used. This behavior is more pronounced when MgCl$_2$ is used as electrolyte. The presence of electrolytes causes an enhanced sejour probability of the surfactant upon the membrane surface as well as in the pore volume of the membrane. This results in a higher surfactant permeability.

On the other hand, the adsorbed cationic surfactant and its higher concentration in the pore volume of the ultrafiltration membrane, increases the charge of this phase. Therefore, the cations (sodium or magnesium) are more hindered from entering the pores by ionic exclusion. The rejection of the corresponding counterion chloride increases, as is shown in Fig. 2. Lower

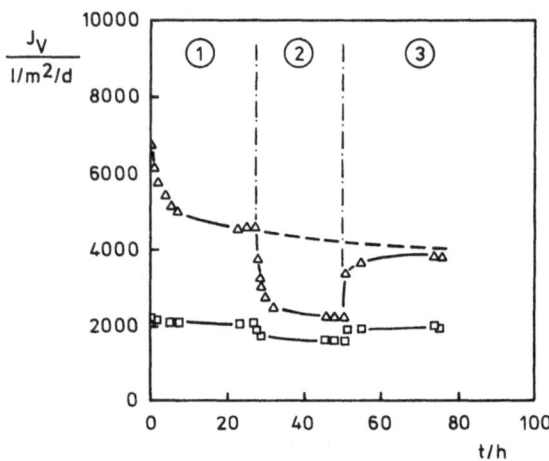

Fig. 3. Volume flow J_v as function of time t. Membranes: CA-10 $\triangle$, CA-75 $\square$. Solutions: distilled water **1**, FT 248 $1.59 \cdot 10^{-4}$ mol/L **2**, distilled water **3**. Dashed line: calculated flow reduction caused by compression; pressure 40 bars

concentrations are more effective than higher concentrations.

Anionic surfactant: The FT 248 is an extremely hydrophobic surfactant; it is a small molecule compared with the cationic surfactant employed in these investigations. Because of its chemical properties it is better adsorbed onto the membrane surface than the other surfactant. A marked influence on the volume flow should result. This can be taken from Fig. 3 which shows the volume flow as a function of time,

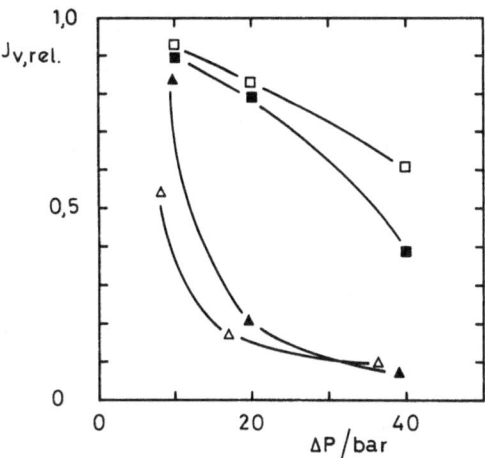

Fig. 4. Relative volume flow $J_{v,rel.}$ as function of effective pressure difference ΔP. Membranes: 10-CA filled symbols, CA-75 open symbols. Solutions: FT 248 $1.11 \cdot 10^{-4}$ mol/L with 0.005 mol/L NaCl □, with 0.1 mol/L NaCl △, 20 °C

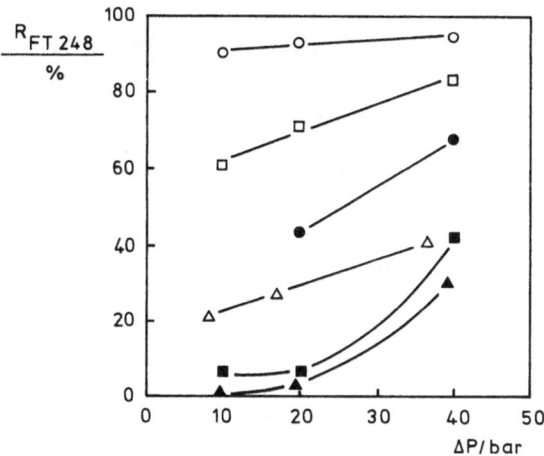

Fig. 5. Rejection of anionic surfactant R_{FT248} as function of effective pressure difference ΔP. Membranes: CA-10 filled symbols, CA-75 open symbols. Solutions: $1.11 \cdot 10^{-4}$ mol/L FT 248 ○, with 0.005 mol/L NaCl □, with 0.1 mol/L NaCl △, 20 °C

and from Fig. 4 wherein the relative volume flow is plotted against the effective pressure difference for different membranes used. The flow reduction caused by the surfactant is reversible. An extended period of rinsing with water can reestablish the former flow properties of the membranes (see Fig. 3). At the higher sodium chloride concentration of 0.1 mol, an additional characteristic can be observed (see Fig. 4). Unlike the cationic surfactant, even at the concentration of 0.11 mmol a gel layer is formed on the membrane which increases when the applied pressure rises. For the CA-10 membrane an amount of 1.0 g/m^2

membrane area was measured at 10 bars. It was 4.2 g/m^2 at 40 bars. The corresponding values for the CA-75 membrane were 2.0 g/m^2, and 3.5 g/m^2, respectively. According to the gel model of Blatt et al. [12], the concentration difference between the membrane surface and the bulk phase reaches its maximum value when the gel layer is formed. If it is assumed that the diffusion coefficient and the streaming conditions in the system are constant, then the back transport of the molecules from the layer adjacent to the membrane into the bulk phase is likewise constant. With increased drive forces the convective transport to the membrane surface and hence the thickness of the gel layer increase because the back diffusion remains constant. Therefore the volume flow decreases.

The surfactant rejection is negatively influenced in the presence of electrolytes as can be seen from Fig. 5 for NaCl using both the CA-10 and the CA-75 membrane. A better presentation of the surfactant permeation can be obtained when the substrate flow is used. This value is calculated with the following formula

$$J_s = c_s \cdot J_v \cdot (1 - R_s) \; . \tag{1}$$

J_s is the substrate flow (mol/m^2/d), c_s is the concentration (mol/L), and R_s is the measured rejection. Since the volume flow decreases with increasing electrolyte concentration, the permeation also becomes lower (Table 3). In some cases various cations form complexes with surfactants which have a low solubility product. Magnesium is one of these cations [11]. Therefore, the transport resistance of the surfactant rises at higher magnesium concentrations. It has been found that the complex precipitates at magnesium concentrations of 2 mol/L and at FT 248 concentrations of $4.77 \cdot 10^{-4}$ mol/L. In contrast, sodium does not form a complex; the normal behavior is observed. The different effects of the electrolytes can also be seen in the chloride rejection. In Table 4 the experimental values using a CA-10 membrane at 10 bars are listed. They indicate a pronounced influence of the magnesium-surfactant interaction at higher magnesium concentrations. From Table 4 it can be seen that the influence of the magnesium is the converse of that of sodium.

Hydronium ion permeation: The different chemical properties of the two charged surfactants are also manifest in the hydronium ion rejection. The pH of the solutions was 6.0. Besides the CA-10 membrane the highly rejecting membrane CA-90 was used in order to perform the investigations, partly at high cation rejection. The most mobile cation in the systems used is the hydronium ion. The corresponding

Table 3. Influence of different sodium chloride concentrations upon solute flow of FT 248

ΔP bar	$J_s \, mol/m^2/d$			
	CA-10		*CA-75*	
	0.005 m NaCl	*0.1 m NaCl*	*0.005 m NaCl*	*0.1 m NaCl*
10	0.23	0.19	0.027	0.023
20	0.30	0.07	0.030	0.011
40	0.13	0.03	0.017	0.009

Table 4. Chloride rejection behavior of a CA-10 membrane using different electrolytes

ΔP bar	Chloride rejection R_{Cl}-%			
	$1.11 \cdot 10^{-4}$ m FT 248		$6.63 \cdot 10^{-5}$ m FT 248	
	0.005 m NaCl	0.1 m NaCl	0.005 m $MgCl_2$	0.1 m $MgCl_2$
10	30.8	17.1	21.0	27.5
20	45.8	17.3	34.8	47.1
30	–	–	43.0	60.2
40	77.9	33.8	52.2	67.0

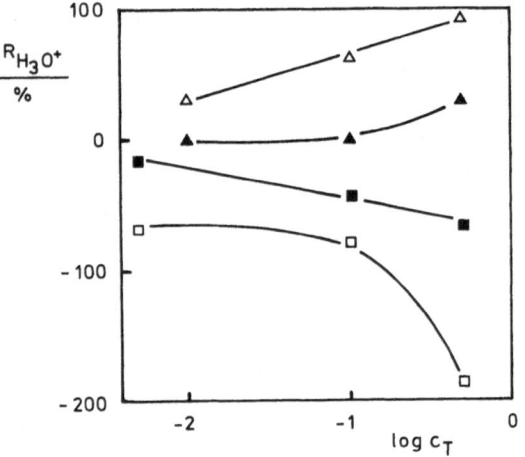

Fig. 6. Rejection of hydronium ion $R_{H_3O^+}$ as function of surfactant concentration c_T. Membranes: CA-10 filled symbols, CA-93 open symbols. Solutions: 0.005 mol/L NaCl with BAC □, with FT 248 △, 20 °C

counterions are either chloride or the surfactant anion FT 248. When the reverse osmosis membrane is used in the solution containing the cationic surfactant, the permeation of the smaller cation should be increased at rising surfactant concentration. This is observed for the hydronium ion. In both cases an ion exclusion which corresponds to a better rejection occurs since the adsorbed surfactants increase the fixed charge of

the membranes. At higher BAC concentration the layer formed adjacent to the membrane hinders the BAC permeation. The same is valid when the anionic surfactant is employed. In Fig. 6 the hydronium ion rejection is shown as a function of the surfactant concentration for the two CA membranes.

Conclusion

The influence of surfactants on the reverse osmosis of electrolyte solutions was investigated. Two ionic-type surfactants were chosen to show an interaction between membrane polymer and substrate; these surfactants were supposed to interfere with the electrolyte permeability. Various cellulose acetate membranes were used for the study. Of these, two types did not possess high rejecting qualities because the interactions could be better observed when the membranes were less rejecting of the substrates. In addition, with more effective membranes the concentration polarization is more marked than with porous membranes. This would lead to difficulties in differentiating IF the influence upon the volume flow and the rejection stems from either the concentration polarization or the surface active agent itself.

The CA-75 membrane is a reverse osmosis membrane with moderate reverse osmosis properties. In addition, this membrane has a low cation-exchange capacity. Therefore, when the pure surfactant solution is applied, the solute is better rejected than the sodium chloride which normally is used for membrane characterization (see Table 1). This depends on the dimension of the molecule. The ion exclusion is also effective because the concentration of the surfactant is low. Electrolytes change the situation. The surfactant permeability is increased and the volume flow is reduced. The anionic surfactant is more hydrophobic. With regard to this property there is no difference compared with the other surface active agents when sodium chloride is added. However, magnesium chloride exerts an additional effect because of its interaction with the surfactant. Thus, the low volume flow results from this surfactant type itself and in addition from the developing gel layer at higher electrolyte concentrations. Contrary to this, the permeability is less influenced by using the cationic surfactant.

The CA-10 membrane shows a more pronounced influence of the electrolytes on the surfactant permeability through them. The most significant effect is the zero or even negative surfactant rejection. This cannot be explained by a preferred solubility of organic within the membrane phase such as is known for phenol, e.g., [13]. However, this is only measured at neutral conditions for a phenolic solution whereas the surfactants are ionic. Apparently, a salt effect, especially at higher salt concentrations, is responsible for the low rejection. Hence, the membrane phase is the favored phase, which may explain the high permeability. This can be derived from the stronger pressure dependence of the electrolyte rejection compared with that of the surfactant. According to the following formula which results from equation (1)

$$R = 1 - J_s/(c_s \cdot J_v) \qquad (2)$$

which gives the ratio of the substrate flow to the volume flow. An increase of the pressure speeds up the BAC permeation compared with that of the electrolytes. Therefore the behavior of the BAC is intermediate between that of organic substances with a high affinity for the membrane polymer and inorganic salts. An additional effect results from adsorption. In agreement with the volume flow reduction at increasing electrolyte concentration a more intense adsorption of the surfactant occurs at the pore walls within the membrane. The local high concentration of this substance involves a lower rejection.

The results of these investigations cannot be transferred to other membranes without some restrictions. One of them is the chemical property of the membrane polymer. Because of the different isoelectric points of the membranes, which also vary for differently annealed CA membranes, the surfactant-polymer interactions can be influenced. The same is valid for the adsorption. Of course, when using highly effective reverse osmosis membranes the surfactant influence is not remarkable. Nevertheless, surfactants will play a role in reverse osmosis investigations.

References

1. Michaels AS, Bixler HJ, Hodges RM (1965) J Colloid Interface Sci 20:1034
2. Sourirajan S, Sirianni AF (1966) Ind Eng Chem Prod Res Dev 5:30
3. Markly R, Cross R, Bixler HJ (1967) OSW Res Develop Rept 281
4. Kesting RE, Subcasky WJ, Paton JD (1968) J Colloid Interface Sci 28:156
5. Kamizawa C, Ishizaka S (1972) Bull Chem Soc Jpn 45:2967
6. Van der Velden PM, Smolders CA (1977) J Colloid Interface Sci 61:446
7. Luppino RJ, Spencer HG (1982) Desalination 41:33
8. Pegoraro M, Penati A, Pizzamiglio A (1974) J Appl Polym Sci 18:379
9. Staude E, Assenmacher W (1984) Desalination 49:321
10. Pusch W (1986) Desalination 59:105
11. Stache H (1981) Tensid Taschenbuch. Hanser Verlag, München, Wien
12. Blatt WF (ed), David A, Michaels AS, Nelsen L (1970) In: Membrane Science and Technology. Plenum Press, New York
13. Pusch W, Burghoff H-G (1976) Proc Int Symp Fresh Water Sea 5th, 4:143–156

Received February 15, 1988; accepted June 7, 1988

Authors' address:

E. Staude
Institut für Technische Chemie
Universität Essen
4300 Essen

Progress in Colloid & Polymer Science

Progr Colloid & Polymer Sci 77:100–108 (1988)

Emulsions and Microemulsions

Measurement of low interfacial tensions by capillary wave spectroscopy. Study of an oil-water-surfactant system near its phase inversion

D. Wielebinski and G. H. Findenegg

Physikalische Chemie II, Ruhr-Universität Bochum, Bochum, F. R. G.

Abstract: Dynamic light scattering by small-amplitude thermal fluctuations at liquid-liquid interfaces (capillary wave spectroscopy) is used to measure interfacial tensions in oil + water + nonionic surfactant systems. Spectra for scattering angles between 0.1° and 0.45° (corresponding to surface modes with wave-numbers between 240 cm^{-1} and 730 cm^{-1}) are detected by the heterodyne technique, using a high-performance signal analyzer. The resulting power spectra are evaluated on the basis of the Herpin-Meunier dispersion relation. Technical details of the experimental setup are given and possible sources of error are discussed. As an illustration of the method, some results for the system decane + water + dodecyl-tetraoxyethylene monoether ($C_{12}E_4$) in a temperature range above the phase inversion are presented. In this temperature range in which the oil-water interfacial tension varies by more than two orders of magnitude, one observes a transition from spectra with a peak maximum at a finite frequency shift (corresponding to slowly propagating capillary waves) to spectra centered at the frequency of the incident light (corresponding to overdamped surface modes). The analysis of the spectra in this transition region is discussed.

Key words: Interfacial tension, capillary wave spectroscopy, light scattering, oil-water-surfactant system

1. Introduction

The interfacial tension of water + alkane (oil) systems can be reduced to very low values (far below 0.1 mN m^{-1}) by either non-ionic or ionic surfactants, and pronounced minima in tension with respect to temperature, salt concentration (salinity), or alkane carbon number have been found in such three-, four-, or five-component systems [1–4]. Such ultralow interfacial tensions cannot be measured by most classical techniques, but two alternative methods can be employed: the spinning drop method, and an optical method based on an analysis of the spectrum of light scattered by a macroscopically flat interface (capillary wave spectroscopy) [5, 6]. On a microscopic level, fluid interfaces are never perfectly flat but a little rough due to thermal excitations. The amplitude of these surface corrugations is very small (only a few Ångstroms) but increases as the interfacial tension decreases. Therefore this method is particularly well suited to studying ultralow tensions.

The hydrodynamic theory of the propagation and dispersion of short-wavelength surface waves (so-called capillary waves or ripples) is well established [7, 8].

Indeed, mechanically generated surface waves of known frequency have long been used as optical diffraction gratings for accurate determinations of surface tension, (see [9] for a recent application of this Bragg diffraction technique). Capillary wave spectroscopy measures the small frequency shifts of light resulting from its interaction with the spontaneous surface modes (quasi-elastic photon-ripplon scattering). This technique requires the coherence properties of laser light sources and the resolution provided by modern light beating methods [10]. Two limiting forms of the spectrum can be considered [11, 12].

If the damping forces due to the viscosity of the fluids are weak, the capillary waves will propagate along the interface and interact with the incident light beam like a moving diffraction grating. Accordingly, the scattered light is frequency shifted by the Doppler effect and exhibits two Lorentz-shaped peaks symmetrically displaced from the incident frequency. In the simplest case, the interfacial tension can then be obtained directly from this frequency shift of the scattered light. The other limiting case is that of highly viscous fluids, when any disturbance of the interface decays exponentially without propagating. In this case the spectrum of the scattered light exhibits no separate peaks but only a broadening of the central line, and the interfacial tension can be obtained from the half-width of that line. For intermediate cases of slowly propagating, strongly damped capillary waves no simple approximate relations are available, but the full dispersion equation must be used for the analysis of the spectra [13].

The present paper describes the application of capillary wave spectroscopy to interfaces in oil + water systems in this intermediate regime. We first repeat the basic theory of light scattering from fluid interfaces and discuss our method of analyzing the spectra. Next, we explain the experimental setup developed in our laboratory [14]. Finally, we present capillary wave spectra for a typical alkane + water + nonionic surfactant system in which the interfacial tension varies by more than two orders of magnitude in a temperature range of only 10 K. The resulting interfacial tensions are then compared with literature data on related systems. Extensive results on interfacial tensions in three-phase regions of oil + water + surfactant systems will be presented in a subsequent paper [15].

2. Theoretical

2.1 Dispersion relation for capillary waves

Thermally excited disturbances on the free surface of liquids or liquid-liquid interfaces can be Fourier analyzed in terms of surface modes. The individual modes (or capillary waves) are characterized by their wave-number q and radian frequency ω. These mea-surable quantities can be related to the material properties of the two-phase system via the appropriate dispersion relation. Here we consider capillary waves of small amplitudes $\zeta_q (q\zeta_q < 10^{-2})$ on an ideal sharp interface which separates two incompressible fluids of mass densities ρ and $\rho' (\rho \geqslant \rho')$ and viscosities η and η'. Following Herpin and Meunier [16] we use the reduced variables

$$Y = \frac{\sigma_q (\rho + \rho')}{4(\eta + \eta')^2 q} \quad \text{with} \quad \sigma_q = \sigma + g(\rho - \rho')/q^2 \quad (1)$$

and

$$S = i\omega\tau_0 \quad \text{with} \quad \tau_0 = \frac{\rho + \rho'}{2(\eta + \eta')q^2} \quad (2)$$

where σ is the interfacial tension and g is the gravitational acceleration. The parameter Y expresses the balance between the driving forces due to the interfacial tension and the gravitational force, and the damping forces which are proportional to the viscosities of the two phases. The parameter τ_0 represents a characteristic decay time of vortex motion. The Herpin-Meunier relation is based on the approximation $\rho/\eta = \rho'/\eta'$ and has the form

$$D(S) = Y + \frac{2\rho\rho'}{(\rho + \rho')^2} S\sqrt{1+2S}(1+\sqrt{1+2S})$$
$$+ \left(\frac{\rho - \rho'}{\rho + \rho'}\right)^2 [(1+S)^2 - \sqrt{1+2S}] = 0. \quad (3)$$

For Y values above a critical value Y_c this quadratic equation yields roots S_1 and S_2 which are complex conjugate, corresponding to propagating capillary waves, i.e.,

$$S_{1/2} = S_r \pm i S_i . \quad (4)$$

Below Y_c two real roots S_1 and S_2 are obtained, corresponding to overdamped (non-propagating) surface motions. Two special cases of Eq. (3) are commonly considered [17, 18]:

(A) For liquid-gas interfaces the gas phase density ρ' is negligible and thus, with $\rho' = 0$:

$$D_g(S) = Y + (1+S)^2 - \sqrt{1+2S} = 0 \quad (5)$$

with $Y_c = 0.145$.

(B) For two fluids of equal mass density $(\Delta\rho = \rho - \rho' = 0)$, which is approximately true for many liquid-liquid systems, Eq. (3) yields

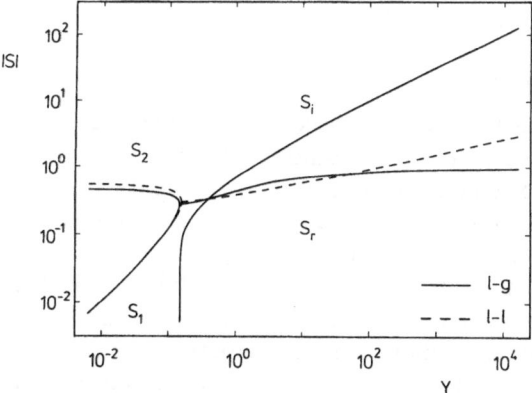

Fig. 1. Roots of the dispersion equation: ———, $D_g(S)$ for a free liquid surface (Eq. 5), – – –, $D_1(S)$ for liquid-liquid interfaces (Eq. 6). The quantities S and Y are defined in the text (Eqs. 1, 2)

$$D_1(S) = Y + \tfrac{1}{2} S \sqrt{1+2S}\,(1+\sqrt{1+2S}) = 0 \qquad (6)$$

with $Y_c = 0.155$.

Figure 1 shows numerical solutions of Eqs. (5) and (6) for a wide range of Y values below and above Y_c.

2.2 Power spectrum of the scattered light

The intensity of light scattered by capillary waves of wave-number q is proportional to the mean square displacement of this surface mode

$$\langle \zeta_q^2 \rangle = \frac{k_B T}{A(\sigma q^2 + g \Delta \rho)} \qquad (7)$$

where A is the surface area, k_B is the Boltzmann constant and T is the temperature. Thus, when the interfacial tension is the leading driving force of the waves ($\sigma q^2 \gg g \Delta \rho$), the intensity of the scattered light will increase with decreasing interfacial tension. At any given q value the scattered intensity is proportional to the spectral density $P_q(\omega)$ of the scattering interface. Here, $\omega = \omega' - \omega_0$ is the frequency shift of the scattered light relative to the frequency of the incident light (ω_0). The spectrum $P_q(\omega)$ was calculated by Bouchiat and Meunier [19] by Fourier transformation of the amplitude autocorrelation function $G_q(t)$. For propagating modes ($Y > Y_c$), the spectrum has the form

$$P_q(\omega) = \langle \zeta_q^2 \rangle \, \tau_0 \, \frac{S_r^2 + S_i^2}{S_i^2} \left[\frac{S_r}{S_r^2 + (\omega \tau_0 + S_i)^2} \right.$$
$$+ \frac{S_r}{S_r^2 + (\omega \tau_0 - S_i)^2} + \frac{S_r}{S_i}\left(\frac{S_i + \omega \tau_0}{S_r^2 + (\omega \tau_0 + S_i)^2} \right.$$
$$\left. \left. + \frac{S_i - \omega \tau_0}{S_r^2 + (\omega \tau_0 - S_i)^2} \right) \right] \qquad (8)$$

while for overdamped modes ($Y < Y_c$) $P_q(\omega)$ becomes

$$P_q(\omega) = \langle \zeta_q^2 \rangle \, \frac{2\tau_0 S_1 S_2}{S_2 - S_1} \left(\frac{1}{\omega^2 \tau_0^2 + S_1^2} - \frac{1}{\omega^2 \tau_0^2 + S_2^2} \right) \qquad (9)$$

where $\langle \zeta_q^2 \rangle$ is given by Eq. (7). In the case of propagating waves (Eq. 8), the spectrum represents a doublet symmetrically disposed about the frequency of the incident light ($\omega = 0$). The spectrum of non-propagating waves (Eq. 9) comprises a single peak centered at $\omega = 0$. Spectra computed on the basis of the Eqs. (8) and (9) for widely varying values of Y are

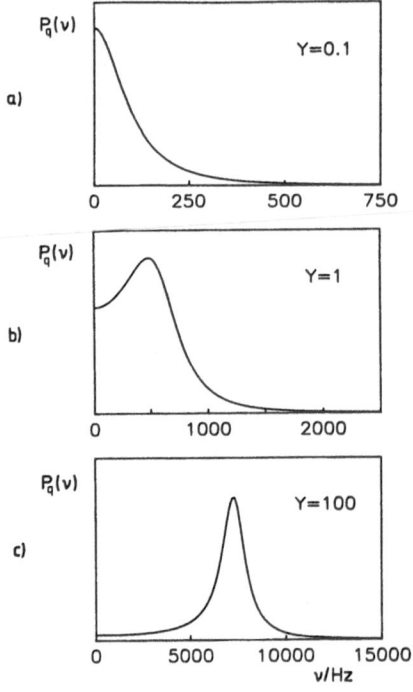

Fig. 2. Theoretical spectra for liquid-liquid interfaces at different values of Y. For $\rho = \rho' = 10^3 \, \text{kg m}^{-3}$, $\eta = \eta' = 10^{-3} \, \text{kg m}^{-1} \text{s}^{-1}$, and $q = 500 \, \text{cm}^{-1}$, these Y values correspond to the following interfacial tensions $\sigma/\text{mN m}^{-1}$: 0.04 (a); 0.4 (b); 40 (c). Graph a: S_1 and S_2 from Eq. (3); graphs b and c: S_r and S_i from Eq. (6)

shown in Fig. 2. For a chosen q value and given values of the densities and viscosities of the two phases, Y is a linear function of the interfacial tension σ; correspondingly, Fig. 2 illustrates the change in the spectrum when the interfacial tension changes by a factor 1 000. In oil + water + surfactant systems such a wide variation of the interfacial tension can indeed occur. Below we explain how the experimental spectra are analyzed in terms of the Eqs. (8) and (9).

2.3 Data analysis

Optical mixing spectroscopy yields the beat signal $\omega = \omega' - \omega_0$, and the resulting photo-current power spectrum is denoted here as $P_q(\nu)$, where $\nu = \omega/2\pi$. Experimental power spectra are either centered at $\nu = 0$ (see Fig. 2a), indicating non-propagating surface fluctuations, or the spectrum exhibits a peak at some frequency ν_{max} (see Figs. 2b, 2c), indicating propagating capillary modes. These two types of spectra correspond to $Y < Y_c$ and $Y > Y_c$, respectively, and the method of the data analysis is somewhat different for these two cases.

(I) Non-propagating fluctuations ($Y < Y_c$): For Y values less than about 0.1, S_1 becomes much smaller

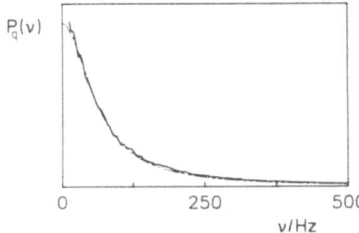

Fig. 3. Spectrum of the oil-water interface of the system decane + water + $C_{12}E_4$ at 22 °C ($q = 340$ cm^{-1}: ——, experimental power spectrum; $- \cdot \cdot - \cdot \cdot -$, two-parameter fit using Eq. (10)

than S_2 (cf. Fig. 1) and thus the expression on the righthand side of Eq. (9) is dominated by the first term in the brackets. Hence the power spectrum reduces to a Lorentzian signal

$$P_q(v) = \frac{A}{v^2 + (\Delta v/2)^2} \quad \text{with} \quad \Delta v = \frac{|S_1|}{\pi \tau_0} \quad (10)$$

where Δv is the full half-width (FWHH) of the signal. The parameters A and Δv were obtained from the experimental spectra by a least-squares fitting procedure. An example of such a fit is shown in Fig. 3. The value of S_1 resulting from the best-fit Δv was inserted into the dispersion relation (Eq. 6) which was solved for Y. The interfacial tension is obtained from the resulting Y when the densities and viscosities of the two co-existing phases are known. In microemulsion systems with low interfacial tensions but relatively large differences in density between the two phases, the gravitational term $g\Delta\rho/q$ cannot be neglected against the interface energy term σq. An approximate relation for the half-width in the limit $Y \ll Y_c$, when Y becomes approximately equal to S_1, has been used in the literature [20] for microemulsion systems with ultralow interfacial tensions

$$\Delta v = \frac{\sigma q + g\Delta\rho/q}{2\pi(\eta + \eta')} \ . \quad (11)$$

In the present study, where the lowest Y value was ca. 0.02, this relation was not generally applicable. In this intermediate regime of Y it was necessary to use the full dispersion equation as outlined above.

(II) Propagating capillary waves ($Y > Y_c$): for sufficiently large values of Y the spectra have a Lorentzian shape and the relevant physical parameters can be obtained from the frequency of the peak maximum v_{max} and the half-width Δv by the Kelvin approximation

$$v_{max} = \frac{1}{2\pi} \left(\frac{\sigma q^3}{\rho + \rho'} + \frac{\varrho - \varrho'}{\varrho + \varrho'} gq \right)^{1/2} \quad (12)$$

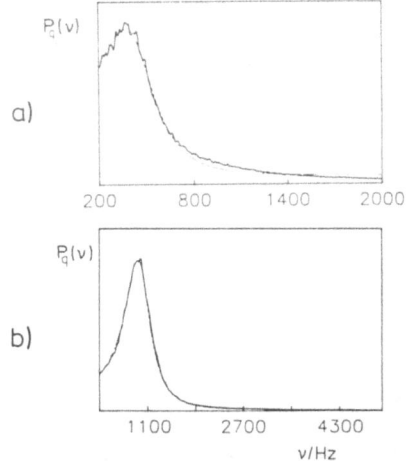

Fig. 4. Spectra of the same sample as in Fig. 3 but at higher temperatures: 25 °C (a) and 31.7 °C (b): ——, experimental; $- \cdot \cdot - \cdot \cdot -$, four-parameter fit based on Eq. (8)

$$\Delta v = \frac{2(\eta + \eta')q^2}{\pi(\rho + \rho')} \quad (13)$$

Our experimental spectra with a peak maximum were mostly in the nearcritical regime of slowly propagating waves and did not conform to this simple approximation. They were analyzed in terms of Eq. (8), which can be written in terms of the parameters $S_i' = S_i/2\pi\tau_0$ and $S_r' = S_r/2\pi\tau_0$. S_i', S_r', a prefactor A and a background term B were treated as adjustable parameters. The experimental spectra are fitted rather well by this relation (except for the low frequency region, which is strongly affected by low frequency noise; see section 3.2). Two examples of such fits are shown in Fig. 4. The resulting best-fit values of S_i and S_r were fed into the dispersion relation $\text{Re}[D_1(S)]$, and this equation was then solved for Y (details of this procedure are given in [14]). Due to the instrumental broadening of the spectra [22–24], the value of S_r obtained by this procedure was greater than the corresponding solution of the complete dispersion equation including $\text{Im}[D_1(S)]$ for given Y values. However, the resulting error in σ was less than 1% on average, which is less than the scatter of the experimental results.

3. Experimental

3.1 Apparatus

A block diagram of the experimental setup is shown in Fig. 5 [14]. The capillary wave spectrometer consists of the optical components, the sample cell and thermostat, the sig-

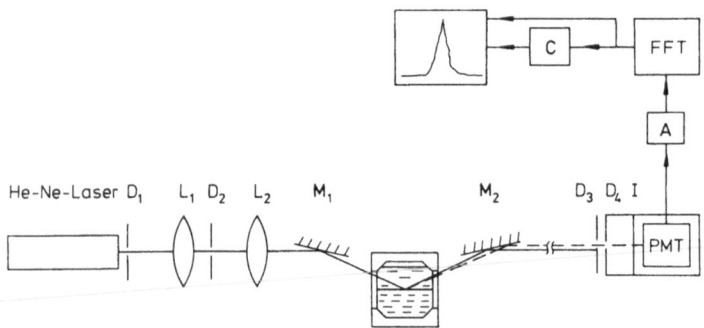

Fig. 5. Schematic of the experimental setup of the capillary wave spectrometer: $D_1 - D_4$ are diaphragms, L_1 and L_2 lenses, M_1 and M_2 mirrors, *PMT* is a photomultiplier tube, *A* an amplifier, *FFT* the signal analyser and *C* a desk computer (see text)

nal analyzer, and a data processing unit. All optical components and the sample cell were placed on a 4×1 meter heavy stone table which was rigidly connected to the floor. On this table the amplitudes of vibrations (mainly in the frequency range around 5 and 15 Hz in horizontal and vertical direction, respectively) were approximately 0.05 μm.

A 1.5 mW He-Ne laser (Spectra Physics, model 102P) was used as a coherent light source ($\lambda = 632.8$ nm). The laser beam was aligned horizontally and was spatially filtered by two diaphragms D_1 and D_2 and a lens L_1. The mirrors M_1 and M_2 can be tilted to adjust the angle of incidence θ_i, and the diaphragm D_4 selects the light scattered in a chosen deflection angle $\Delta\theta$ in the plane of incidence or in the horizontal plane. The lens L_2 ($f = 200$ cm) is used to focus the reflected beam on the photocathode of the photomultiplier (PMT). The diaphragm D_3 and the interference filter I reduce light from other sources.

The sample cell used in the present study was a cylindrical stainless-steel container with a sample volume of 20 cm^3 and with optical windows at the two ends sealed by Teflon O-rings. These windows were of fused silica (Suprasil), 20 mm diam. and 3 mm thick, flat to $\lambda/20$. The liquid-liquid interface was nearly square with an area of 9 cm^2. The cell was thermostated to ± 0.02 K, and the temperature was measured in the sample cell using a calibrated thermistor. The cell was mounted on a translation stage and could be moved up and down by 12 cm. In this way it was possible to reflect the laser beam at the interface from either above or below.

The photomultiplier with the diaphragms D_3 and D_4 was mounted on a linear translation stage operated by a step motor. In the present work the light scattered from the interface was detected mainly in the horizontal plane. In this case the scattering angle $\Delta\theta$ is related to the wave-number q of the detected capillary waves and to the wave-number of the scattered light, $k_i = 2\pi/\lambda$, by

$$q = k_i \tan(\Delta\theta) \quad \text{with} \quad \tan(\Delta\theta) = \Delta x/l$$

where Δx is the distance from the center of the reflected beam to the center of the aperture of the detector, and l is the optical path length from the scattering center to the detector. Scattering angles in the range $0.1° - 0.45°$ were used in this work. The above relation for q in the horizontal detection plane holds, independent of the angle of incidence θ_i and of the refractive indexes of the two phases.

In studies of free liquid surfaces a diffraction grating is often employed to define the scattering angle $\Delta\theta$ and also to generate an intense reference beam of unshifted frequency [20–25]. This "local oscillator" mixes with the frequency-shifted signal at the photomultiplier and yields a heterodyne

signal. In the present study it was not necessary to use a diffraction grating, as explained in the next section. The heterodyne signal produced by the photomultiplier (EMI, model 9658) was fed to an amplifier A (PAR 113 by EG & G) which had adjustable low-pass (DC − 1 kHz) and high-pass filters (3 Hz − 300 kHz). The resulting power spectrum was obtained by an electronic signal analyzer (Iwatsu, model SM-2100 B) which operates in frequency ranges $0 - 0.2$ Hz, or up to 100 kHz with a resolution of up to 1024 data points. It also provides a variety of auxiliary computation routines including fast Fourier transformation (FFT). Determination of the spectra in the frequency domain (rather than in time domain) was preferred because this detection mode makes it easier to identify and eliminate low-frequency noise and interfering signals [26].

3.2 Accuracy

The accuracy of this method of measuring interfacial tensions depends on an accurate determination of the wave-number q of the scattered light and the densities and viscosities of the two bulk phases, as well as on the elimination of faults and noise from the measured spectra. It is also affected by systematic errors in the data analysis.

The photo-detector is positioned at the chosen scattering angle $\Delta\theta$ with a resolution of 6 μm. The zero position of the detector can be determined with an accuracy of only ± 100 μm, by matching the reflected laser beam with the aperture of the detector. This uncertainty causes an error in q of about 1% for $q = 600$ cm^{-1}, but nearly 2% for $q = 300$ cm^{-1}. To reduce this error in the determination of Δx, spectra were taken at equal settings of the detector on either side of the nominal zero. The distance l from the scattering center to the detector was 2040 ± 5 mm.

The viscosity of the equilibrated liquids was measured using an Ubbelohde-type micro-viscosimeter (Viscoboy 2 by MGW Lauda). Repeated measurements agreed within ± 1%. The densities were measured in pyknometers, or using a vibrating-tube densitometer with an accuracy much better than 1%. For the present system the densities of the equilibrated phases differed from those of pure water and decane by no more than 0.1%.

Several factors which can cause systematic errors in the analysis of capillary spectra have been discussed in the literature: instrumental broadening, a homodyne contribution and low-frequency noise in the observed spectra, curvature effects of the interface, etc. The experimental spectrum always represents a superposition of contributions from several wave vectors. This instrumental broadening mostly affects the evaluation of elastic and viscous properties of monolayer films on liquid surfaces (large values of

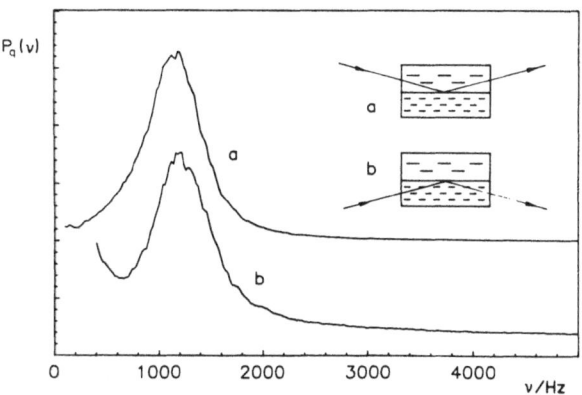

Fig. 6. Comparison of experimental spectra for (a) light incident from the upper phase (i.e., the optically denser oil phase), (b) light incident from below (aqueous phase); $q = 390 \, cm^{-1}$

Y). Corrections for instrumental broadening become less important for strongly damped waves (low Y values) as in the present oil + water systems. We estimate the error in the interfacial tension caused by this effect to be always less than 5% [14].

The observed spectra generally contain a contribution due to homodyne detection, which yields Lorentzian signals with a half-width twice as large as that of the heterodyne signal. Suppression of this homodyne contribution is most important in those cases in which the interfacial tension is determined from the width of the spectrum (i.e., for $Y < Y_c$). The homodyne contribution to the resulting signal is inversely proportional to I_0/I_q, the ratio of the intensities of the elastically and quasi-elastically scattered fields at the detector. In the present work, this ratio I_0/I_q was varied by changing the angle of incidence θ_i: I_0 is largest at the limiting angle of total reflection θ_i^0, while I_q decreases with increasing θ_i. Hence the homodyne contribution decreases as the angle of incidence from the optically denser phase is increased. It was found that the signal half-width $\Delta\theta$ decreases as θ_i approaches θ_i^0 from smaller angles, but reaches a constant value at $\theta_i > \theta_i^0$ (at ca. 75° in the present study were θ_i^0 was ca. 71°). It was assumed that the homodyne contribution is negligible in this regime. The influence of a given homodyne contribution on the resulting value of the interfacial tension decreases as the parameter Y (or σ) becomes smaller. Figure 6 shows that the spectra obtained for light reflected from the interface from above and below exhibit their peak maxima at nearly the same frequency, as expected. However, the signal-to-noise ratio is significantly better when the light is reflected from the optically denser (upper) phase.

The experimental spectra always exhibit low-frequency noise which can be suppressed by a suitable setting of the low-frequency cut-off of the high-pass filter. Obviously, this procedure can cause systematic errors like a shift of the peak maximum and a change of the half-width, and it will affect the determination of the interfacial tension if the chosen cutoff frequency comes too close to the peak. On the other hand, the signal-to-noise ratio generally becomes more favorable in systems of low interfacial tension (low frequencies ν_{max}).

In the present study, power spectra were taken for a series of q values in the range $240 - 730 \, cm^{-1}$ for each temperature of the sample. The individual values of the tension σ obtained from these spectra had a standard deviation of ca. 5% for σ in the range $2 - 0.2 \, mN \, m^{-1}$, and ca. 10% in the range $40 - 8 \, \mu N \, m^{-1}$. A similar precision is reported by other authors [20, 21] who used a diffraction grating as a local oscillator.

For tensions down to ca. $0.2 \, mN \, m^{-1}$ the results obtained in this way could be compared with measurements made by the Wilhelmy method. The two sets of results agreed nearly within the reproducibility of the capillary wave measurements, even in those cases where Y was close to the critical value Y_c. No well-defined test system with ultralow tensions was available to compare the results for the overdamped regime ($Y < Y_c$) with results obtained by the spinning drop technique.

3.3 Materials and sample preparation

Fluka "Puriss." grade decane (99% GC), and n-Dodecyl-tetraoxyethylene monoether (designated $C_{12}E_4$) supplied by the Nikko Chemicals Co. (stated isomeric purity 99%) were used as received. Water used in this study was deionized and double distilled.

A larger sample of an overall decane-to-water mass ratio 1 : 1 with 1.1 mass percent $C_{12}E_4$ was prepared. Mixing of the sample by shaking, stirring, or by treatment in an ultrasonic bath inevitably caused emulsion formation in both the aqueous phase and the oil phase. It was possible to break these emulsions by heating the sample to 40 °C for several hours. Cooling to 18 °C caused a turbid layer of a third phase to appear between the two transparent phases. This turbid layer disappeared when the temperature was raised above 20 °C, but a narrow layer of emulsion remained at the oil-water interface and did not dissolve over long periods of time. Samples of $7 - 10 \, cm^3$ of the two equilibrated transparent phases were then transferred into the optical cell at constant temperature. In this way it was possible to prepare an optically clear oil-water interface at temperatures above the appearance of the inhomogeneous turbid layer.

4. Results and discussion

The system decane + water + $C_{12}E_4$ exhibits a rich phase behavior. At temperatures above ca. 18 °C, dilute $C_{12}E_4$ + water mixtures form a dispersion of lamellar liquid crystal particles (L_a phase) [27] which extends deep into the composition triangle of the ternary alkane + water + surfactant system. The temperature interval of the phase inversion, in which a middle-phase microemulsion coexists with the aqueous phase and the oil phase, is reported to extend from $17 - 20 °C$ for the present system [28]. The possibility of a coexistence of this microemulsion and L_a phase results in a phase diagram with up to three different three-phase regions [29, 30]. Microscopic observations on the present system [29] show that layers of L_a and microemulsion phase are formed between the aqueous and oil phases when 1 wt% aqueous $C_{12}E_4$ is contacted with decane. Above the temperature range of the phase inversion, the L_a crystals dissolve in the oil phase and this dissolution of water-rich L_a phase causes a spontaneous emulsification of water in the oil.

The turbid inhomogeneous layer observed in the present study in a temperature interval from ca. 17 to 20 °C probably represents a composite of middle-phase microemulsion and L_a phase. The interfaces of this layer against the upper and lower phases were not sufficiently clear to take capillary wave spectra in this temperature range, which will be referred to as the three-phase region of the phase inversion (despite the possible presence of L_a phase as a further phase).

4.1 Temperature dependence of the interfacial tension

Figure 7 shows the interfacial tension of our decane + water samples with 1.1 wt% $C_{12}E_4$ in the temperature range from 20 to 31 °C. The numerical results are summarized in Table 1. The interfacial tension decreases from ca. 2 mN m^{-1} at 31 °C to values below 0.01 mN m^{-1} near 20 °C. This pronounced temperature dependence of σ causes a transition of the capillary wave spectrum from the type with a peak maximum at a finite frequency ν_{max} (corresponding to propagating capillary waves) at higher temperatures to spectra with the maximum at the origin (corresponding to overdamped surface disturbances) near 20 °C. No results were obtained for temperatures between 22 and 24 °C, where the interfacial tensions correspond to Y values in the range $0.14 < Y < 0.6$ (depending on the chosen q value). In this transition region the experimental spectra could not be clearly identified as belonging to either $Y < Y_c$ or $Y > Y_c$ (Eqs. 8 or 9).

The oil-water interfacial tension σ is expected to exhibit a minimum at or near the mean temperature of the three-phase interval, $T_m = (T_l + T_u)/2$, if T_l and T_u, the temperatures of the lower and upper critical end-point of three-phase coexistence (oil + water + microemulsion), are not too far apart. If σ is a symmetrical analytical function about this minimum at T_m, then we can write

$$\sigma = \sigma_m + \sigma'(T - T_m)^2 + \ldots \qquad (14)$$

Our experimental data for the temperature range closest to the three-phase region conform to this simple behavior. A three-parameter least-squares fit of our data by Eq. (14) in the temperature range 20–22 °C yields $\sigma_m = 2.0 \, \mu N \, m^{-1}$, $\sigma' = 3.2 \, \mu N \, m^{-1} \, K^{-2}$, and $T_m = 18.6$ °C. This extrapolated

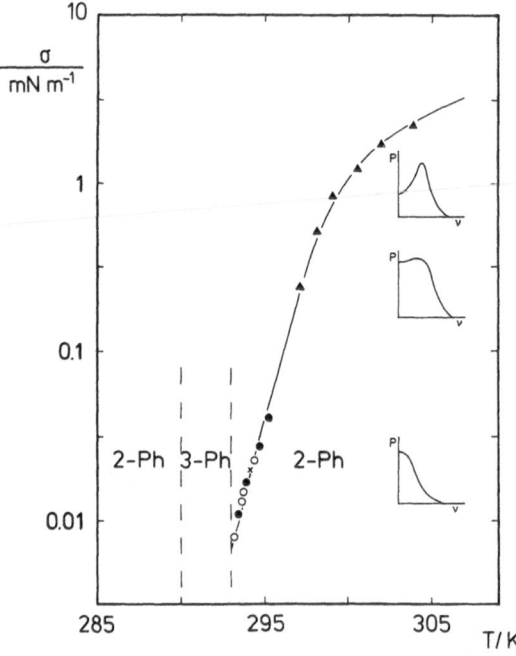

Fig. 7. Equilibrium interfacial tension σ (on a logarithmic scale) as a function of temperature T for decane + water + 1.1 wt% $C_{12}E_4$. Four different samples of the same composition are shown ($\bullet$, $\times$, $\circ$, $\blacktriangle$). Two-phase regions (2-Ph) and the oil + water + microemulsion three-phase region (3-Ph) are indicated. The changing form of the power spectrum caused by the variation of the interfacial tension is also shown

Table 1. Interfacial tension (σ) of the oil-water interface in the system decane + water + 1.1 mass% $C_{12}E_4$ as a function of temperature (T) above the phase inversion interval. The numbers in brackets indicate different samples of the same composition

T/K	$\sigma/\mu N \, m^{-1}$	T/K	$\sigma/\mu N \, m^{-1}$
293.21 (1)	8 ± 1	297.10	240 ± 10
293.45 (2)	11 ± 1	298.09	520 ± 20
293.69 (1)	13 ± 1	299.05	850 ± 40
293.79 (1)	15 ± 1	300.54	1240 ± 50
293.93 (2)	17 ± 2	301.94	1720 ± 50
294.14 (3)	20 ± 2	303.89	2230 ± 60
294.38 (1)	23 ± 2		
294.71 (2)	28 ± 2		
295.28 (2)	41 ± 3		

value of T_m is consistent with the reported mean temperature of the three-phase interval, $T_m = (17 + 20)/2 = 18.5$ °C [28].

A symmetrical behavior of the oil-water interfacial tension in a narrow three-phase region is implied by a simple mean-field theory of interfaces [31]. This theory predicts that the tensions of the three interfaces

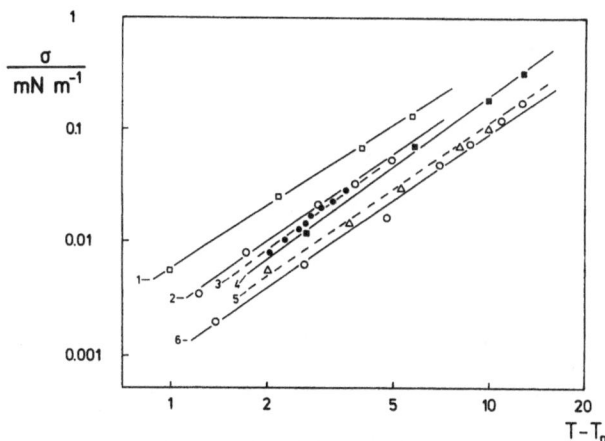

Fig. 8. Comparison of the present results for the system decane + water + $C_{12}E_4$ (labeled as line 3, data points ●) with data on related systems taken from the literature [34], at temperatures above the interfacial tension minimum (T_{min}): 1, hexane + brine + $C_{12}E_4$ (□); 2, octane + brine + $C_{12}E_4$ (○); 4, dodecane + brine + $C_{12}E_4$ (■); 5, dodecane + brine + $C_{12}E_6$ (△); 6, heptane + brine + $C_{12}E_6$ (○). (brine = 0.085 M aqueous NaCl)

obey Antonow's sum rule and that the minimum oil-water interfacial tension σ_m is related to σ_0, the tension at the two end-points, by $\sigma_m = 2\sigma_0/3$. For the present system it appears that this rule is not obeyed: with $T_u = 20\,°C$ we have $\sigma_0 = 8\,\mu N\,m^{-1}$, and thus our extrapolated value of σ_m is closer to $\sigma_0/3$ than to $2\sigma_0/3$. Such a deviation from the "2/3 rule" is expected if the middle phase does not wet the oil-water interface, as has indeed been found for other oil + water + surfactant systems [3, 15, 32, 33].

For narrow three-phase regions (when σ_m becomes very small) an empirical correlation for interfacial tension data near the phase inversion temperature is obtained from Eq. (14) by setting $\sigma_m = 0$, viz.

$$\sigma \simeq \sigma^0 (T - T_{min})^n \qquad (15)$$

where T_{min} is the temperature of the interfacial tension minimum ($T_{min} = T_m$ for symmetrical behavior) and $n \simeq 2$ is expected for sufficiently small σ_m values. When our experimental data for the temperature range $20-22\,°C$ are fitted by this relation, assuming $T_{min} = T_m = 18.6\,°C$, we obtain $n = 1.8$ and $\sigma^0 = 4.17$ in units of $\mu N\,m^{-1}$ and K.

Interfacial tension data exhibiting a pronounced minimum as a function of temperature were published recently for a series of alkane + water (+ NaCl) + $C_{12}E_n$ systems ($n = 4, 5, 6$) by Aveyard and Lawless [34]. A comparison of the present results with these data on the basis of Eq. (15), for $T > T_{min}$, is shown in Fig. 8. The data for all of these systems can be represented by nearly parallel lines with slopes $n = 2 \pm 0.2$.

The present results for the system decane + water + $C_{12}E_4$ fall between the lines for the systems octane + water (+ NaCl) and dodecane + water (+ NaCl) with the same surfactant, as expected.

4.2 *"Dynamic" interfacial tension during temperature variations*

In separate experiments the interface was studied during heating and cooling cycles in a temperature interval from 24 to 31 °C, in which the interfacial tension of the equilibrated system varies by a factor of 10 (cf. Fig. 7). These dynamic measurements were made at a constant wave-number q, and the frequency of the peak maximum (ν_{max}) was taken as an indicator of variations of the interfacial tension. Figure 9 shows three spectra obtained before and during such a cycle: spectrum a was taken at the initial temperature (24 °C), and spectra b and c represent averages of 250 scans taken about 5 and 11 minutes later during continuous heating at a rate of 0.6 K min^{-1}; the time needed for taking these 250 scans was about one minute. Spectrum c corresponds to a temperature of 31 °C and exhibits the same ν_{max} as the equilibrated system at that temperature, indicating that the interface follows changes in temperature rather quickly. This is shown more directly in Fig. 10, wherein the variation of temperature (measured in the liquid near the interface) and the corresponding variation of ν_{max} are plotted as functions of time during a cooling period.

The compositions of the two bulk phases cannot relax to the new (temperature dependent) equilibrium during the time of these measurements, and thus we expect a steady mass transfer through the oil-water in-

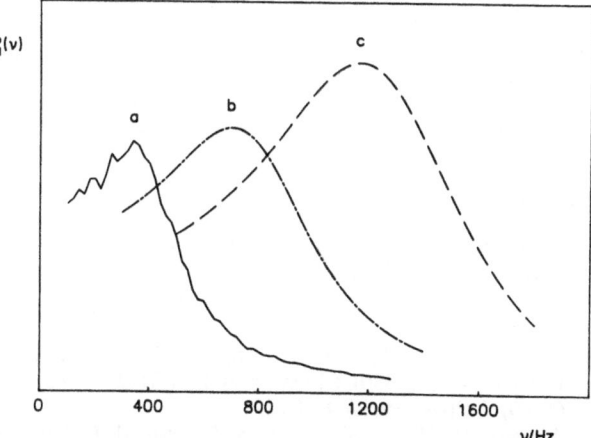

Fig. 9. Power spectra of the oil-water interface taken before (a) and during steady heating of the sample (b and c) at a rate of 0.6 K min^{-1}. The spectra b and c represent averages of 250 scans which were taken during ca. 1 min

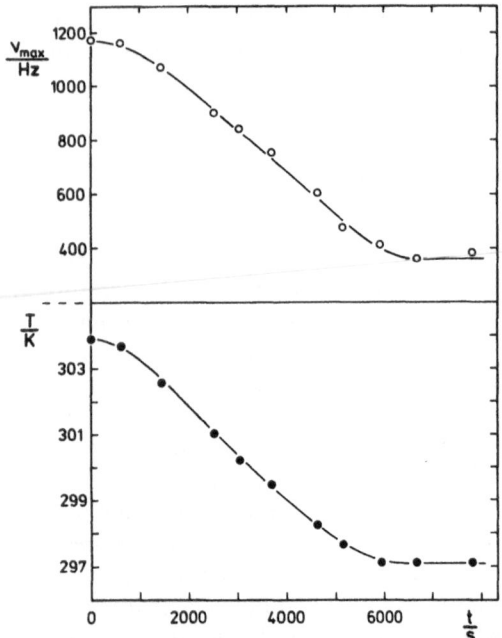

Fig. 10. Shift of the peak maximum ν_{max} of the spectrum and the corresponding variation of the sample temperature during a cooling experiment from 31 °C to 24 °C

terface. The steady-state adsorption of the surfactant should then be lower, and the corresponding dynamic interfacial tension should be greater than the corresponding values at equilibrium [35]. The present measurements were not sufficiently accurate to detect any small deviations from the equilibrium tensions, but we are planning to study such dynamic effects in more detail by this nonperturbative method.

Acknowledgement

The authors wish to thak Dr. D. Langevin and her colleagues for helpful advice during a visit by one of us (D. W.) to the Ecole Normale Supérieure, Paris, in the early stages of this work. We also wish to thank Mr. J. Masuch for help in the construction of parts of the apparatus. Financial support contributed to this work by the Minister für Wissenschaft und Forschung des Landes Nordrhein-Westfalen (Az. IV B 4 103 085 85) is gratefully acknowledged.

References

1. Wade WH, Morgan JC, Schechter RS, Jacobson JK, Salager JL (1978) Soc Petrol Eng J 242
2. Kunieda H, Shinoda K (1982) Bull Chem Soc Jpn 55:1777
3. Pouchelon A, Meunier J, Langevin D, Chatenay D, Cazabat AM (1980) Chem Phys Lett 76:277; see also (1980) J Phys (Paris) 41:L239
4. Aveyard R, Binks BP, Lawless TA, Mead J (1985) J Chem Soc, Faraday Trans I 81:2155
5. Chatenay D, Langevin D, Meunier J, Bourbon D, Lalanne P, Bellocq AM (1982) J Dispersion Sci Technol 3:245
6. Langevin D (1985) In: Physics of Amphiphiles: Micelles, Vesicles and Microemulsions, XC Corso, Soc Ital di Fisica, Bologna, p 181
7. Levich VG (1962) Physicochemical Hydrodynamics. Prentice-Hall, Englewood Cliffs, NJ, p 603
8. Lucassen-Reynders EH, Lucassen J (1970) Adv Coll Interf Sci 2:347
9. Nagarajan N, Webb WW, Widom B (1982) J Chem Phys 77:5771
10. Cummins HZ, Pike ER (1974) Photon Correlation and Light Beating Spectroscopy, Plenum, NY
11. Bouchiat MA, Meunier J (1968) CR Acad Sci (Paris) 266 B:301; (1972) J Phys (Paris) 33:C1–141
12. Papoular M (1968) J Phys (Paris) 29:81
13. Earnshaw JC, McGivern RC (1987) J Phys D, Appl Phys 20:82
14. Wielebinski D (1987) Dissertation, Ruhr-Universität Bochum
15. Bonkhoff K, Hirtz A, Findenegg GH (1988) in preparation
16. Herpin JC, Meunier J (1974) J Phys (Paris) 35:847
17. Huang JS, Webb WW (1969) Phys Rev Lett 23:160
18. Byrne D, Earnshaw JC (1979) J Phys D, Appl Phys 12:1133
19. Bouchiat MA, Meunier J (1971) J Phys (Paris) 32:561
20. Guest D, Langevin D (1986) J Coll Interf Sci 112:208
21. Jon DI, Rosano HL, Cummins HZ (1986) J Coll Interf Sci 114:330
22. Hård S, Hamnerius Y, Nilsson O (1976) J Appl Phys 47:2433
23. Byrne D, Earnshaw JC (1977) J Phys D, Appl Phys 10:L207; (1979) ibid 12:1145
24. Edwards RV, Sirohi RS, Mann JA, Shih LB, Lading L (1982) Appl Optics 21:3555
25. Hård S, Neumann RD (1981) J Coll Interf Sci 83:315
26. Langevin D, Meunier J, Chatenay D (1984) In: Surfactants in Solution, Vol 3, p 1991, Mittal KL, Lindman B (eds). Plenum, NY
27. Mitchell DJ, Tiddy GJT, Waring L, Bostock T, McDonald MP (1983) J Chem Soc, Faraday Trans I 79:975
28. Kahlweit M, Strey R (1985) Angew Chem 97:655
29. Benton WJ, Raney KH, Miller CA (1986) J Coll Interf Sci 110:363
30. Raney KH, Benton WJ, Miller CA (1987) J Coll Interf Sci 117:282
31. Lang JC, Lim PK, Widom B (1976) J Phys Chem 80:1719
32. Bellocq AM, Bourbon D, Lemanceau B, Fourche G (1982) J Coll Interf Sci 89:427
33. Kahlweit M, Strey R, Haase D, et al (1987) J Coll Interf Sci 118:436
34. Aveyard R, Lawless TA (1986) J Chem Soc, Faraday Trans I 82:2951
35. Van hunsel J, Joos P (1987) Langmuir 3:1069

Received February 22, 1988;
accepted June 5, 1988

Authors' address:

Prof. Dr. G. H. Findenegg
Physikalische Chemie II
Ruhr-Universität Bochum
Postfach 102148
4630 Bochum 1, FRG

Progress in Colloid & Polymer Science　　　　　　　Progr Colloid & Polymer Sci 77:109–114 (1988)

Dielectric spectroscopy
– a method of investigating the stability of water-oil emulsions

B.-M. Sax, G. Schön, S. Paasch*) and M. J. Schwuger*)

Essen University – GHS – FB 8 – Institute for Inorganic Chemistry

Abstract: Emulsions are thermodynamically unstable systems. Therefore it is of great importance for technical applications to characterize their physical stability. Unfortunately, the most reliable methods of determining the stability of emulsions require long periods of time.

A new method based on dielectric spectroscopy is presented, which permits distinction between stable and unstable water-oil (w/o) emulsions. These measurements register the volume fraction of the inner phase (water) and its changes with time at different locations in the sample.

Simple model emulsions were investigated using this method. Different mechanisms of destroying the emulsion such as sedimentation, flocculation, and coalescence can be identified and these processes can be recorded time-dependently. The results of such measurements allow the prediction of the physical stability of w/o emulsions.

Key words: Stability of w/o-emulsions, time-dependence, dielectric spectroscopy

1. Introduction

Emulsions are of great importance in both the industrial and consumer sectors. They have been used since the earliest food industry (e.g., milk, butter). Some examples of technological applications are: cooling lubricants (drilling and cutting oils), corrosion-inhibiting emulsions, and emulsion polymerization. One of the best known applications is in cosmetic and pharmaceutical products such as ointments, creams, and lotions.

The stability of such systems plays an important role in the development of emulsions for industrial and cosmetic applications. Cooling emulsions must not show signs of separation in storage tanks, otherwise there is no guarantee that the lubricating components will reach the areas where they are needed during the processing of a workpiece. Cosmetic products must not show signs of aging. Since such products are often stored for a period of months, sometimes for years, the manufacturers must be able to assess and guarantee their stability.

Various methods for testing the stability of emulsions and predicting their behavior over long periods

of time have been developed and described [1–3]. The simplest method, which also gives the most direct results regarding long-term stability, is storage under the conditions found in practice, or under even more extreme conditions (temperature increase, temperature cycles). However, this is very time consuming and the investigations can take several years.

Appreciably quicker results on the stability of disperse systems are given by rheological tests. However, the rheological characteristics can change in the period following sample manufacture so that extrapolation over a longer period of time is not always reliable. For this reason, new methods that will enable early recognition of instabilities are being sought.

In o/w emulsions and suspensions in which water is the continuous phase, it is possible to determine the stability by measuring the conductivity at various places in the sample [2]. Because the oil phase makes no contribution, the conductivity in the upper part of the sample diminishes if the oil component creams, and it increases in the lower part as the proportion of water increases. The difference between the conductivities in the upper and lower parts of the sample, and the time-related changes in conductivity, make it possible to draw conclusions about the stability of the emulsion.

*) Henkel KGaA, Düsseldorf

This principle should apply to all other parameters which depend on the volumetric ratio of the aqueous phase to the oil phase. For w/o emulsions, for which the conductivity method proves unsuitable because of the insulating properties of the oil phase, the dielectric constant can be measured. A method of predicting the stability of w/o emulsions based on capacitance measurements is described in the following.

2. Basis of the method

The dielectric constant is a measure of the interaction between a medium and an electric field. This interaction results in macroscopic polarization. The investigation of how this polarization varies with frequency is called dielectric spectroscopy. Such polarization can be brought about by various mechanisms.

2.1 Polarization mechanisms

Distortion polarization

Electric fields bring about a displacement of the negatively charged electron shells and the positively charged nuclei of the atoms of the molecules in the field. Each molecule becomes a dipole.

The induced moment μ_i is proportional to the strength E of the electric field:

$$\mu_i = a_i \cdot E$$

a_i is the molecular polarizability.

Orientation polarization

If polar molecules, i.e., electric dipoles, are already present in a substance, they line up under the influence of the electric field, whereby the resulting polarization depends on the magnitude of the dipole moment, the strength of the field, and the temperature.

Because distortion polarization also takes place in the presence of permanent dipoles, the sum of both effects is always measured. The relationship between the two molecular effects and the macroscopically measurable dielectric constant is expressed by the Clausius-Mosotti equation.

Surface polarization

An additional peculiarity of disperse systems is the Maxwell-Wagner-Sillars effect [4, 5], which concerns the so-called surface polarization. Charged particles (ions) in drops of water in a w/o emulsion move in the direction of the potential gradient, collecting at the phase boundary. Since the conductivity of the oil phase is much lower, these charges cannot flow away. This results in additional polarization of the drops of water. If flocculation brings the water drops near to each other, the surface polarization increases, causing the macroscopic dielectric constant to increase too.

If the electric field is switched off, the polarization fades away [6]. The analysis of such relaxation processes is generally called "dielectric analysis". As the phenomena are frequency dependent, the term "dielectric spectroscopy" is also used.

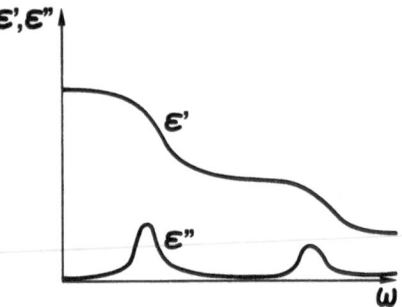

Fig. 1. Schematic representation of the frequency dependence of the dielectric constants

2.2 Frequency dependence of polarization

At low frequencies all charged particles and dipoles can still follow the field. With rising frequency, first the contribution made by orientation polarization is lost, then at even higher frequencies surface polarization disappears. The "optical" dielectric constant left at very high frequencies is due to distortion polarization. The frequency ranges in which the individual types of polarization disappear are known as relaxation frequency domains [7].

Almost all dielectrics display electronic or ionic conductivity as well as polarization effects. Polarization and conductivity cause energy to be absorbed, whereby the absorbed energy of the alternating field is dissipated as heat. To describe the characteristics of dielectric behavior a complex permittivity $\varepsilon^*(w)$ was introduced.

$$\varepsilon^*(w) = \varepsilon'(w) - i\varepsilon''(w)$$

The real part $\varepsilon'(w)$ describes dielectric dispersion phenomena and $\varepsilon''(w)$ dielectric absorption phenomena, both functions generally are interconnected. The frequency dependence of the two parts varies (Fig. 1). The real part of the dielectric permittivity $\varepsilon'(w)$ is almost constant at low temperatures. In the dispersion zones $\varepsilon'(w)$ decreases, and at high frequencies it reaches the so-called optical dielectric constant. The imaginary part $\varepsilon''(w)$ reaches peak values in the relaxation frequency domains.

Besides the frequency dependence of the dielectric constant, it must be remembered that because of the geometry of the problem, the dielectric constant of the system is not simply the sum of the dielectric constants of the aqueous and the oil phases [8–10]. Since we are not concerned here with the quantitative interpretation of the individual measurements, we will not go further into that aspect.

3. Assembling of the apparatus

The block diagram of the measuring apparatus is shown in Fig. 2. The central part of the apparatus is an impedance analyzer which is able to measure all characteristic impedance values in the 5 Hz–13 MHz range. The complex dielectric constant can be calculated from the changes in the phase and amplitude of the measured signal in the sample. A total of five

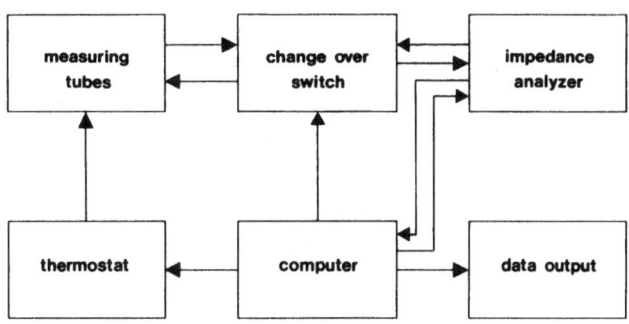

Fig. 2. Block diagram of the measuring apparatus

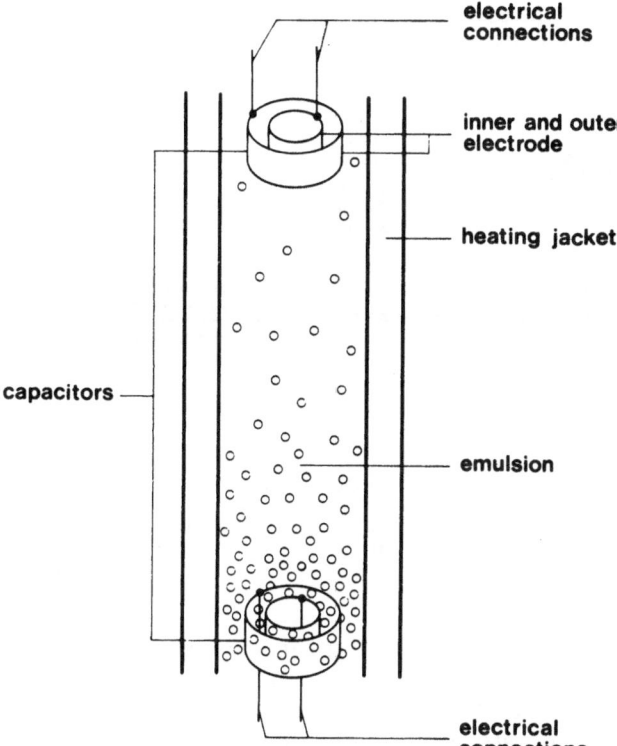

Fig. 3. Measuring cell for dielectric measurements (schematic)

measuring cells are connected to the impedance analyzer through a measuring position change-over switch. A microcomputer looks after data collection and controls the change-over switch.

A schematic representation of the measuring cells used is given in Fig. 3. The emulsions form the dielectric between the electrodes of two annular capacitors. The total volume of a measuring cell is 150 mL, the double-walled cells are thermostatically regulated. The difference in the real parts of the dielectric permittivity determined in the upper and lower sections of the capacitor is taken as a measure of the stability of the

emulsions. The measurements must *not* be made in a relaxation frequency domain, as this generally changes with the composition of the emulsion. For this reason, the complex dielectric constant must be determined for the range 10 Hz – 13 MHz, and the frequency at which measurements are to be made must be specified, before the actual stability measurements are made. To measure the actual stability, the dielectric constant is registered at given time intervals and the change in the difference between the dielectric permittivity of the upper and lower parts of the samples is taken as a criterion of the stability.

Independent of the dielectric measurements, the water content was determined at various places in the emulsion, according to the Karl Fischer method, using specially prepared sedimentation cells. The results were compared with the dielectric values obtained.

4. Substances used and sample preparation

Two w/o emulsion systems were tested in which the emulsifier content was always 5% by weight and the water content varied from 5–65% by weight. The following substances were used:

Double-distilled water
Thin-bodied liquid paraffin DAB 8, $\varrho = 0.85$ g/mL
Emulsion system I : sorbitan mono-oleate (Arlacel 80, Atlas Chemie)
Emulsion system II: POE glycerol sorbitan oleostearate (Arlacel 581, Atlas Chemie)

The emulsions were manufactured from their components at 75 °C under reduced pressure with an ultra-turrax. They were then cooled down to 30 °C in an ice bath, while being subjected to shearing, and transferred to the measuring cells.

5. Results and discussion

The two emulsion systems showed very different stabilities. The emulsions containing Arlacel 80 were considerably less stable than those stabilized with Arlacel 581.

The dielectric measurements were represented as follows: the initial, identical values obtained at both the upper and lower capacitors were taken to be 100%, and the subsequent values obtained were related to this. The curves in Figs. 4–8 show how the dielectric constant of the Arlacel 80 emulsion system varies with

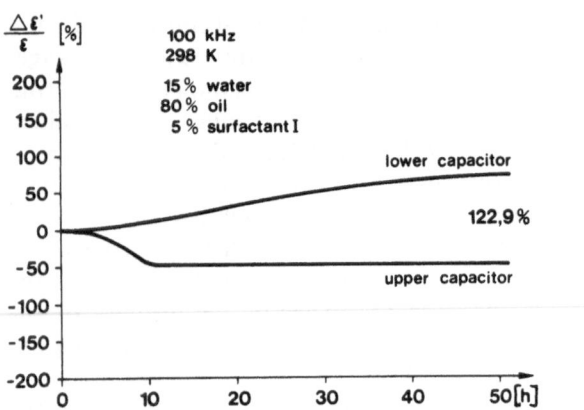

Fig. 4. Measurement curve for Emulsion I/1, 15% water, 80% oil, 5% Arlacel 80

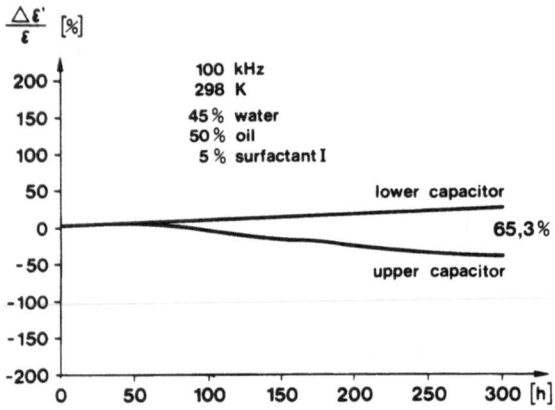

Fig. 7. Measurement curve for Emulsion I/4, 45% water, 50% oil, 5% Arlacel 80

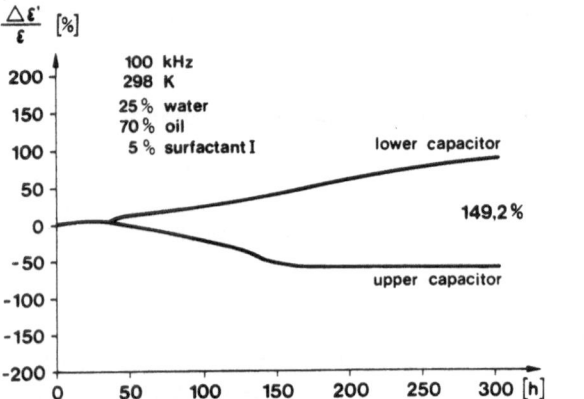

Fig. 5. Measurement curve for Emulsion I/2, 25% water, 70% oil, 5% Arlacel 80

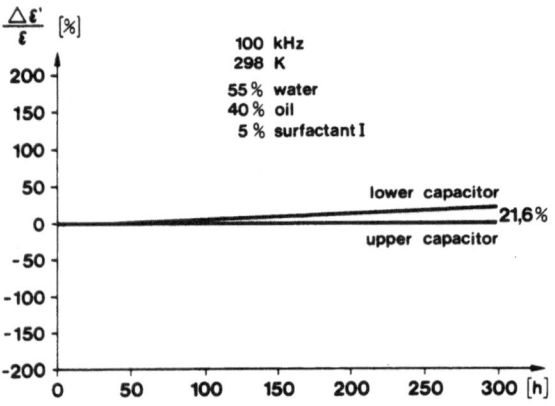

Fig. 8. Measurement curve for Emulsion I/5, 55% water, 40% oil, 5% Arlacel 80

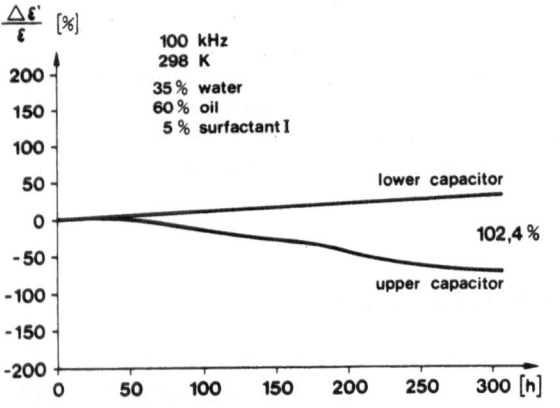

Fig. 6. Measurement curve for Emulsion I/3, 35% water, 60% oil, 5% Arlacel 80

time. In these unstable emulsions, the influence of the volume of the dispersed water phase on the divergence from the initial value can be seen clearly. In the emulsions containing low water volumes, sedimentation of the water drops occurs so quickly that the gap between the electrodes of the upper capacitor is filled with pure oil after only 10 hours.

The stability of the emulsions increases with increased water content. The time taken before only the pure oil phase is present in the upper capacitor increases as the proportion of water to oil increases. For the samples with very high water content this situation had not been reached after 300 hours.

In Fig. 9 the dielectric constant measurements are shown alongside the water content determined using the Karl Fischer method. The two curves show a good correspondence.

This result shows clearly that the divergence of the measurements correlates with the composition of the emulsion at the upper and lower electrodes.

In Fig. 10 the results for the appreciably more stable emulsion system II are shown. Only after two months did the sample show a slight oil separation. The difference in the values of the dielectric constant measured at the upper and lower capacitors was clearly

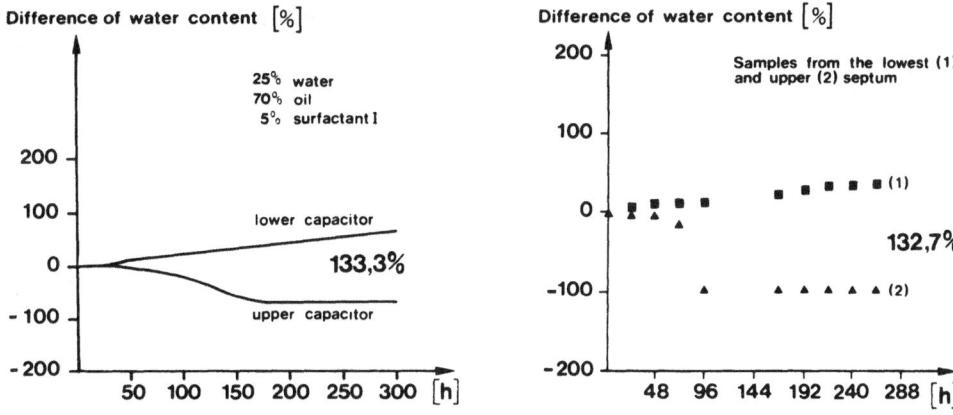

Fig. 9. Comparison dielectric constant — water determination

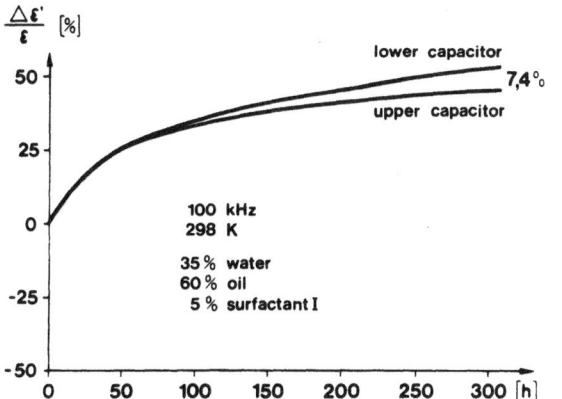

Fig. 10. Measurement curve for Emulsion II, 35% water, 60% oil, 5% Arlacel 581

less than for the first unstable, emulsifier system. Thus the visual observations agree with the small difference between the dielectric constant values at the upper and lower electrodes.

Although the difference is small in absolute terms it shows the onset of the separation after only a few days, whereas it could first be visually recognized after two months, a much longer period. In this respect, the dielectric method permits much quicker conclusions to be drawn regarding the long-term stability of w/o emulsions.

It is interesting that the dielectric constant initially increases at both electrodes. Under the microscope it can be seen that aggregates form in the freshly prepared emulsions, without the droplets coalescing. During this process the droplets approach each other more closely, which leads to increased surface polarization and thus an increase in the value of the dielectric constant [10]. In unstable emulsions such as system I, aggregate formation occurs too quickly to be registered by dielectric measurements. aggregate formation is thus a second destabilizing mechanism alongside sedimentation and creaming.

6. Conclusion

Measurements of two model w/o emulsions show that dielectric spectroscopy is a suitable method for the early recognition of instabilities. In particular it is possible to continually monitor aging processes such as sedimentation and aggregation of the droplets. Dielectric spectroscopy can be used to investigate the effect of systematic variation of one parameter (water to oil proportion, emulsifier concentration) on these processes and thus on the general stability characteristics.

Quantitative conclusions about the longterm stability of an emulsion cannot yet be made by means of dielectric spectroscopy. It is however possible to make a qualitative estimate of the stability. In any case, it enables the onset of separation in emulsions to be recognized much earlier than by visual methods. This means that the user can considerably shorten the time needed for longterm storage tests.

References

1. Howe AM, Mackie AR, Robins MM (1986) J Dispersion Science and Technology 7:231
2. Quack JM, Reng AK, Skrypzak W (1975) Parfümerie und Kosmetik 56:309
3. Cannell JS (1985) Int J Cosmetic Science 7:291
4. Maxwell J (1892) A Treatise on Electricity and Magnetism, Vol 1 (2nd edn). Clarendon Press, Oxford

5. Wagner K, Schering H (1929) Isolierstoffe der Elektrotechnik, Berlin
6. Kronig R (1938) Z techn Physik 19:509
7. Oehme F (1959) Erdöl und Kohle 12:623
8. Wiener O (1912) Abh Sächs Akad Wiss 32:509
9. Bruggman D (1935) Ann Phys Lpz 24:636
10. Hanai T (1960) Kolloid-Z 171:23
    Hanai T (1961) Ibid 175:61

Received November 27, 1987;
accepted June 4, 1988

Authors' address:

M. J. Schwuger
Henkel KGaA
4000 Düsseldorf

Progress in Colloid & Polymer Science   Progr Colloid & Polymer Sci 77:115–119 (1988)

Porod's limit of small angle x-ray scattering from AOT-H$_2$O isooctane micro-emulsions

C. Robertus[1]), J. G. H. Joosten[2]), and Y. K. Levine[1])

[1]) Department of Molecular Biophysics, State University Utrecht, Physics Laboratory, Utrecht, The Netherlands;
[2]) DSM Research, Section FA-GF, Geleen, The Netherlands

Abstract: Water-oil micro-emulsions consisting of AOT-H$_2$O isooctane of different water-surfactant, W_0, and oil-surfactant, S_0, ratios, have been studied by SAXS over a wide range of scattering vectors. The shape of the high angle portion of the scattered intensity is found to be independent of the micro-emulsion composition and furthermore, does not follow the classical $1/q^4$ Porod Law. Rather an additional $1/q^2$ contribution to the scattering curve is observed. These results are explained in terms of the distance correlation function for a pseudo two-phase system, water and oil, with a diffuse interface formed by a monomolecular layer of surfactant molecules. The observed $1/q^2$ behavior is shown to arise from a $1/r$ contribution to the correlation function. A quantitative analysis shows that the total interfacial area of the micro-emulsion is proportional to the total number of surfactant molecules in solution. Thus the average area per surfactant molecule is independent of the micro-emulsion composition.

Key words: X-ray scattering, micro-emulsion, interfacial area, surfactants, aerosol OT

Introduction

Micro-emulsions are mixtures of oil and water stabilized by surfactant molecules [1]. The mixtures form transparent, isotropic, fluid systems that are stable at room temperature for long periods. These systems are homogeneous dispersions of large amounts of two otherwise immiscible liquids. Variation of the water-oil ratio can induce dramatic changes in the macroscopic properties of the micro-emulsion similar to critical phase transitions in single component systems. For example, a conductivity threshold and a maximum in the permittivity as a function of water concentration or temperature, or both, is observed in the Aerosol OT-H$_2$O-iso-octane micro-emulsion [2, 3]. The behavior of the dielectric properties of this Aerosol OT-H$_2$O-iso-octane micro-emulsion can be well understood within the framework of percolation theory [2]. This, however, gives a description of the micro-emulsion only on a macroscopic scale.

On a microscopic scale it has been shown that at low water-oil ratio the water is dispersed in small spherical droplets ($R = 5-70$ Å) in the oil phase [4–6]. The average droplet size, R, is determined by the area per surfactant molecule, Σ_s, and the molar ratio of water to surfactant, W_0. The molar ratio of oil to surfactant, S_0, then determines the droplet concentration or micro-emulsion concentration. A concentration increase can induce different types of structural transitions [4–8]. The micro-emulsion can change from a droplet structure at low concentration to a bicontinuous structure with irregular regions of water interspersed in oil regions of a similar arbitrary shape. But the micro-emulsion may retain its droplet nature at higher concentrations where large aggregates of droplets exist. The type of transition will most probably be governed by the geometry of the surfactant molecule and the way it is incorporated in the water-oil interface [9].

Techniques such as light, neutron, and x-ray scattering provide information concerning the shape, size, polydispersity, and particle interactions in inhomogeneous systems [10, 11]. Particularly interesting is the high-angle portion of the scattered intensity of x-rays or neutrons. This part, governed by Porod's Law, contains information about variations in the electron density (x-ray scattering) or scattering-length density (neutron scattering) over small distances. Importantly, fluctuations across the water-oil interface contribute to the scattered intensity. Porod formulated a $1/q^4$ dependence of the scattering curve at high angles for a two-phase system with a sharp interface (vide Eq. 3) [12–14]. This behavior, however, is not observed in the x-ray scattering of the Aerosol OT-H_2O-iso-octane micro-emulsion.

We shall consider here the origin of this deviation from Porod's Law and show that it can be explained using a model of a pseudo two-phase system, oil and water, with a diffuse interface formed by a monomolecular layer of surfactants. It is shown that quantitative information about the surfactant structure in the interface as a function of micro-emulsion concentration can be obtained from the experimental data.

Theory

X-rays are scattered at small angles by fluctuations in electron density of the medium. The fluctuations can be described using the distance correlation function [11, 15]

$$\gamma(\vec{r}) = \langle \eta(\vec{x}+\vec{r})\eta(\vec{x}) \rangle / \overline{\eta^2} \tag{1}$$

with $\eta(\vec{x}) = \varrho(\vec{x}) - \bar{\varrho}$, where $\varrho(\vec{x})$ is the electron density at point $\vec{x}$, and $\overline{\eta^2}$ is the average of the squared electron density about some average value $\bar{\varrho}$. For solutions and other centro-symmetric systems $\gamma(\vec{r})$ is a function of r only. If no long-range order exists, $\gamma(r)$ will decay to zero for a large r. The scattered intensity, can then be given in terms of $\gamma(r)$ as

$$I(q) = I_e V \overline{\eta^2} \int 4\pi r^2 \gamma(r) \frac{\sin(qr)}{qr} dr \tag{2}$$

with $q = 4\pi/\lambda \sin(\theta/2)$, V the volume irradiated and

$$I_e = 7.9 \cdot 10^{-26} I_0 \frac{(\cos^2\theta + 1)}{2R_s^2}$$

where I_0 is the incident radiation intensity of wavelength λ, R_s the sample to detector distance, and θ the scattering angle. $I_e V$ can be taken as arbitrarily equal to unity. We note that $qI(q)$ and $r\gamma(r)$ form a sine Fourier pair [10, 16].

Porod has derived an expression for $I(q)$ in the limit for large q, known as Porod's Law, for a system composed of two phases with a difference in electron density

$$I(q) = 4\pi\overline{\eta^2} \frac{S}{q^4} \tag{3}$$

with S the total interface between phases. Equation (3) is, however, valid only for two-phase systems with an ideal interface at which the electron density changes discontinuously.

In the case of AOT micro-emulsions it is doubtful whether this condition is satisfied. The AOT surfactant is made up of an SO_4 headgroup attached to two hydrocarbon chains. These chains will have approximately the same electron density as iso-octane and therefore will not contribute to an electron density fluctuation. The sulfate polar headgroup however has a relatively high electron density compared with either water or iso-octane. There will thus be a strong variation in electron density across the oil-water interface over a distance of approximately $1-5$ Å. Consequently we do not expect to find the q^4 decay of the scattering curve at high angles as described by Eq. (3) [17].

An expression for the scattered intensity at high q can nevertheless be obtained for a general pseudo two-phase system. In this system arbitrarily shaped regions of electron densities ϱ_1 and ϱ_2 are separated by a small layer, ∂, of electron density ϱ_0. The value of $\gamma(r)$ can now be calculated using Eq. (1) analogous to the way of Weigel et al. [18]. For a two-phase system with a radius of curvature, R, we find

$$\overline{\eta^2}\gamma(r) = a'r^3 - b'r + c' \quad \text{for} \quad r < \partial \tag{4a}$$

$$\overline{\eta^2}\gamma(r) = ar^3 - br + c + \frac{d}{r} \quad \text{for} \quad \partial < r < 2R \tag{4b}$$

with

$$a' = \overline{\eta^2}$$

$$b' = \frac{S}{4V}((\varrho_0-\varrho_2)^2 + (\varrho_0-\varrho_1)^2)$$

$$\times \left[1 + \frac{\partial}{R}\frac{(\varrho_0-\varrho_2)^2 - (\varrho_0-\varrho_1)^2}{(\varrho_0-\varrho_2)^2 + (\varrho_0-\varrho_1)^2} + \frac{\partial^2}{4R^2} \right]$$

$$c' = \frac{S}{4V}\frac{(\varrho_0-\varrho_2)^2 + (\varrho_0-\varrho_1)^2}{12R^2}$$

and

$$a = \overline{\eta^2}$$

$$b = \frac{S}{4V}(\varrho_2-\varrho_1)^2 \left[1 + \frac{\partial}{R}\frac{\varrho_1+\varrho_2-2\varrho_0}{\varrho_2-\varrho_1} + \frac{\partial^2}{4R^2} \right]$$

$$c = \frac{S}{4V} \frac{(\varrho_2 - \varrho_1)^2}{12R^2}$$

$$d = \frac{S}{4V} (\varrho_0 - \varrho_1)(\varrho_0 - \varrho_2) 2 \partial^2 .$$

It can be shown that the scattered intensity at high angles can be expressed as a function of b', b, and d only

$$q^4 I(q) = b' \qquad \text{for} \qquad 1/q < \partial \qquad (5\,a)$$

$$q^4 I(q) = dq^2 + 2b \quad \text{for} \quad \partial < 1/q < 2R . \qquad (5\,b)$$

Note that Eq. (5 b) is invariant to a simultaneous interchange of electron densities ϱ_1, ϱ_2 and reversal of the sign of R. In fact Eqs. (4a, b) are identical to the distance correlation function for a sphere of radius R and electron density ϱ_1, coated with a layer of electron density ϱ_0 and suspended in a medium with electron density ϱ_2. However, it is also possible to show that the Eqs. (4a, b) provide a useful approximation for the distance correlation function for a pseudo two-phase system not containing finite closed volumes.

In the limit of $\partial/R \ll 1$ the coefficient of the term linear in r of $\gamma(r)$ is given by

$$b = \frac{S}{4V} (\varrho_2 - \varrho_1)^2 \qquad \partial/R \ll 1 . \qquad (6)$$

This is identical to the expression found by Auvray et al. [17], who calculated the scattered intensity from lamellar systems. However this approximation is only justified when ϱ_1 and ϱ_2 are considerably different. If this is not the case then the term in ∂/R can contribute significantly to b and the finite curvature of the surface will influence the observed intensity.

The expression for $\gamma(r)$, Eq. (4b), differs from the expression for an ideal two phase system only in the appearance of the $1/r$ term. It is this term that introduces the $1/q^2$ dependence in the high-angle scattering limit, Eq. (5b). It can be seen from Eqs. (4) and (5) that for the pseudo two-phase system considered here, the high-angle limit of the scattered intensity can be written in terms of the total area of interface between components, the relative electron densities, and the radius of curvature of the interface. These quantities can be extracted from the experimental data by plotting $q^4 I$ vs q^2 [17]. The slope, d, and intercept, $2b$, in Eq. (5b) are both linear functions of the area of interface, S, and their ratio is characteristic of the diffuse interface. For $\partial/R \ll 1$ this ratio is given by

$$\frac{d}{2b} = \frac{(\varrho_0 - \varrho_1)(\varrho_0 - \varrho_2)}{(\varrho_2 - \varrho_1)^2} \partial^2 \qquad \partial/R \ll 1 . \qquad (7)$$

Unfortunately, the total area of interface can only be obtained from the measurements on the calibration of the intensity. Consequently an absolute determination of the area per surfactant molecule is not possible; only relative values can be determined. This is not the case for the ratio $d/2b$ which is independent of the experimental factors characteristic of the apparatus.

Materials and methods

The surfactant Aerosol OT (AOT, sodium-di-2-ethylhexyl-sulfosuccinate) was obtained from Fluka Chemie AG, purity >98%, and was used without further purification. Iso-octane (2,2,4-trimethyl pentane) was obtained from Merck. The water used was deionized and quadruple distilled in a quartz still.

Three concentration series of samples with constant water-surfactant ratio W_0 (26, 35, 45) and increasing S_0 were prepared. In a single series the micro-emulsion with the lowest S_0 was prepared by weighing AOT and dissolving the material in the appropiate amount of iso-octane. Water was then added to establish W_0. The S_0 values were increased by consecutive dilution. Small angle x-ray scattering (SAXS) experiments were performed at Daresbury Laboratory (SRS), UK, using station 8.2, the small angle diffraction beamline. The samples were measured in a 2 mm diameter glass capillary (Mark-Röhrchen 1/100 mm) purchased from W. Müller, Berlin. The capillary was placed in a thermostatted (22 °C) sample holder, identical to those used in a classical Kratky camera.

Radiation ($\lambda = 1.5$Å), scattered at angles between 3 mrad and 100 mrad was detected with a single-wire gasfilled linear-position detector, situated at approximately 2.0 m from the sample. The total scattered radiation was monitored using an ionization chamber placed behind the sample. Data could then be easily corrected for sample absorption, incident beam intensity fluctuations, and the measuring time. Intensities, corrected for detector response, can be compared on an absolute scale, so that spectra recorded with an identical setup differ only by a single multiplication factor. The scattering from pure iso-octane and capillary were recorded and used for buffer corrections. A single measurement took between 5 and 35 min depending on the number of counts registered at the higher angles.

For the calculation of the volume fractions of water, iso-octane, and AOT, the bulk densities for water and iso-octane were used. The density for AOT in solution was taken as 1.138 g/ml [19].

Results

Figure 1 gives a typical example of SAXS data for a micro-emulsion concentration series with constant W_0. In the low-angle region ($q < 0.1$ Å) interference effects appear when the water content is increased, giving rise to a broad peak at approximately 0.053 Å, which is typical of micro-emulsion scattering. In contrast the shape of the scattered intensity for $q > 0.1$ Å appears to be the same over the concentration range used. The scattered intensity for different concentrations differs here only by a multiplication factor. This is manifested as a vertical shift on the logarithmic plot of Fig. 1. The modified Porod plots for a high

(61 vol% AOT + water) and a low (3 vol% AOT + water) concentration of micro-emulsion are shown in Fig. 2.

Note that the intensity is scaled according to the concentration of AOT. It can be clearly seen that the plot of $q^4 I$ vs q^2 closely follows the modified Porod Law given by Eq. (5b). This behavior extends over the entire high-angle region observed in our experiments. The simple $1/q^4$ dependence does not describe the experimental data. The slope and intercepts, d and $2b$, Eq. (5b), of the modified Porod plots are shown in Fig. 3 and Fig. 4 as a function of AOT concentration, for the three different concentration series. Both the slope and the intercept are found to be linearly dependent on the surfactant concentration. No absolute

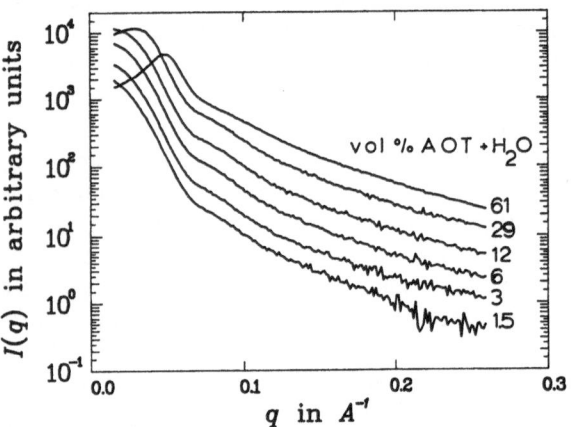

Fig. 1. SAXS from an AOT-H$_2$O-iso-octane micro-emulsion at $T = 22\,^\circ$C, with constant water/surfactant ratio $W_0 = 35$ and vol% AOT + H$_2$O increasing from 1.5 to 61%

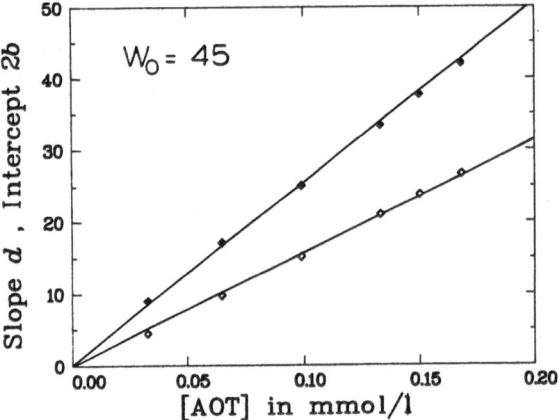

Fig. 3. Intercepts d ($\diamond$) and $2b$ ($\blacklozenge$) of Eq. (5b) at different AOT concentrations and constant water/surfactant ratio $W_0 = 45$. The quantities d and $2b$ in arbitrary units are measures of the total area of the interface

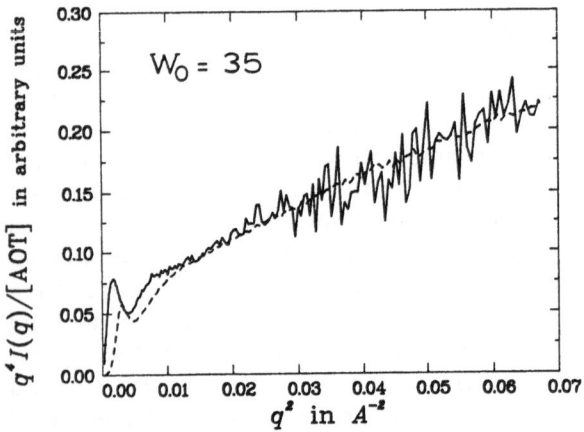

Fig. 2. Modified Porod plot, Eq. (5b), of the 3 (———) and 61 (– – –) vol% AOT + H$_2$O micro-emulsion of Fig. 1, $W_0 = 35$

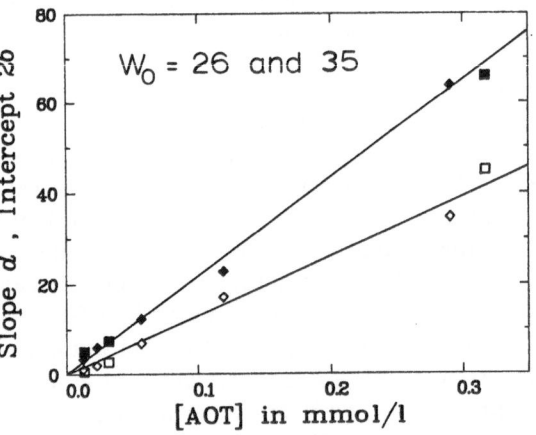

Fig. 4. Coefficients d and $2b$, Eq. (5b) for a $W_0 = 26$ ($\square$, $\blacksquare$) and $W_0 = 35$ ($\diamond$, $\blacklozenge$) micro-emulsion at different AOT concentrations. Data recorded with different experimental setup than data from Fig. 3

determination of intensity was undertaken in order to calibrate these quantities. The calibration independent variable, $d/2b$, was found to be 69 ± 3, independent of the concentration and water/surfactant ratio.

Discussion

We have shown here that the SAXS from the AOT-H_2O-iso-octane micro-emulsion does not follow the classical Porod Law. This deviation can, however, be readily understood in terms of the pseudo two-phase system described above.

The linear dependence of the intercepts d and $2b$ of Eq. (5b), on the surfactant concentration, shown in Fig. 3 demonstrates that the total area of interface is proportional to the total number of surfactant molecules in solution. If all the surfactants are localized in the water-oil interface, then the average area per surfactant, Σ_s, is independent of the micro-emulsion concentration. Furthermore, we find that Σ_s is independent of W_0, as the increase of the total interfacial area for every additional AOT molecule is the same for $W_0 = 35$ and $W_0 = 26$ micro-emulsions, Fig. 4.

The invariance of $d/2b$, Eq. (5b), of the modified Porod plot to changes in concentration or in W_0, indicates that the manner in which the AOT is incorporated in the oil-water interface is independent of the composition of the micro-emulsion. The observation of a constant Σ_s and $d/2b$ does not preclude structural changes in the micro-emulsion when the concentration is varied. However, the structural changes that can take place are limited. For example, a description of the micro-emulsion using a polydisperse spherical droplet model will have the restriction imposed that $\overline{R^2}$ is constant.

Further information about the micro-emulsion system can only be obtained from analysis of the entire scattering curve. To this end we have fitted the SAXS data of AOT-H_2O-iso-octane to a polydisperse spherical droplet model where the droplets are surrounded by a layer of high electron density. It was found that in dilute micro-emulsions (< 2 vol% AOT $+ H_2O$) the average droplet size R is proportional to W_0 and the relative variance in R is independent of W_0. From this analysis a value of Σ_s equal to $55 \pm 2 Å^2$ is obtained. This is smaller than the value of $68 Å^2$ calculated by Kotlarchyk et al. from neutron scattering on the AOT-H_2O-decane micro-emulsions. Although part of the difference can be ascribed to a bigger spread (30%) in R found here than was reported from neutron scattering experiments, its origin is not yet clear.

The finding of a W_0 independent Σ_s from these fits is consistent with the results from the pseudo two-phase model for the scattering at high angles. However, analysis of concentrated micro-emulsions using the droplet model is possible only when a model for interparticle interactions is included. This work is now in progress in our laboratory.

Acknowledgement

The use of the SRS was made possible by the Netherlands Organization for Scientific Research (NWO) in cooperation with the SERC. Special thanks go to Dr. H. Gerritsen for valuable assistance at line 8.

References

1. Hoar TP, Schulman JH (1943) Nature 152:102
2. Van Dijk MA, Casteleijn G, Joosten JGH, Levine YK (1986) J Chem Phys 85:1
3. Van Dijk MA, Broekman E, Joosten JGH, Bedeaux D (1986) J Physique 47:727
4. Kotlarchyk M, Chen SW (1983) J Chem Phys 79:2461
5. Kotlarchyk M, Chen SW (1982) J Phys Chem 86:3273
6. Kotlarchyk M, Chen SW, Huang JS, Kim MW (1984) Phys Rev A 29:2054
7. Kaler EW, Davis HT, Scriven LE (1983) J Chem Phys 79:5685
8. Kaler EW, Bennet KE, Davis HT, Scriven LE (1983) J Chem Phys 79:5673
9. Mittal KL, Lindman B (eds) (1977) Surfactants in Solution. Plenum Press, London New York
10. Glatter O, Kratky O (1982) Small Angle X-ray scattering. Academic Press, London
11. Kerker M (1969) The Scattering of Light and other Electromagnetic Radiation. Academic Press, New York London
12. Porod G (1951) Kolloid Z 124:83
13. Porod G (1952) Kolloid Z 125:51
14. Porod G (1952) Kolloid Z 125:109
15. Debye P, Bueche AM (1949) J Appl Phys 20:51
16. Cabane B (1987) Surfactant Solutions, New Methods of Investigation, Surfactant Science Series, vol. 22, R. Zana (ed). Marcel Dekker, New York
17. Auvray L, Cotton JP, Ober R, Taupin C (1984) J Physique 45:913
18. Weigel D, Renouprez A, Imelik B (1965) J Chim Phys 62:125
19. Ekwall P, Mandell L, Fontell K (1970) J Coll Int Sci 33:574

Received February 15, 1988;
accepted June 3, 1988

Authors' address:

C. Robertus
Department of Molecular Biophysics
State University, Utrecht, Physics Laboratory
P.O.Box 80000
3508 TA Utrecht
The Netherlands

Progress in Colloid & Polymer Science

Progr Colloid & Polymer Sci 77:120–122 (1988)

Kinetics of phase transformations in a fluorinated micro-emulsion system

C. Tondre and C. Burger-Guerrisi

Laboratoire d'Etude des Solutions Organiques et Colloïdales (L.E.S.O.C.), Université de Nancy I. Faculté des Sciences. Vandoeuvre-les-Nancy, France

Abstract: Structural transitions have been induced by fast temperature jumps in organized liquid systems including water, a fluorinated oil, and a nonionic fluorinated surfactant. The rate at which the structural rearrangements occur is shown to be very dependent on the type of transition involved. For this particular system, the characteristic times have been found to be on the order of 25 ms for liquid crystal formation, 0.3 s for the formation of an excess phase in the biphasic region, and $<15 \, \mu s$ for transition between two disconnected isotropic phases, respectively.

Key words: Fluorinated micro-emulsions, phase transitions (kinetics), liquid crystal, temperature jump

Introduction

The rate at which structural rearrangements occur when a phase transition is induced by a change of pressure or temperature in organized liquid systems can be a parameter useful to determine how close the structures are that characterize the initial and the final state. If a very fast phase transformation occurs there will be a strong supposition that closely related structures are involved. Structural relations are, for instance, expected to exist between the liquid crystalline and the isotropic liquid L_2 phases found in many ternary systems [1–3]. Unfortunately, adequate experimental conditions which permit measurement of the rate of a structural rearrangement when crossing a phase limit are rarely met. Relaxation methods have previously been used for similar purposes. For example, the formation and breakdown of lyotropic crystalline phases in binary water-surfactant systems have been studied with the temperature-jump technique [4], whereas pressure-jump experiments have been performed to obtain information on the kinetics of phase separation in fluid mixtures [5].

We have investigated the kinetics of phase transitions induced by fast temperature jumps in fluorinated micro-emulsion systems rich in both water and fluorocarbon [6, 7]. Three different situations have been examined here: the rate of formation of a liquid crystalline phase, the transition between two disconnected isotropic phases, and the rate of formation of an excess phase when penetrating the two-phase region.

Experimental

The monodisperse nonionic polyoxyethyleneperfluoroalkyl surfactant $C_6F_{13}CH_2(EO)_5$ was synthesized in our laboratory [8]. The fluorinated oil 1H,1H,2H-perfluorodecene $(C_8F_{17}CH=CH_2)$ was donated by Atochem (France). Doubly distilled water was used for the preparation of the samples to which NaCl was added so that a 0.1 M concentration was obtained.

The Joule-heating temperature-jump apparatus was purchased from Meßanlagen Studienges. mbH, Göttingen (F. R. G.). A schematic diagram of the setup, represented in Fig. 1, recalls the principle of the method [9]: the energy stored in the high-voltage capacitor H can be suddenly dissipated in the sample contained in the observation cell C, by closing the contact F. The time constant of the temperature rise was on the order of 15 µs in the present experimental conditions. A monochromatic light beam ($\lambda = 515$ nm) goes through the sample and is received by the photomultiplier D. Either turbidity of birefringence changes can be used to monitor the kinetics. In the latter case, crossed polarizers are placed on both sides of the observation cell. The data can be sampled in a transient recorder (Biomation 805) and transferred to a computer (Computer Automation NM 4/30). A non-linear least-squares program was available to fit the kinetic curves with exponential functions. The temperature change ΔT imposed on the sample is proportional to the square of the capacitor voltage, which can be adjusted as desired.

Results and discussion

The temperature-composition phase diagram of the system investigated, according to Shinoda's representation [10] was drawn in Fig. 2. It was obtained with the aid of an automated setup [11]. From one sample

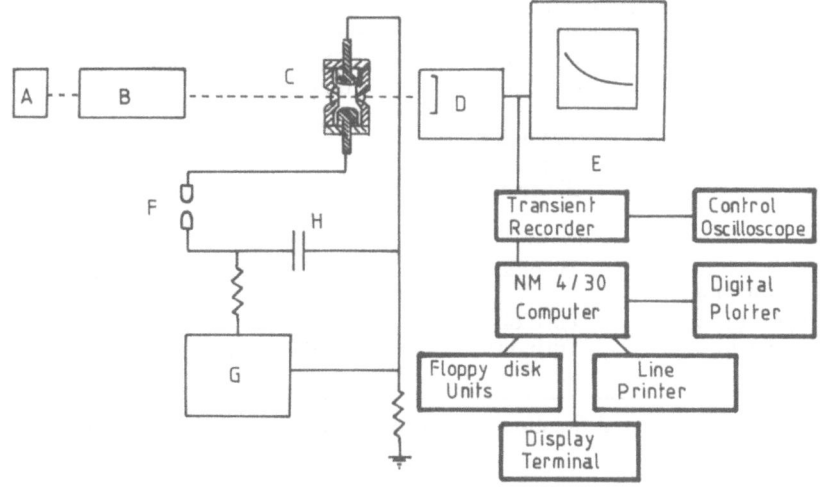

Fig. 1. Schematic diagram of the temperature-jump setup (adapted from [12]). *A*, light source; *B*, monochromator; *C*, observation cell; *D*, photomultiplier; *E*, storage oscilloscope; *F*, spark gap; *G*, high-voltage generator; *H*, capacitor

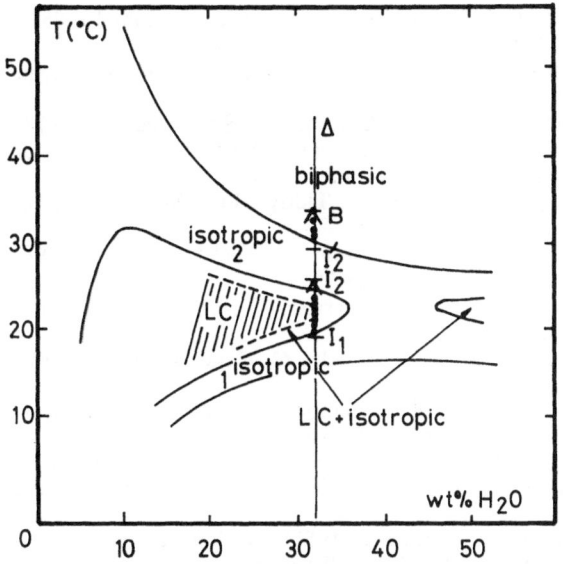

Fig. 2. Temperature-composition phase diagram obtained when salted water is added to a binary system of initial composition (wt/wt): 24.7% $C_6F_{13}CH_2(EO)_5$ and 75.3% $C_8F_{17}CH=CH_2$. Temperature jumps have been performed between I_1 and I_2 and between I'_2 and B (final states are indicated by dots)

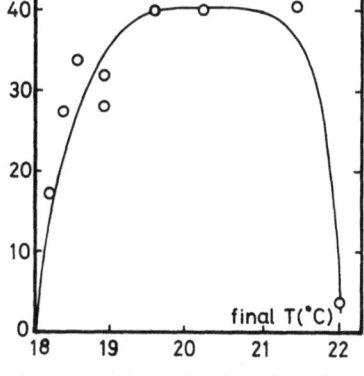

Fig. 3. Reciprocal relaxation time vs final temperature for temperature-jump experiments between I_1 and I_2 in Fig. 2

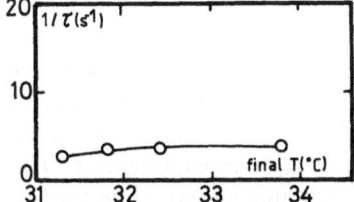

Fig. 4. Reciprocal relaxation time vs final temperature for temperature-jump experiments between I'_2 and B in Fig. 2

preparation to another one the phase limits have been found to be reproducible to within ±2 °C. This uncertainty arises from tiny differences in the surfactant or fluorocarbon batches, that are otherwise undetectable.

The measurements described in the present work have been performed along line Δ in Fig. 2, wherein the sample composition (in wt/wt) is 31.9% water, 51.3% fluorocarbon, 16.8% surfactant, 0.1 M NaCl.

The change in the transmitted light was recorded when the system was brought from the initial state represented by I_1 to different final states between I_1 and I_2 (dots in Fig. 2). The system was allowed to return to its initial state (T_{init} = 17.6 °C) after each temperature jump. A quasi-exponential decrease in the transmitted light was generally observed after a short induction period. The best fits with an exponential function gave the values of the relaxation times plotted vs the final temperature in Fig. 3.

Table 1. Type of transitions performed and characteristic times involved

Composition of the micro-emulsion system	Type of transition	Mode of detection	Characteristic time[a]	Ref.
Water 25.1 $C_8F_{17}CH=CH_2$ 56.4 $C_6F_{13}CH_2(EO)_5$ 18.5 0.1 M NaCl	isotropic $1 \rightarrow L.C.$	Turbidity and Birefringence	2.5 ms 25 – 30 ms	[7] [6, 7]
Water 31.9 $C_8F_{13}CH=CH_2$ 51.3 $C_6F_{13}CH_2(EO)_5$ 16.8 0.1 M NaCl	isotropic $1 \rightarrow (LC$ $\qquad\qquad + isotropic)$ isotropic $1 \rightarrow$ isotropic 2 isotropic $2 \rightarrow$ biphasic	Turbidity Turbidity Turbidity	25 – 50 ms $<15\ \mu s$[b] 0.25 – 0.5 s	this work [6, 7] this work and [7]

[a]) Relaxation times obtained from the best fit with an exponential function, except for [b]) (no signal detectable, see [6, 7]).

The plateau observed at $1/\tau \sim 40\ s^{-1}$ corresponds to the limit of the existence domain of the liquid crystal (Fig. 2). The value of the corresponding relaxation time (25 ms) is in perfect agreement with that previously obtained with a birefringence detection [6] when the final state is right in the middle of the LC domain (at 25.1% water in Fig. 2).

We have also investigated the case where the initial state (I_1) is in the lower isotropic channel and the final state (I_2) in the upper one. This particular situation has been discussed in detail elsewhere [6, 7]. The structural rearrangement involved in the transition appears to be quasi-instantaneous in this case, being completed within the heating time of the solution.

Finally, the rate of appearance of the excess phase when the biphasic system forms has been obtained from the changes of turbidity accompanying temperature jumps, which bring the system from the initial state I'_2 to final states in the biphasic region (dots between I'_2 and B in Fig. 2). The kinetic curves obtained have in this case a marked sigmoïdal shape due to the fact that the nuclei of the excess phase need to have a certain size before they become detectable. We nevertheless made use of a least squares exponential fit to obtain an evaluation of the characteristic time for this process. Values in the range 0.25 – 0.5 s were obtained, at least ten times larger than for the formation of a liquid crystalline phase.

A summary of the kinetic features observed for the different types of transition is presented in Table 1. Further work is needed to determine whether the time scales involved here are specific for compounds having a fluorocarbon nature.

References

1. Fontell K, Hernqvist L, Larsson K, Sjöblom J (1983) J Coll Interf Sci 93:453
2. Ravey JC, Stébé M-J (1987) Progr Colloïd & Colymer Sci 73:127
3. Tondre C, Xenakis A, Robert A, Serratrice G (1986) In: Mittal KL, Bothorel P (eds) Surfactants in solution. Plenum, NY, Vol 6:1345
4. Knight P, Wyn-Jones E, Tiddy GJT (1985) J Phys Chem 89:3447
5. Schneider GM, Dittmann M, Metz U, Wenzel J (1987) Pure & Appl Chem 59:79
6. Tondre C, Burger-Guerrisi C (1987) J Phys Chem 91:4055
7. Burger-Guerrisi C, Tondre C, Canet D (1988) J Phys Chem 92:4974
8. Selve C, Castro B, Leempoel P, Mathis G, Gartiser T, Delpuech JJ (1983) Tetrahedron 39:1313
9. Eigen M, De Maeyer L (1963) In: Techniques of Organic Chemistry. Wiley-Interscience, New York, Vol VIII, Part 2
10. Shinoda K, Ogawa T (1967) J Coll Interf Sci 24:56
Shinoda K (1967) J Coll Interf Sci 24:4
11. Tondre C, Robert A, Burger C (1986) J Disp Sci & Technl 7:581
12. Bernasconi CF (1976) In: Relaxation Kinetics. Academic Press, New York

Received February 4, 1988;
accepted June 7, 1988

Authors' address:

Dr. C. Tondre
LESOC – UA CNRS 406
Faculté des Sciences
Université de Nancy I
B.P. 239
F-54506 Vandoeuvre-les-Nancy Cedex
France

Petroleum emulsions, micro-emulsions, and micellar solutions

H.-J. Neumann and B. Paczyńska-Lahme

Institut für Erdölforschung, Clausthal-Zellerfeld

Abstract: Petroleum is a micellar solution of asphaltenes and resins. Crude oils mostly are recovered as water in oil emulsions. Micro-emulsions in the state of phase inversion are very important for an enhanced oil recovery by surfactant flooding.

The results of investigations on the composition and the properties of asphaltenes and petroleum resins are discussed along with the importance of micro-emulsions for petroleum recovery and their formation and properties. The stabilization and the demulsifying of petroleum emulsions are also discussed. The phenomenon of the phase inversion is especially important for petroleum emulsions and micro-emulsions.

Key words: Petroleum emulsions, micro-emulsions, micellar solutions, enhanced oil recovery, petroleum treatment

1. Introduction

It seems that Nature itself has a special affinity for colloidal-dispersed systems. Petroleum is such a colloid system. It is a micellar solution, and it will most often be recovered as water in oil emulsions. Surfactants and micro-emulsions are important in modern processes for enhanced oil recovery.

The idea that crude oils exist as colloidally dispersed systems dates back to the beginning of this century, but it was not proved until the early 1960s.

Crude oils can be separated by ultrafiltration of their colloidal compounds. The asphaltenes and the petroleum resins are separated out of their dispersion medium, the "oily phase".

2. Petroleum as a micellar solution

Petroleums are micellar solutions of asphaltenes and petroleum resins in an oil phase. In the dispersed asphaltenes and resins, the hetero compounds of the petroleum are enriched; in the asphaltenes there are the acidic oxygen and sulfur compounds, the anionic interfacially active components, and the majority of metal organic compounds; in the petroleum resins, the basic nitrogen and the cationic interfacial active components. The micellar solution crude oil is thermodynamically stable.

Asphaltenes and petroleum resins differ from one another in a characteristic way. They differ in their particle size and also in their composition. Both the asphaltenes and the petroleum resins are spherical, oleophilic, resoluble, polydispersed micellar colloids [1]. The asphaltenes are insoluble in *n*-pentane, the resins are soluble. The asphaltenes are hard and brittle, the resins soft and greasy. The asphaltenes are not meltable, because there is a thermal decomposition before they reach the melting point. The petroleum resins are meltable [2–7].

The micelles of the asphaltenes are formed of several various molecules by dipole forces, but the micelles of the petroleum resins are formed of several other different molecules by dispersion forces. Petroleum resin particles are smaller than those of the asphaltenes. The asphaltenes have in colloidal solution a medium relative particle mass on the order of 5 000 to 8 000; for petroleum resins it is about 1 000 to 1 500.

The interfacial activity at oil-water interfaces of the asphaltenes are high, and highly dependent on the pH. This interfacial activity of the petroleum resins is medium, and only slightly dependent on the pH.

The medium relative particle mass in solution both of the asphaltenes and of the petroleum resins depends on the temperature. It increases with decreasing temperature. Both of the petroleum colloids, the

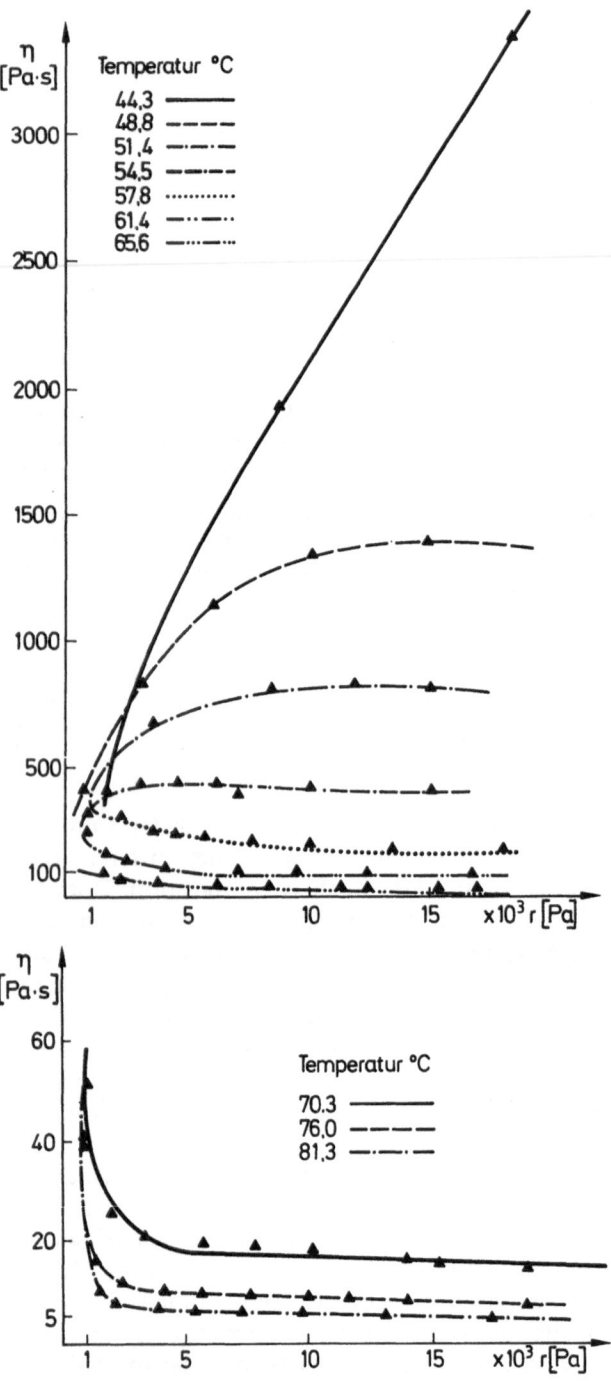

Fig. 1. The viscosity η of petroleum resins as a function of the shear stress τ

with increasing shear stress, for higher temperatures it decreases with the shear stress.

3. Micro-emulsions

There are micro-emulsions [8] in quasi-ternary systems of crude oil, brine, and surfactant with the crude oil as a micellar solution. The brine is a solution of inorganic salts up to a concentration of 30% in water, and the surfactant is a mixture of interfacially active substances [9]. Depending on the temperature and on the salinity of the brine equilibriums exist with two liquid phases of a micro-emulsion w-o and water, and of a micro-emulsion o-w and oil. Another equilibrium has three liquid phases, a middle phase micro-emulsion in the state of phase inversion and a water and an oil phase. Oil and water become quasi miscible in such a micro-emulsion.

The middle phase micro-emulsion is rich in interfacially active substances, and the petroleum resins are

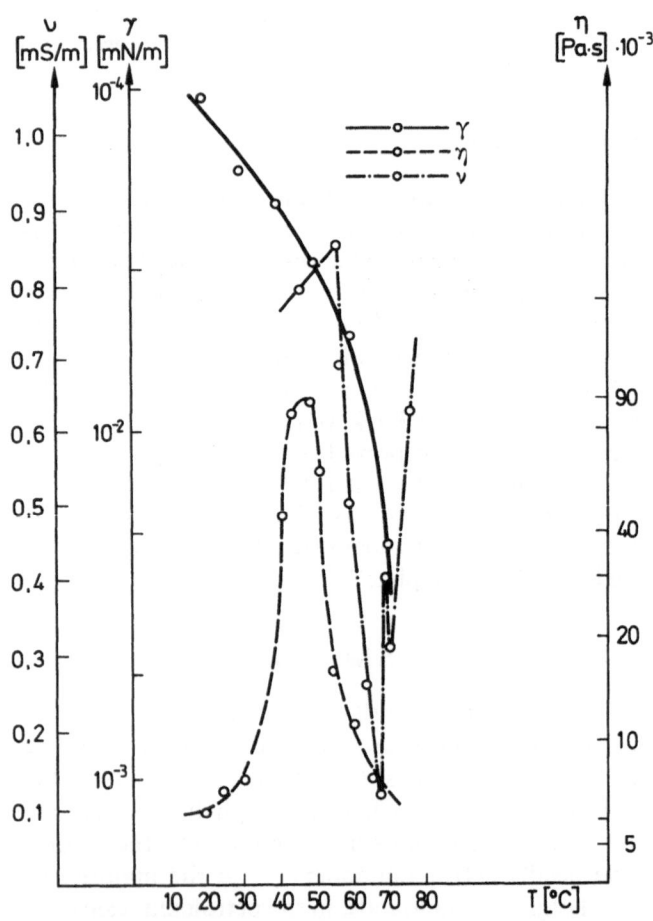

Fig. 2. The three physical properties of a middle phase micro-emulsion interfacial tension γ, viscosity η, and conductivity ν as functions of the temperature T

asphaltenes and the petroleum resins have a non-Newtonian viscosity behavior. The substances themselves and their concentrated solutions are pseudoplastic. Figure 1 shows the viscosity η of petroleum resins for several temperatures as a function of the shear stress τ. For temperatures below 55 °C the viscosity increases

enriched in this phase, while the asphaltenes are enriched in the oil phase [9, 10]. The oil and the brine in the middle phase have other compositions than the oil of the oil phase and the brine of the water phase. The polar compounds are enriched in the middle phase.

Several physical properties go through a minimum or a maximum with the temperature or with the salinity of the brine. Figure 2 shows three of these as a function of the temperature [11]. The curve for the interfacial tension γ has a minimum, a maximum the curve for the viscosity η, a minimum and a second additional minimum the curve for the electrical conductivity ν. The extreme values are not exactly at the same temperature.

The enhanced oil recovery by surfactant flooding needs surfactants with wide ranges of temperature and salinity. We have discovered in laboratory tests mixtures of polyglycolethers with ethoxylated sulfonates without any co-tenside that are suitable compounds for a surfactant flooding to recover more oil out of a reservoir than the 30% averaged today [12]. The ratio of the components makes possible an adaption to different temperatures and salinities.

4. Emulsions

Most of the crude oil is recovered as a water in oil emulsion. These emulsions are thermodynamically instable, but often constant for a long time.

The brine in crude oil emulsions are stabilized by asphaltenes and petroleum resins, which are adsorbed at the oil-water interfaces. They are stabilized against coalescence, but not against flocculation. Therefore a demulsifier has to cause a coalescence by change the wetting of the interfacial films from oil wetted to water wetted [13].

Often there is a small quantity of a micro-emulsion besides the emulsion in the recovered crude oil.

In these cases we got only 80% of the emulgated water out of the emulsion using a demulsifier and heat. But with the use of ultrasound all the water was demulsified out of the emulsion.

5. Phase inversion

The phenomenon of the phase inversion from oil in water to water in oil and vice versa exists for micro-emulsions and for (macro)emulsions. For emulsions it is visible under the microscope. Many phase inversions we have photographed with crude oils of different densities and with brines of different salinities [14]. While the w-o and the o-w emulsions show spherical

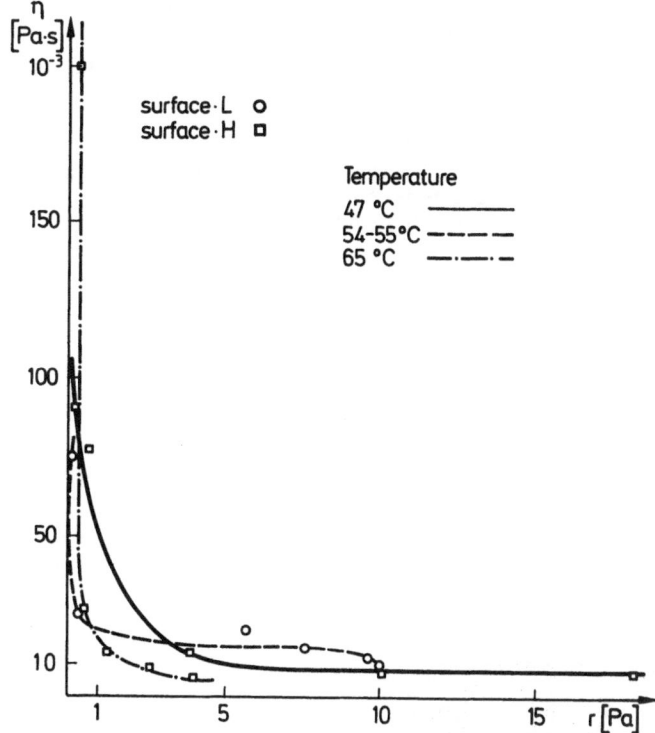

Fig. 3. The viscosity of middle phase emulsions with two different surfactants as a function of the shear stress τ

droplets, the emulsion in the state of phase inversion shows neither spherical volumes nor features distinguishable between oil in water and water in oil.

Emulsions in the state of a phase inversion have higher viscosities than the non-emulgated oil and water. Their viscosity decreases with increasing shear stress. For two different surfactants Fig. 3 shows the viscosity of such emulsions in the state of phase inversion as a function of the shear stress τ.

References

1. Neumann HJ (1967) Habilitationsschrift Braunschweig
2. Neumann HJ, Paczyńska-Lahme B (1981) Composition and Properties of Petroleum. Ferd. Enke Publishers, Stuttgart
3. Neumann HJ (1965) Erdöl Kohle-Erdgas-Petrochemie 18:865−870
4. Neumann HJ, Rahimian I, Taghizadeh D (1967) Brennstoff-Chemie 48:66−69
5. Neumann HJ (1970) Erdöl Kohle-Erdgas-Petrochemie 23:496−499
6. Paczyńska-Lahme B, Neumann HJ (1978/79) DGMK Compendium 2:1533−1549
7. Paczyńska-Lahme B (1979) Dissertation Clausthal
8. Prince LM (1977) Micro-emulsions, Theory and Practice. Academic Press Inc, New York, San Francisco, London

9. Neumann HJ, Paczyńska-Lahme B (1986) Erdöl Erdgas Kohle 162:397–401
10. Tunç S (1982) Bundesministerium für Forschung und Technologie, Bonn, Forschungsbericht T 82-093
11. Gade B (1986) Dissertation Clausthal
12. Wachsmann V (1987) Dissertation Clausthal
13. Neumann HJ, Paczyńska-Lahme B, Rahimian I (1987) Chem-Ing-Tech 59:966–967
14. Bomka HD (1988) Diplomarbeit Clausthal

Received February 11, 1988;
accepted June 7, 1988

Authors' address:

Prof. Dr. H. J. Neumann
Institut für Erdölforschung
Walther-Nernst-Straße 7
3392 Clausthal-Zellerfeld

Progress in Colloid & Polymer Science

Progr Colloid & Polymer Sci 77:127–130 (1988)

Phase equilibria in the systems
used for vinyl acetate micro-emulsion polymerization

D. F. Anghel, M. Balcan, and D. Donescu*)

Central Institute of Chemical Research, Department of Physical Chemistry and *Department of Research for Plastic Materials, Bucharest, Romania

Abstract: Phase behavior of vinyl acetate-water system in the presence of various surfactant-cosurfactant mixtures was studied. The surfactants used were sodium dodecyl sulfate and ethoxylated nonylphenol with 3 and 30 moles of ethylene oxide, respectively. Alcohol cosurfactants from methanol to 2-ethylhexanol were employed. The effect of salinity (i.e., 1% wt. NaCl) was also considered. The most efficient surfactant in reducing the multiphase zone was sodium dodecyl sulfate. For NP-30 surfactant the micro-emulsion increases with surfactant concentration and cosurfactant hydrophilicity. The electrolyte increases the multiphase area. Structure investigation showed a cosolubilized system rather than a micro-emulsion to be present in the isotropic phases.

Key words: Vinyl acetate, phase behavior, surfactants, cosurfactants, cosolubilized systems

Introduction

During the past three decades much study has been devoted to micro-emulsion systems. These systems are transparent and thermodynamically stable dispersions of two immiscible liquids, stabilized by a suitable surfactant or surfactant-cosurfactant mixture. The dispersed phase is either organic or aqueous and consists of spherical droplets whose diameter range is 10–100 nm. Because of these properties, such systems have many applications in the detergent, cosmetic, and pharmaceutical industries, in the chemical industry as special catalysts, in biotechnology, in solvent extraction and heat generating processes, and in enhanced oil recovery. Micro-emulsions of perfluorocarbons have a also been successfully used in medicine as blood substitutes.

Meanwhile micro-emulsions have proved useful in polymer chemistry; the lipophilic styrene is by far the most intensively studied monomer [1–7]. Monomers of low polarity such as methylacrylate and methylmethacrylate [8, 9] and of high polarity like acryl amide [10] were also investigated.

In a previous work [11], we described the micro-emulsion polymerization of a monomer of intermediate polarity (i.e., vinyl acetate). The system used consisted of water, sodium dodecyl sulfate and n-propanol. This study is devoted to phase equilibria in the above mentioned system, as well as in the systems containing nonionic surfactants and various short-chain alcohol cosurfactants.

Experimental

The sodium dodecyl sulfate (SDS) used was reagent grade supplied by MERCK. The nonionic surfactants were polydisperse ethoxylated nonylphenols with 3 (NP-3) and 30 (NP-30) moles of ethylene oxide, respectively. The real mean ethoxylation degrees found by high performance liquid chromatography (HPLC) were 4.39 and 29.00 for NP-3 and NP-30, respectively.

Alcohol cosurfactants such as methanol, ethanol, n-propanol, iso-propanol, and iso-butanol were reagent grade REACTIVUL (Bucharest). They were used without purification. The 2-ethylhexanol, technical grade, was freshly distilled before use.

Reagent grade sodium chloride REACTIVUL (Bucharest) was used.

Distilled water and vinyl acetate (VAc) stabilized with 10 ppm hydroquinone were employed.

Viscosity was determined by means of an Ubbelohde viscometer ($K = 0.01001$). A pycnometer was employed for density measurements.

Electrical conductivity was measured with the aid of a radiometer conductivity meter fitted with a CDC 314 bright platinum electrodes cell. The nominal constant of the cell was 0.316 cm.

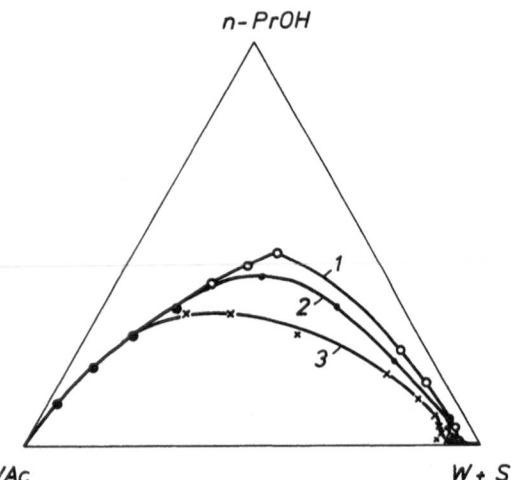

Fig. 1. Phase behavior of water (W)–vinyl acetate (VAc) – *n*-propanol system in the absence (curve 1), in the presence of SDS (curve 2), and in the presence of NP-30 (curve 3). Water-surfactant: 12:1 (wt.)

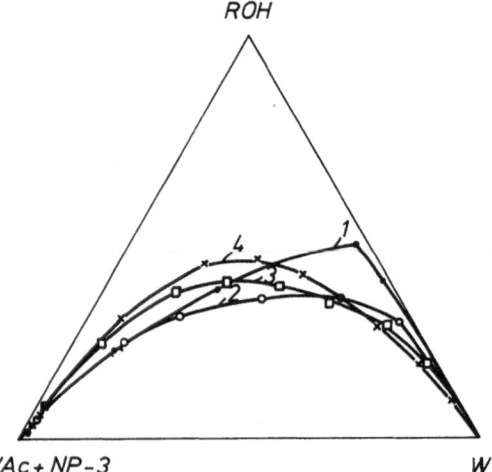

Fig. 3. Effect of cosurfactant on NP-3 containing systems: 1) MeOH, 2) EtOH, 3) i-PrOH, 4) n-PrOH; Vac-NP-3: 10:1 (wt.)

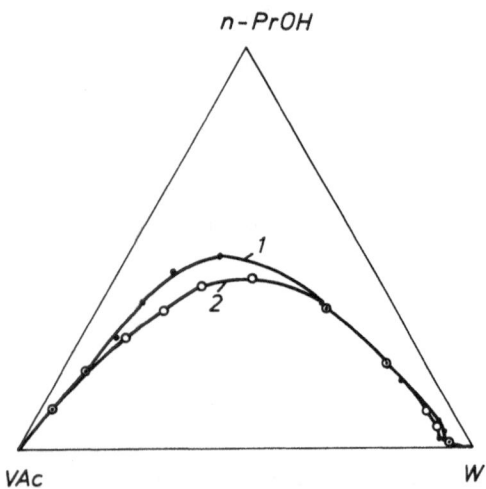

Fig. 2. Influence of NP-30 concentration on phase behavior: 1) 5.9% (wt.); 2) 7.7% (wt.)

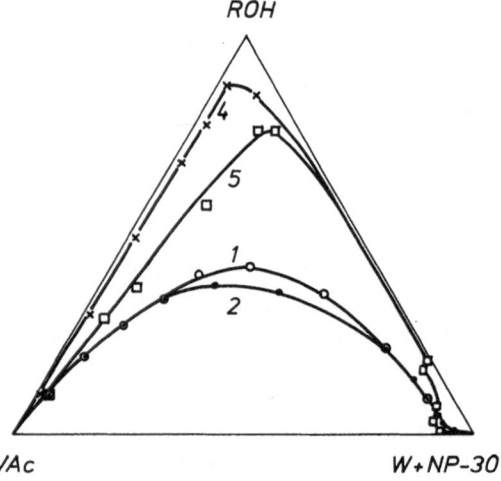

Fig. 4. Influence of cosurfactant on NP-30 based systems: 1) n-PrOH, 2) i-PrOH, 3) i-BuOH, 4) i-C_8H_{17}OH. Water-NP-30: 12:1 (wt.)

Results

Phase equilibria

The phase diagrams of the following systems were studied:

1) Water-SDS-vinyl acetate-*n*-propanol
2) Water-NP-30-vinyl acetate-cosurfactant
3) Water-vinyl acetate-NP-3-cosurfactant.

The weight ratios of VAc-NP-3 were 10:1 and 12:1; those for water-NP-30, 12:1 and 16:1, while for water-SDS 12:1.

Figure 1 shows the phase behavior of the water-vinyl acetate-*n*-propanol system in the absence (curve 1) and in the presence of surfactant (curve 2 = SDS; curve

3 = NP-30). The presence of the surfactant in the system diminished the emulsion zone, sodium dodecyl sulfate being more efficient than the nonionic surfactant. The transparent phase obtained when the system contains no surfactant corresponds to a moleculary dispersed solution.

The influence of the surfactant concentration on phase behavior was tested, and the results obtained for the system containing NP-30 are presented in Fig. 2. Increasing the surfactant concentration entails an increase in the single phase area.

The effect of cosurfactant in systems with hydrophobic (NP-3) as well as hydrophilic (NP-30) nonionic surfactants was also considered. The monomer-surfac-

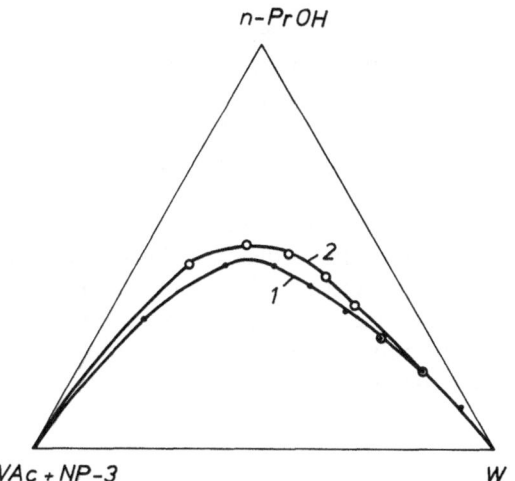

Fig. 5. Influence of electrolyte on phase behavior of NP-3 containing system. 1) no electrolyte, 2) 1% NaCl solution. VAc-NP-3: 10:1 (wt.)

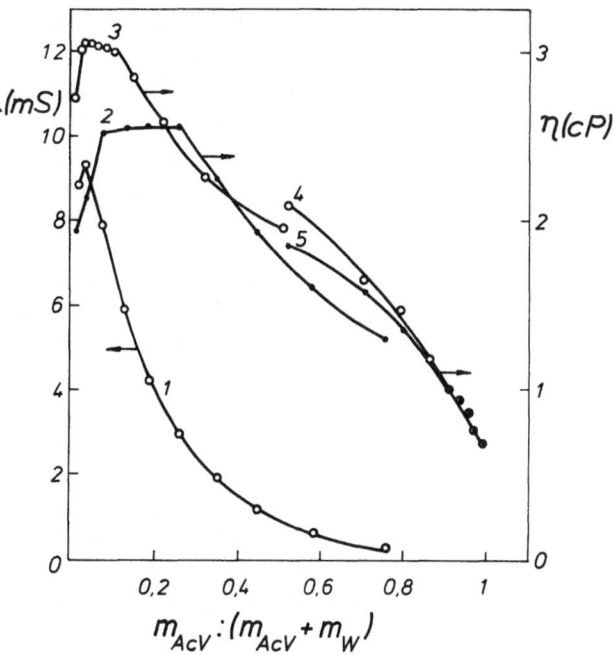

Fig. 6. Electrical conductivity and viscosity on monophasic systems: water-SDS (12:1 wt.)–VAc–n-PrOH (curves 1 and 2); water-NP-30 (10:1 wt.)–VAc–i-PrOH (curve 3); water –VAc-NP-3 (10:1 wt.)–n-PrOH (curve 4) and 1% NaCl solution–VAc-NP-3 (10:1 wt.)–n-PrOH (curve 5)

tant ratio was 10:1 (wt.) and the water-surfactant ratio 12:1 (wt.). The results obtained are shown in Figs. 3 and 4. In the case of hydrophobic surfactant the following sequence of the single-phase micro-emulsion was found: methanol < *n*-propanol < *iso*-propanol < ethanol, while for NP-30 the micro-emulsion phase increases with cosurfactant hydrophilicity: *iso*-octanol < *iso*-butanol < *n*-propanol ≪ *iso*-propanol.

The effect of the electrolyte on phase behavior was investigated for the system containing the NP-3 surfactant. Addition of the electrolyte to the system decreases the micro-emulsion zone, as can be seen in Fig. 5.

Physico-chemical properties

The phase diagrams described above indicate that either vinyl acetate or water can be solubilized by the surfactant-cosurfactant mixture employed to an extent depending on the relative composition of the system. This result implies that the continuous phase should reverse for a definite composition, and we have attempted to detect such an inversion. In this respect, we have studied the evolution of the continuous phase by viscosimetry and conductometry. The results obtained for some of the studied systems are presented in Fig. 6.

Electrical conductivity of the SDS-containing systems decreased steadily with the increase of monomer:(monomer + water) ratio. Except for the zones very rich in water, this holds true for the whole concentration range investigated. This behavior does not evidence any inversion of the micro-emulsion continuous phase.

Viscosity data are more relevant. They depend on the surfactant used. In hydrophilic surfactant systems, upon increasing the amount of water, the viscosity first increases, then levels off and finally decreases. Hydrophobic surfactant systems show a continuous increase in the viscosity. This is not affected by the presence of electrolyte for small amounts of water. On increasing the water content, the electrolyte produces a small decrease in viscosity.

Discussion

The following discussion deals mainly with the system based on short-chain alcohols because they yield more extended monophasic zones than the systems containing long-chain alcohols and afford greater possibilities to study their properties.

From the results presented above, it appears that hydrophobic and hydrophilic surfactants and cosurfactants behave differently in micro-emulsifying vinyl acetate. For example, in the system based on hydrophobic surfactants (i.e., NP-3) the addition of a short-chain alcohol (*n*-propanol) can produce micro-emulsions. In systems with hydrophilic surfactants (i.e., NP-30) the long-chain alcohols (i.e., *iso*-octanol) are too hydrophobic to form a water-soluble mixture with hydrophilic surfactants and drastically reduce the micro-emulsion zone (see Figs. 3, 4).

The problem that arises now is whether the surfactant present in the monomer-cosurfactant-water mixture can aggregate or not. An analogy with emulsions suggests that an oil-in-water emulsion is more electrically conducting than a water-in-oil emulsion. A sharp change in conductivity is recorded when passing from one type to the other. The electrical conductivity of the SDS-containing systems does not exhibit such a behavior; it increases continuously with the amount of water. This may suggest that the continuous phase is a ternary solution containing appreciable amounts of the three components of which vinyl acetate or water predominate.

Since the system contains a high polarity alcohol, a structure of a cosolubilized system seems to be more realistic. This structure was proposed by Shah et al. [12] for the system containing water-hexadecane-potassium oleate-*n*-pentanol, where the surfactant and cosurfactant form a liquid medium which can dissolve both water and oil as a molecular dispersion. Small aggregates of water molecules or of surfactant and cosurfactant molecules may also be present. By increasing the amount of water in the system, it becomes more electrical conducting as shown by the data presented in Fig. 6.

The low viscosities obtained in our systems demonstrate that there are no extended aggregates with ordered interfaces. The low viscosity rules out extended ordered bicontinuous structures. It should be recalled that cubic mesophases having such structures are extremely viscous.

On the other hand, no liquid crystalline phases were detected in the monophase systems by viewing them directly in polarized light. This is consistent with the fact that the cosurfactants are used in micro-emulsion formulations either to preclude the formation or to reduce the viscosity of liquid crystalline phases.

At the same time, electron spin resonance investigations of the monophase systems showed that they are practically nonstructured and agree with the previously reported NMR data [13]. Based on those data, Lindman and Stilbs have concluded that in micro-emulsions containing medium-chain alcohols (buta-

nol, pentanol) both hydrocarbon and water continuous solutions can be found.

Taking account these results, one may conclude that isotropic phases studied for vinyl acetate polymerization are cosolubilized systems and not micro-emulsions.

Finally, one must point out that this research allowed us to develop screening procedures for optimal polymerization of vinyl acetate. The kinetics of polymerization in cosolubilized systems always differs from that of emulsion polymerization and the properties of the resultant polymers are also different.

References

1. Gan LM, Chew CH, Friberg SE, Higashimura T (1981) J Polym Sci 19:1585
2. Atik SS, Thomas JK (1981) J Am Chem Soc 103:4279
3. Lianos P (1982) J Phys Chem 86:1935
4. Gan LM, Chew CH, Friberg SE (1983) J Polym Sci Polym Chem Ed 21:513
5. Jayakrishnan A, Shah DO (1984) J Polym Sci Polym Lett Ed 22:31
6. Friberg SE, Liang P (1984) J Polym Sci Polym Chem Ed 22:1699
7. Johnson PL, Gulari E (1984) J Polym Sci Polym Chem Ed 22:3967
8. Stoffer JO, Bone T (1980) J Polym Sci 18:2641
9. Stoffer JO, Bone T (1980) J Dispersion Sci Technol 1:37
10. Leong YS, Candau F (1982) J Phys Chem 86:2269
11. Donescu D, Anghel DF, Goşa K, Balcan M (1986) Paper presented at "The IInd National Symposium on Colloid and Interface Chemistry", Cluj-Napoca, Romania, 8–10 September 1986
12. Shah DO, Walker RD Jr, Hsieh WC, Shah NJ, Dwivedi S, Nelander J, Pepinsky R, Deamer DW (1976) SPE 5815 presented at Improved Oil Recovery Symposium
13. Lindman B, Stilbs P (1984) In: Mittal KL, Lindman B (eds) Surfactants in Solution. Plenum Press, New York, Vol 1, p 1651

Received September 16, 1987;
accepted June 9, 1988

Authors' address:

D.F. Anghel
Central Institute of Chemical Research
Dept. of Physical Chemistry
Spl. Independenţei 202
79611 Bucharest, Romania

Progress in Colloid & Polymer Science   Progr Colloid & Polymer Sci 77:131–135 (1988)

In vitro drug release from liquid crystalline creams; cream structure dependence

H. L. G. M. Tiemessen, H. E. Boddé, C. van Mourik, and H. E. Junginger

Division of Pharmaceutical Technology, Center for Bio-Pharmaceutical Sciences, Leiden University, Leiden, The Netherlands

Abstract: The general aim of this study is to develop skin-compliant liquid crystalline creams for controlled transdermal drug delivery. To achieve this aim, we try to use the colloidal gel structures inside the creams to control the drug release rate. This study deals specifically with the relationship between cream structures and drug release in vitro. Creams have been prepared by mixing Brij 96^R (polyoxyethylene(10)-oleylether, a nonionic surfactant) and water. Depending on the mixing ratio, they can adopt a lamellar, a viscous isotropic, or a hexagonal gel structure. The model drug benzocaine (ethyl p-aminobenzoate) was solubilized (1%) in the creams without disturbing the gel structures. Because the creams are able to swell quickly upon contact with water, an in vitro drug release model was developed using a water-impermeable membrane (dialysis membrane impregnated with silicone adhesive) to separate the cream from the aqueous acceptor phase (0.1 M HCl). In the lamellar and hexagonal creams the effective diffusion coefficient of benzocaine increases gradually with the water content due to a volume increase of the hydrophilic domains of the creams that form the continuous phase. In the viscous isotropic cream the diffusion of benzocaine is relatively high, most likely because of the low tortuosity of the cream.

Key words: Drug release, liquid crystalline creams, lyomesophases, polyoxyethylene alkylethers, nonionic surfactants

Introduction

In modern pharmaceutics there is a strong interest in transdermal drug delivery because it offers an opportunity to avoid the hepatic first-pass metabolism and the gastrointestinal side-effects often associated with oral dosing and because a more constant plasma concentration of the drug can be maintained for longer periods of time [1]. The rate of drug delivery from a transdermal dosage form is often controlled not by the system itself but by the stratum corneum, the upper layer of the human skin which has a very low permeability for drugs and therefore acts as the main diffusional barrier. As a consequence, transdermal drug absorption rates are low and show considerable inter- and intra-individual variations. Ideally, for controlled transdermal drug delivery the diffusional resistance inside the vehicle should be higher than the one in the stratum corneum. This requirement may be fulfilled by using the following double strategy:

1) by increasing the drug permeability of the stratum corneum with the help of so-called penetra-

tion enhancers; these enhance the transdermal flux by interfering with stratum corneum structures,

2) by using vehicles which contain microstructures functioning as diffusional barriers which can be manipulated in such a way that the drug permeability of the vehicle can be "set" at a desired level. The general aim of this study is to develop skin compliant liquid crystalline creams for controlled transdermal drug delivery, following the above-mentioned strategy. The creams studied are based upon nonionic surfactants which may function as penetration enhancers [2, 3], but which are the least irritating among the surfactants.

The specific aim of this study is to investigate the relationship between the colloidal gel structures inside the creams and the kinetics of drug release in vitro. The influence of the creams on the skin permeability will be dealt with in further studies.

Mixtures of water and the nonionic surfactant Brij 96^R (polyoxyethylene(10) oleylether) can be considered as simplified versions of creams; henceforth they will be referred to as creams. Depending on the surfactant-water mixing ratio, these creams can have various liquid crystalline gel structures (lamellar, hexagonal or viscous isotropic) which have been studied extensively [4, 5]. Benzocaine was chosen as a model drug for release studies. Since the creams tend to exchange water with an aqueous acceptor phase, it was necessary to design a new in vitro drug release model. The most critical part of the drug release model is the membrane which separates the cream compartment from the acceptor phase. The membrane has to be impermeable for cream components as well as for the acceptor phase to prevent changes in cream composition during the drug release experiment. In order to make sure that the membrane did not control the drug release rate, thereby hiding information about the diffusional behavior of the model drug in the cream, a membrane with a high permeability for the model drug was chosen so that the release profiles yielded accurate values for the effective diffusion coefficient of the drug inside the cream. For this purpose a special dialysis membrane impregnated with a silicone polymer was developed.

Materials and methods

Materials

Brij 96[R], a polyoxyethyleneoleylether with, according to specifications, an average of 10 ethyleneoxide (EO) units per molecule was a gift of Atlas Chemie (Essen, FRG). The two batches of the surfactant Brij 96[R] used were investigated with ^{1}H NMR [6] to check the average number of ethyleneoxide (EO) units per molecule. The two products appeared to have a lower EO-number than specified: 8.5 ± 0.2 and 9.3 ± 0.2. In the following the two batches will be referred to as Brij 96[R] (EO 8.5) and Brij 96[R] (EO 9.3).

Reagent grade water was produced by a Milli-Q filtration system, the final resistivity of the water was greater then $10 \ M\Omega \cdot cm$. Benzocaine, ethyl-p-aminobenzoate, obtained from Brocacef, (Maarsen, NL) purity > 99%, was used as received.

The silicone-impregnated membrane for the drug release experiments was made by pouring out a solution of a 2% w/v silicone polymer in 1,1,2-trichlorotrifluoroethane (Merck, Darmstadt, FRG) on a dialysis membrane (MWCO 5000 Diachema, Zürich, CH). As a soluble silicone polymer Medical Adhesive 355 or X7-2920 (a gift from Dow Corning, Midland, Michigan, USA) was used. The membrane finally contained 2.6 mg silicone per cm^2 and had a thickness of $20 \ \mu m$.

Methods

Preparation of the creams: Brij 96[R], water and benzocaine (1%) were weighed into glass vials, sealed, heated to 80°C, homogenized by shaking, and cooled down to room temperature. Water which had evaporated during the preparation procedure was not replaced and the final water content was determined gravimetrically, so that the final weight fractions Brij 96[R] (B), water (W) and benzocaine (F)

could be calculated. The water contents of the creams ranged from 15 up to 60% w/w.

Characterization of the creams was performed with an Olympus polarizing microscope (Tokyo, Japan) and small angle x-ray diffraction according to a method described elsewhere [4].

In vitro drug release

The in vitro drug release experiments were performed at a temperature of 21 °C using a stainless steel donor compartment in which the cream layer (2.5 mm) was separated from the aqueous acceptor phase (0.1 N HCl) by the silicone-impregnated dialysis membrane with a surface area of $3.14 \ cm^2$. The membrane was impermeable to the surfactant while the change in the water content of the creams was less than 2% during the first 600 min.

Since benzocaine molecules that diffuse from the cream into the acceptor phase (0.1 N HCl) were protonated, the concentration of the unprotonated drug was very low and the acceptor phase may be considered as a perfect sink.

The drug concentration in the acceptor phase was determined continuously by pumping it through a spectrophotometer (Ultrospec 4052 TDS, LKB Biochrom Ltd, Cambridge, UK) using flow-through cells in which the absorbance at 227 nm was measured at 10 minute intervals.

Results and discussion

Cream characterization

Small-angle x-ray diffraction was used to characterize the gel structures of the creams and to determine the repeat distances inside these gel structures [7]. Figure 1 shows the determined gel structures and repeat distances of the creams. Creams having a water content between 15 and 30% have a lamellar gel structure (L) which consists of extended layers containing the lipophilic moieties of the surfactant molecules, separated by hydrophilic layers which contain the water and the hydrophilic moieties of the surfactant molecules. Creams with a relatively high water content have hexagonal gel structures (H) composed of cylindrical surfactant aggregates arranged in a hexagonal structure and surrounded by water. The hydrophobic surfactant moieties are on the inside of the cylinders; the hydrophilic parts are on the outside and contribute to the continuous hydrophilic phase. The distances d between the centers of the cylinders can be calculated from the small-angle x-ray diffractograms and are given in Fig. 1.

Using formulae described elsewhere [4] d_1, the thickness of the hydrophobic layers in the lamellar creams or the diameter of the hydrophobic cores of the cylinders in the hexagonal creams could be calculated and was found to be $2.00 \pm 0.05 \ nm$ in both cases. The observed increase of the repeat distances d with an increasing water content of the cream is therefore caused

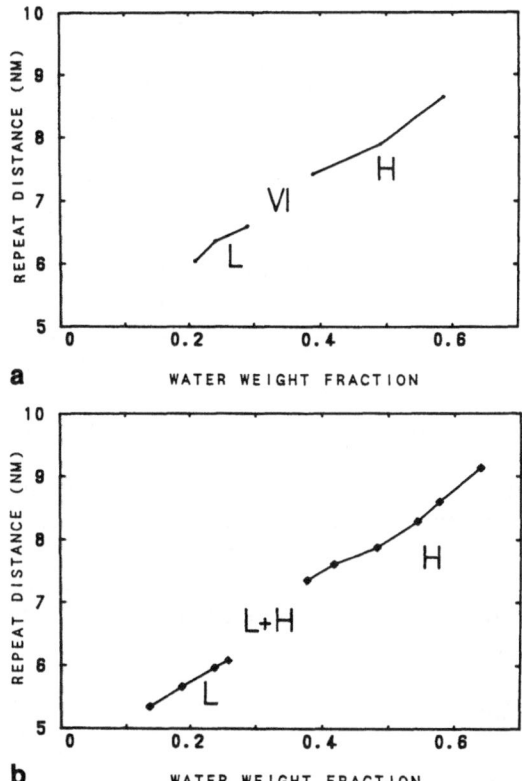

Fig. 1. Gel structures and characteristic repeat distances of the creams containing 1% w/w benzocaine, as function of the water weight fraction, water/(water+Brij 96^R), at 21 °C; L = lamellar, H = hexagonal, VI = viscous isotropic. a) creams containing Brij 96^R (EO 9.3); b) creams containing Brij 96^R (EO 8.5)

by an increase of d_h ($= d-d_l$), the thickness of the hydrophilic layers in the lamellar creams or the thickness of the hydrophilic domains between two cylinders in the hexagonal creams.

Depending on the batch of the surfactant used, the creams with water contents of about 30% may exhibit a mixture of lamellar and hexagonal gel structures (Fig. 1b) [4] or a viscous isotropic gel structure (Fig. 1a). A viscous isotropic gel structure in Brij 96^R-water mixtures has also been found by others [8]. The exact structure of the viscous isotropic phase is not yet fully understood but is supposed to be continuous in three axial directions, with respect to both the hydrophilic and hydrophobic regions [9]. The possible occurrence of the viscous isotropic phase depends on differences in the quality of the surfactant batch.

The solubilization of 1% w/w benzocaine did not influence the gel structure and repeat distance as could be determined with polarized light microscopy and SAXD: the difference in repeat distance between a cream with 1% w/w benzocaine and a cream with the same Brij 96^R/water ratio but without benzocaine was smaller than 0.05 nm in all cases.

Performance of the water-resistant membrane

When using an unmodified dialysis membrane as described in [10] we found that these liquid crystalline creams swell very quickly in an uncontrolled way; the water content of the studied creams increased by more than 20% during the first hour of the drug release experiment, so that no reliable information about the diffusional behavior of the drug could be obtained. However, the water uptake by the creams through the silicone-impregnated dialysis membrane was less than 2% throughout a 10-hour experiment, so that the gel structures in the creams remained unchanged. Hence it was quite feasible to correlate kinetic data with structural data.

The silicone-impregnated membrane may also be useful in the in vitro evaluation of hydrophilic transdermal therapeutic systems under controlled swelling conditions, which in an unmodified USP paddle setup [11] would not be possible.

In vitro benzocaine release from creams

Typical examples of single release experiments are given in Fig. 2. The release of the drug was slow: only a fraction of the drug content of the cream (2.5 mg/cm^2) was released during the first 10 h. After

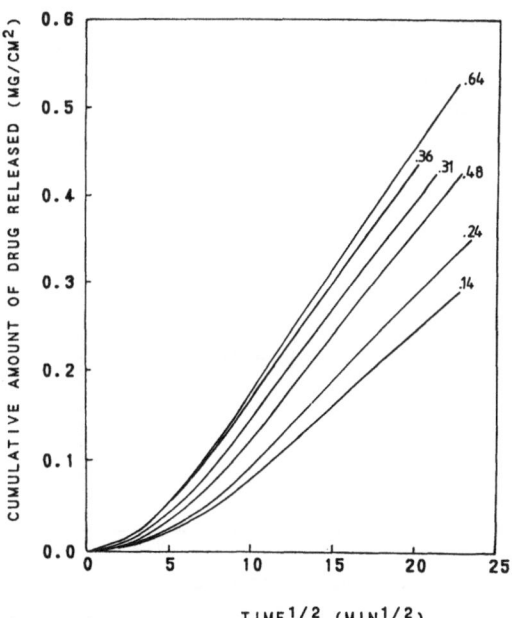

Fig. 2. Release of benzocaine from Brij 96^R (EO 9.3)/water creams at 21 °C. The weight fraction water, water/(water+ Brij 96^R) is given at the righthand side

an initial time lag which takes 25–100 min, complete linearity is obtained when the amount of drug released is plotted versus the square root of time. The slopes of the plots in the time interval 100–500 min were determined using linear regression. A correlation coefficient higher than 0.999 was obtained in every experiment.

Using the Higuchi [12] equation values of the effective diffusion coefficients, D could be obtained.

$$Q = 2AC_0 \left(\frac{Dt}{\pi}\right)^{1/2} \tag{1}$$

where

Q = cumulative amount of drug released (mg)
A = membrane area (cm^2)
C_0 = initial concentration of the drug in the cream (mg cm^{-3})
D = effective diffusion coefficient (cm^2 s^{-1})
t = time (s)

In Fig. 3 the calculated effective diffusion coefficients are plotted vs the weight fraction of water in the creams. Every given effective diffusion coefficient is a mean obtained from three or four separate experiments.

First consider the lamellar (W = water weight fraction between 0.14 and 0.26 using Brij 96^R (EO 9.3), and W between 0.15 and 0.34 using Brij 96^R (EO 8.5)) and hexagonal (W between 0.38 and 0.64 using Brij 96^R (EO 9.3)) creams. In the lamellar and hexagonal Brij 96^R (EO 9.3) creams the effective diffusion coefficient of benzocaine increases gradually with an increasing water content of the cream. In case of the lamellar Brij 96^R (EO 8.5) creams an almost linear in-

crease of the effective diffusion coefficient of benzocaine with the water content of the cream was observed. These differences in the effective diffusion coefficient cannot be explained in terms of (macroscopic) viscosity because the highest effective diffusion coefficients are observed in the hexagonal creams which have a viscosity more than 1 000 times higher than lamellar creams [13, 14].

A possible explanation for the observed phenomena may be based upon the following reasoning: in the creams there are distinct hydrophobic and hydrophilic domains. As has been shown above, the thickness of the hydrophobic layers in the lamellar cream and the diameter of the hydrophobic cores of the cylinders in the hexagonal cream remain constant with increasing water content. Therefore the volume fraction of the hydrophobic domains decreases when the water content of the creams increases. When the benzocaine is not equally distributed among the two cream domains and the diffusion can only take place through one of the two, then [15]

$$D_{eff} = \frac{D_0 \times \varepsilon}{\tau} \tag{2}$$

D_{eff} = effective diffusion coefficient
D_0 = intrinsic diffusion coefficient
ε = porosity, in this case the volume fraction of the accessible domains
τ = tortuosity

If the benzocaine were able to diffuse only through the hydrophobic domains, then the effective diffusion coefficient would decrease with an increasing water content, which is obviously not the case. The experimental data may now be explained very logically by assuming that benzocaine diffuses through the hydrophilic domains of the creams which form a continuous phase, and by taking ε (Eq. 2) as the volume fraction of the hydrophilic domains.

Other indications that the benzocaine is located mainly in the hydrophilic domains are taken from the literature. The hexane-water partition coefficient of benzocaine is 1.25 at 37 °C [16]. Furthermore the maximal solubility of benzocaine in the aqueous phase is highly increased by the presence of polyethylene glycol(400): from 5 times (34 vol%) to >300 times (78 vol%) at 25 °C [17]. Since the hydrophilic domains consist of a concentrated solution of the polyoxyethylene fragments of the surfactant molecules in water (from 24 to 78 vol% in creams containing 36 and 86% w/w surfactant, respectively) the solubility of benzocaine in the hydrophilic domains will be much higher than in the hydrophobic domains and therefore it is

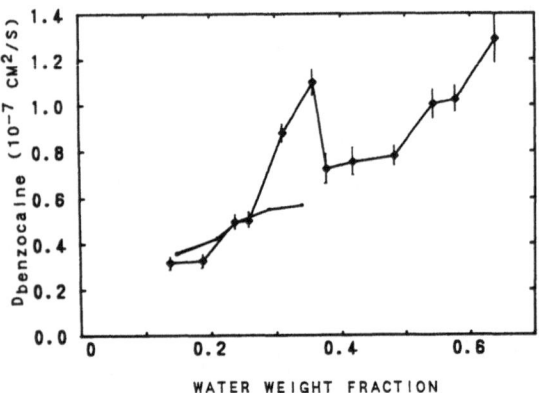

Fig. 3. Effective diffusion coefficients of benzocaine vs the water weight fraction, water/(water + Brij 96^R) in the creams at 21 °C. ♦) creams with Brij 96^R (EO 9.3), the standard deviation is indicated; ●) creams with Brij 96^R (EO 8.5)

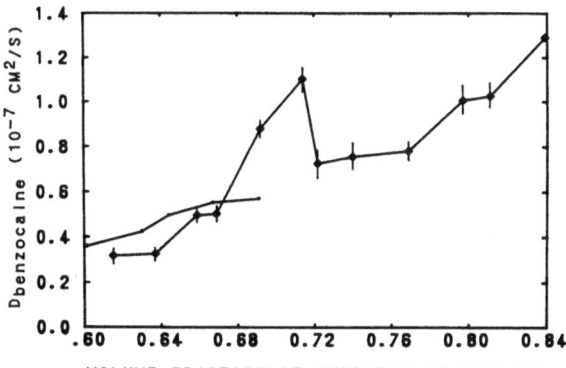

Fig. 4. Effective diffusion coefficients of benzocaine vs the volume fraction of the hydrophilic domains at 21 °C as calculated with Eq. (3). ◆) creams with Brij 96^R (EO 9.3), the standard deviation is indicated; ●) creams with Brij 96^R (EO 8.5)

likely that the benzocaine will partition mainly into the hydrophilic domains.

It is interesting to see to what extent the effective diffusion coefficient depends on the volume fraction of the hydrophilic domains (Q_h). Q_h can be calculated from

$$Q_h = \frac{(\mu B / \delta_{eo}) + (W / \delta_w)}{[(1-\mu)B/\delta_{hc}] + (\mu B / \delta_{eo}) + (W / \delta_w)} \qquad (3)$$

where, B and W are the weight fractions of surfactant and water, respectively and the quantity μ is the (average) molecular weight fraction of the oxyethylene part of the surfactant molecules; δ is the specific density (g cm^{-3}) of the oxyethylene chain (*eo*), the hydrocarbon chains (*hc*) and the water (*w*), respectively. The densities are considered to be invariant with a changing surfactant content of the cream (ideal behavior). The theoretical values δ_{hc} and δ_{eo} are derived from literature data in [4].

In Fig. 4 the effective diffusion coefficient is plotted versus the volume fraction of the hydrophilic domains. Here again we observe an almost proportional increase in the effective diffusion coefficient with an increasing volume fraction of the hydrophilic domains for the lamellar and hexagonal creams, confirming the aforementioned explanation.

Now consider the viscous isotropic creams ($W = 0.31$ and $W = 0.36$ using Brij 96^R (EO 9.3) in Fig. 3). Their effective diffusion coefficients do not "fit" into the regular pattern observed for the lamellar and hexagonal creams; their effective diffusion coefficients are relatively high. Effects of tortuosity and interfacial adsorption may be involved here.

According to Eq. (2) the tortuosity plays an important role in the final value of the effective diffusion coefficient. The high effective diffusion coefficient of benzocaine in the viscous isotropic creams may be explained by the low tortuosity of the hydrophilic pathway in the bicontinuous gel structure.

Another factor which may influence the diffusion of benzocaine in these creams may be the adsorption of the permeant to the interface between the hydrophilic and hydrophobic domains. It is worth noticing that benzocaine, being a local anaesthetic agent, has surface-active properties. The high effective diffusion coefficient of benzocaine in the viscous isotropic creams should then partly be explained by the low tortuosity of the hydrophobic pathway which is followed if the drug is adsorbed to the interfaces between the hydrophilic and hydrophobic domains.

Further studies are now being undertaken that deal with the relationship between the partition of the drug inside liquid crystalline creams and the drug lipophilicity, as well as its diffusivity in the creams.

References

1. Guy RH, Hadgraft J (1987) J Contr Rel 4:237
2. Ashton P, Hadgraft J (1986) Pharm Acta Helv 61:228
3. Shen WW, Danti AG, Bruscato FN (1976) J Pharm Sci 65:1781
4. Jousma H, Joosten JGH, Junginger HE (1988) Coll & Polym Sci, submitted
5. Felger M (1977) Thesis, Techn Univ Braunschweig, FRG
6. Ludwig FJ (1968) Anal Chem 40:1620
7. Luzatti V, Mustacchi H, Skoulios A, Husson F (1960) Acta Cryst 13:660
8. Hoffmann HN, Paulus EF (1969) Fette–Seifen–Anstrichmittel 71:399
9. Lindblom G, Larsson K, Johansson L, Fontell K, Forsén S (1979) J Amer Chem Soc 101:5465
10. Müller-Goymann CC, Frank SG (1986) Int J Pharm 29:147
11. Shah VP, Tymes NW, Yamamoto LA, Skelly JP (1986) Int J Pharm 32:243
12. Higuchi WI (1962) J Pharm Sci 51:802
13. Unpublished results
14. Nurnberg E, Pohler W (1983) Dtsch Apoth Ztg 42:1993
15. Desai SJ, Simonelli AP, Higuchi WI (1965) J Pharm Sci 54:1459
16. Yalkowski SH, Flynn GL, Slunick TG (1972) J Pharm Sci 61:853
17. Rubino JT, Yalkowsky SH (1985) J Parent Sci Technol 39:106

Received December 9, 1987;
accepted June 1, 1988

Authors' address:

H. L. G. M. Tiemessen
Division of Pharmaceutical Technology
Center for Bio-Pharmaceutical Sciences
Leiden University
P.O. Box 9502
2300 RA Leiden, The Netherlands

Progress in Colloid & Polymer Science   Progr Colloid & Polymer Sci 77:136–142 (1988)

Dispersed Systems

Physico-chemical properties of small bubbles in liquids

M. Sakai

Department of Industrial Chemistry, Faculty of Engineering, Kyushu Sangyo University, Fukuoka, Japan

Abstract: The object of this investigation is to explore the physico-chemical properties of ions and their structures in an electrical double layer on the surface of 4.92 µM MgX (X = Cl$^-$, Br$^-$, NO$_3^-$, ClO$_4^-$, and SO$_4^{2-}$) or 4.92 µM MCl (M = Li$^+$, Na$^+$, K$^+$, NH$_4^+$, Rb$^+$, and Cs$^+$) aqueous salt solutions which were investigated by measuring the surface electric charges (σ_0) at a Stern layer and the ion distributions at a gas-liquid interface by means of (I) the rotating cylindrical cell technique and (II) the jet drop method, respectively.

Key words: Electric double layer, surface salt solution, surface electric charges, Stern layer, ion distributions

1. Introduction

When a small bubble is first introduced into 4.92 µM aqueous solutions of MgX or MCl salt, the arrangement of water molecules strongly polarized at the liquid surfaces is formed and immobilized to be completely oriented at its surfaces. Oxygen ions (O$^{\delta-}$) in the oriented water dipoles are directed to the gas side of the surfaces, whereas hydrogen ions (H$^{\delta+}$) point to the body side of liquid to produce an electric field. The positively charged H$^+$ ions on the liquid surface attract the hydrolytic OH$^-$ and the negatively charged other anionic species (X, Cl$^-$) (potential determining ions) by electric and Van der Waals forces and hence form a negatively charged plane (IHP; cf, Fig. 6). Then, the oppositely charged cationic species (Mg^{2+}, M) near the body side of liquid are not so dense, and form positively charged planes (OHP). Under these conditions, an electrical double layer consisting of a Stern layer and a diffusion double layer on the basis of the Gouy-Chapman model is generated at neutrality. It is also possible that the H$^{\delta+}$ of polar water makes anions adsorb on the surface to draw cations into bulk microlayer on the surface. Frumkin [1] has observed that most of the salts impart a negative charge to the surface at the air-water interface and that the magnitude of this charge depends on the nature of the anion. Water in aqueous solutions of inorganic salts plays the role of a surface-active substance. In this way the H$^{\delta+}$ ions on polar water in aqueous solutions of MgX or MCl salt are covered with adsorbed negative ions (X or Cl$^-$). At the same time, cations (Mg^{2+} or M) from the bulk liquid are bound close to some of the negative adsorbed ions. The mean electric charge per unit area of a small bubble (σ_0 µC cm^{-2}) are equal to the charges on the uncovered negative ions on its surface. The bound cations are held less firmly than the adsorbed negative ions and are removed by collisions due to the general motion of thermal agitation in a liquid bulk. In attempting to explain the electric charges of a small bubble, the selective adsorption of X ions on the H$^{\delta+}$ portion of polar water is covered in turn by Mg^{2+} ions from the bulk liquid of the MgX salt. Since X and Mg^{2+} ions are larger than OH$^-$ and H$^+$, they are unable to approach each other so closely and are more readily removed by thermal agitation of the surrounding liquid. Furthermore, the diffusion velocity of cations is generally different from that of anions in the neighborhood of these dipole groups, the distribution of electric charges deviates with surface excess of anions and, in the absence of cations, causes an electric field in the diffuse double layer. The deviated electric charges (σ_0) created by the difference between the water dipoles-anions and the their anion-cation interactions, and the effects of the electric fields on the water dipoles play a major role in determining the magnitude of zetapotential (ζ mV), based on the Boltzmann distribution for an ion concentration, Poisson's equation for the potential, and the local ion or apparent charge distributions at the gas-liquid interface.

In this study more detailed analyses of the electric charges per area on the surfaces of a small nitrogen bubble suspended in 4.92 μM aqueous solutions of MgX or MCl salt are presented which were performed by using the rotating cylindrical cell technique and by measuring the electric mobilities of small bubbles in 4.92 μM aqueous solutions of MgX or MCl salt under an applied potential gradient. The bubbles were maintained in the center of a horizontal cylindrical tube in a manner similar to the technique described in [2–5].

The rotating cylindrical cell technique: a) surface electric charges (σ_0) at the Stern layer for the hydrated anionic radius r_h(Å) of various chemical ionic species; b) from the result the structure of polarized water molecules and the behavior for cationic species Mg^{2+} and M in aqueous MgX or MCl salts solutions; c) the interactions between hydrated anionic species X and Mg^{2+} in the electrical double layer of 4.92 μM aqueous MgX salt solutions; and d) the interactions between cationic species M and Cl^+ in the Stern layer of 4.92 μM aqueous MCl salt solutions are mainly discussed. The effect of the hydration force on the surface negative electric charges ($-\sigma_0$) around a small bubble is investigated using the excellent method of Whybrew et al. [4]. An examination was also successfully carried out to determine the interactions between anionic species X and Mg^{2+} ions for aqueous solutions of MgX salt and these between Cl^- and cationic species M for aqueous solutions of MCl salt in terms of the surface negative electric charges ($-\sigma_0$) and of bubble size, which was correlated with the effect of added electrolytes on surface potential difference determined using the surface chemical technique by Jarvis and Scheiman [5].

The jet drop method (MacIntyre's onion-like shell model [6]): a) the ion concentrations and surface electric charges (σ_0), respectively, in the jet drops that jump into the air, tearing off the thin film of bubble surface region when a small bubble bursts on liquid surface are analyzed; b) from the result an electric charge distribution pattern at the electric double layer around a small bubble in 1.50–112.3 μM aqueous $MgCl_2 \cdot 6H_2O$ solutions; c) the ion or potential distributions; and d) the empirical Mg^{2+} ion distribution at the gas-liquid interface of 4.92 mM aqueous $MgCl_2 \cdot 6H_2O$ solution are discussed.

2. Theory

The surface electric charges (σ_0) can be obtained from the force balance on a small bubble under the

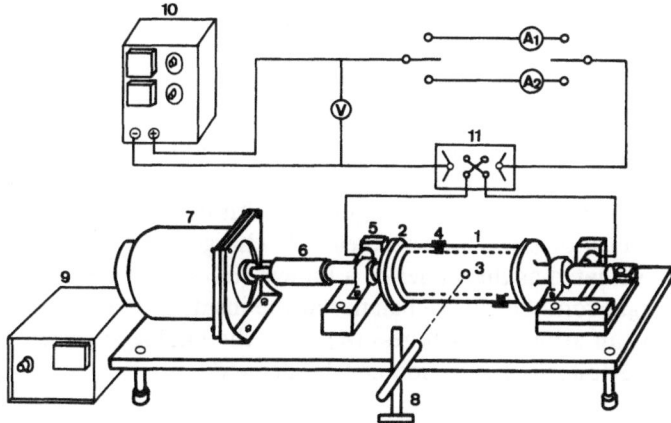

1	Rotating acrylic cylinder	7	Moter
2	Electrodes	8	Traveling microscope
3	Small bubble (N₂ gas)	9	Constant voltage
4	Bubble entrance	10	D.C. power supply
5	Carbon brush		(the force of an electric field)
6	Rubber tube	11	Switch

Fig. 1. Diagram of a small rotating cylindrical acrylic cell ($30 \oslash \times 680$ mm²) filled with 4.92 μM aqueous MgX or MCl salt solutions

constant electric field (E Volt cm^{-1}) as shown in Fig. 1. By using the similar technique investigated by Whybrew et al. [4], the following equation becomes

$$(2/3)d_B^3 \rho \Omega v k_1 + 3\pi\mu d_B v k_2 = (\pi d_B^2 \sigma_0 E)/300. \quad (1)$$

Dynamic	Viscous	Electric force
(F_1)	(F_2)	(F_3)

The electric mobilities $U(= v/E)$ over a wide range and for large value of d_B reduces to

$$U = Cd_2/(Ad_B^2 + B) \simeq (C/A)(1/d_B) \quad (2)$$

where,

$$A = (2/3)\rho\Omega k_1, \quad B = 3\pi\mu k_2, \quad \text{and} \quad C = (\pi\sigma_0/300)$$

v = bubble velocity, cm s^{-1}
d_B = bubble diameter, cm
μ = coefficient of viscosity, g cm^{-1} s^{-1}
Ω = angular speed (= 1 700), rpm
k_1, k_2 = experimental factors (= 0.71, 2.40, respectively) are given by a similar manner as Whybrew [4])
E = electric field (= V/l: 48 Volt/6.8 cm), V cm^{-1}
V, l = voltage applied and length of cell, respectively
σ_0 = apparent surface electric charges, esu cm^{-2}

Equation (2) can be rewritten as

$$\log (d_B/U) \simeq \log d_B^2 + \log (A/C) \qquad (3)$$

where $A/C = (200 \rho \Omega k_1/\pi \sigma_0)$. Equation (3) shows that an empirical plot of the ratio (d_B^2/U) vs d_B^2 on a log-log scale gives a straight line with unit slope. The values of (A/C), surface electric charges (σ_0) can be evaluated as the values of a known (d_B/U) for a fixed d_B. From the charge balance at the inside and outside of an arbitrary thickness (δ) from the bubble surface the following equation can be derived:

$$\sigma_0 \quad + \int_{\delta}^{\infty} \rho(\delta)d\delta = 0 , \qquad (4)$$

Stern   Diffusion layer

where the first (σ_0) and second $\left(\int_{\delta}^{\infty} \rho(\delta)d\delta \right)$ terms represent the electrical double layer (Stern and diffusion layers) in the range of $\delta = 0 - \delta$ and the bulk layer in $\delta = \delta - \infty$, respectively. As for Eq. (4), the boundary condition: $(d\psi/d\delta)|_{\delta = \infty} = 0$ and Poisson's equation is satisfied

$$\rho(\delta) = -\varepsilon(d^2\psi/d\delta^2) . \qquad (5)$$

Also the amount of electric charges per unit area in the diffusion layer is expressed in

$$\sigma_0 = (\sigma_s^- + \sigma_s^+) + \sigma_\delta = -\varepsilon(d\psi/d\delta)|_{\delta = \delta} , \qquad (6)$$

where, the slope of the right side in Eq. (6) is given by

$$(d\psi/d\delta)|_{\delta = \delta} = (2RT/\varepsilon)^{1/2}$$
$$\times \left(\sum_i C_i(\infty)(V^{z_i} - 1) \right)^{1/2} . \qquad (7)$$

Substituting Eq. (7) into Eq. (6), we obtain

$$\sigma_0 = -(2RT/\varepsilon)^{1/2} \left(\sum_i C_i(\infty)(V^{z_i} - 1) \right)^{1/2} , \qquad (8)$$

where $V = \exp(-z_i F \psi(\delta)/RT)$. Symbols are as follows: $(\sigma_s^- + \sigma_s^+ =$ dipolar water molecules, $\varepsilon =$ dielectric constant, $R =$ gas constant, $C_i(\infty) =$ bulk concentration of species i, $F =$ Faraday's constant, $z_i =$ charge number of species i, and $\eta_i = V_i^z =$ relative concentration of species i. From a known amount of $C_i(\infty)$, σ_0 in Eq. (8), and the remaining amount of η_i can be calculated. Further an electric potential

$(= $ zetapotential $\zeta)$ at the Stern layer, $\psi_\delta(= \zeta)$ is represented as

$$\psi_\delta = -(RT/z_i F) \ln (\eta_i) . \qquad (9)$$

The charge density $\rho(\delta)$ is defined by

$$\rho(\delta) = \sum_i C_i z_i F = C(\infty)Fz(\eta_+ - \eta_-) , \qquad (10)$$

wherein it is assumed that $z_+ = z_- = z$ for z valence z valence electrolytes and that $C_+(\infty) = C_-(\infty) = C(\infty)$. From Eqs. (5) and (6), an electric charge density $\rho(\delta)$ is derived as

$$\rho(\delta) = (d\sigma_0/d\delta) \quad \text{or} \quad (d\sigma_\delta/d\delta) . \qquad (11)$$

3. Experiments and conditions

The experimental arrangement is shown in Fig. 1 and is comprised mainly of pure water, 4.92 μM aqueous MgX or MCl salt solutions filled in horizontal cylindrical acrylic cell (30 u × 680 mm² in internal diameter and length) **1** with a pure nitrogen gas bubble entrance **4**. Both circular platinum-plate electrodes **2** are fitted at the ends of cell along the axis of rotation, a switch **11** for reversing the constant voltage **9** between the electrodes connected with a DC bridge **10**, and another traveling microscope **8**. First, a small nitrogen bubble is introduced into 4.92 μM aqueous solutions of MgX or MCl salt contained in a cylindrical cell which rotates about a horizontal axis. A small bubble takes up a position on the axis and its velocity $(v$ cm s$^{-1})$ under an electric field applied between the ends of the cell is measured continuously by the traveling microscope **8** as the bubble is slowly absorbed into the liquid bulk until it disappears. The surface electric charges (σ_0) on a small bubble given from an empirical plot of (d_B/U) vs d_B^2 on a log-log scale are independent of the bubble size and have the diameter above a critical value of about 0.2 mm. After equilibrium is reached, the electric charges on the bubble becomes independent not only of time but also of bubble diameter. The aqueous solutions of inorganic electrolytes used are 4.92 μM MgX (X = Br$^-$, NO$_3^-$, Cl$^-$, ClO$_4^-$, SO$_4^{2-}$), 4.92 μM MCl (M = Li$^+$, Na$^+$, K$^+$, NH$_3^+$, Rb$^+$, Rs$^+$) and ordinary distilled water (conductivity $k = 1.28$ μS cm^{-1}) made by Wako Co., Ltd. The nitrogen gas is saturated with water vapor by passing it through a humidifying bubbler. The bubble diameter is varied over the range of 0.30 – 1.40 mm. The temperature is 26 – 27 °C and the electric field is 48 Volt/6.8 cm^{-1}.

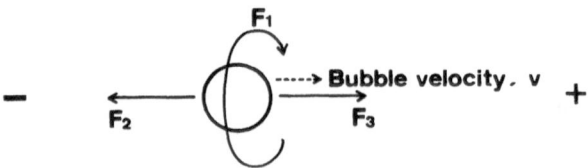

Fig. 2. Force balance on a small bubble under a constant electric field

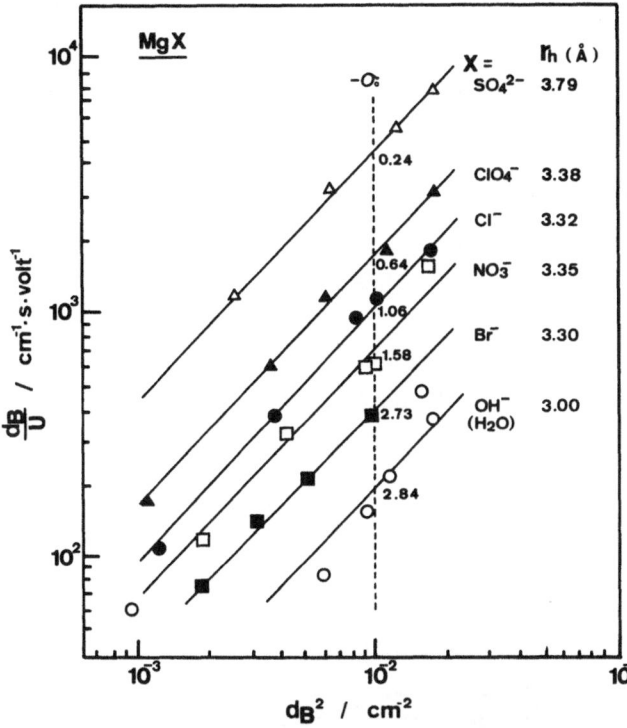

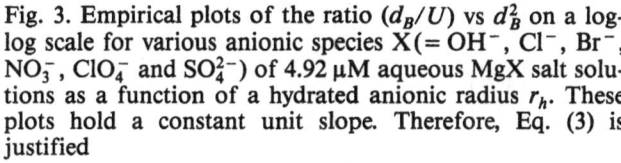

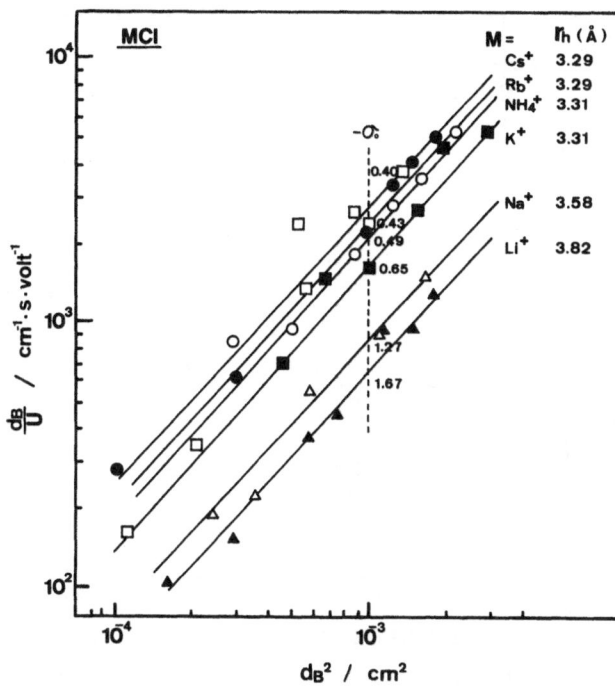

Fig. 3. Empirical plots of the ratio (d_B/U) vs d_B^2 on a log-log scale for various anionic species X(= OH⁻, Cl⁻, Br⁻, NO₃⁻, ClO₄⁻ and SO₄²⁻) of 4.92 μM aqueous MgX salt solutions as a function of a hydrated anionic radius r_h. These plots hold a constant unit slope. Therefore, Eq. (3) is justified

Fig. 4. Empirical plots of the ratio (d_B/U) vs d_B^2 on a log-log scale for various cationic species M(= Li⁺, Na⁺, K⁺, NH₄⁺, Rb⁺ and Cs⁺) of 4.92 μM aqueous MCl salt solutions as a function of a hydrated cationic radius r_h. Eq. (3) is further justified

4. Results and discussions

In order to obtain empirical surface charges (σ_0) at the gas-liquid interface a detailed operational method and the technique are illustrated in Figs. 3 and 4, respectively. An electric charge is an important determinant of adsorbed ions or the collection of particles at an interface. In studying the role of bubble charge, it is necessary to use electrophoresis cell for determining the electric charges of small bubbles of various sizes suspended in bulk solutions.

Figure 3 shows an empirical plots of the ratio (d_B/U) vs d_B^2 on a log-log scale for various anionic species X of 4.92 μM aqueous MgX salt solutions together with the parameters of hydrated anionic radius r_h. These plots are linear over the d_B range of 0.3–1.0 mm, holding a constant unit slope. Therefore, Eq. (3) is justified. Further the values of electrically negative charges $(-\sigma_0)$ are caused by the selective adsorption of ions in the Stern layer and can be computed based on the relation between the known (d_B/U) and d_B^2. These values decrease with increasing

r_h as illustrated in Fig. 3. The presence of an ionic species X for MgX can greatly alter the mobility of a charged bubble when adsorbed in the Stern layer. Even the pure water surface has a surface electric charge characterized by polar water molecules and by selective adsorption of OH⁻ ion.

In the same way as described above, Fig. 4 shows an empirical plot of the ratio (d_B/U) vs d_B^2 on a log-log scale for various cationic species M of 4.92 μM aqueous MCl salts solutions together with the parameters of hydrated cationic radius r_h. These plots also hold a constant unit slope, which means Eq. (3) is justified. The values of the electric negative charges $(-\sigma_0)$ decrease with decreasing r_h. This order of cationic species M agrees with both those of Hofmeister, and the surface potential difference (ΔV) unit is determined by the method of Jarvis and Scheiman [7].

Figure 5 shows the relationships between a apparent negative surface electric charges $(-\sigma_0)$ at the Stern layer and hydrated ionic radius r_h in 4.92 μM aqueous MgX or MCl salt solutions as a function of the weakly or strongly hydrated ionic species X and M. As shown

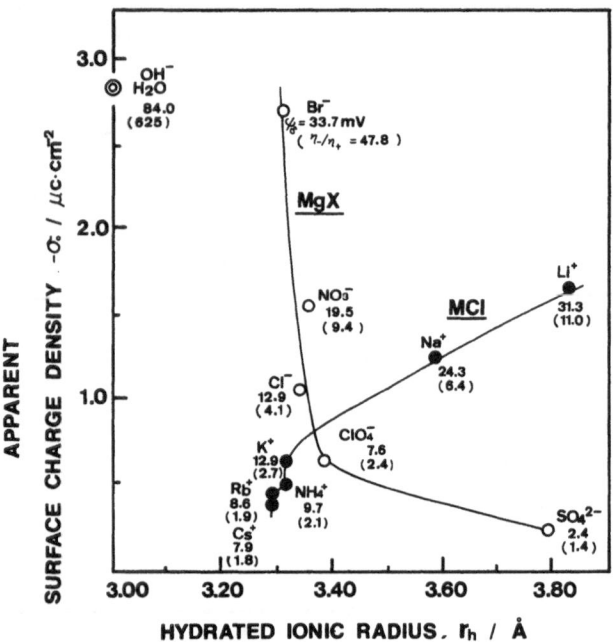

Fig. 5. Relationship between surface electric charges ($-\sigma_0$) at the Stern layer and hydrated ionic radius r_h in 4.92 µM aqueous MgX or MCl salt solutions as a function of the weakly or strongly hydrated ionic species $X (= OH^-, Br^-, NO_3^-, Cl^-, ClO_4^-, SO_4^{2-})$ and $M (= Li^+, Na^+, K^+, NH_4^+, Rb^+$ and $Cs^+)$

in Fig. 5, the values of electric potential $\psi(\delta)$ and separation ratio (η_-/η_+) determined from Eqs. (6), (7) and (8), decrease with increasing r_h in the order: $Br^- > NO_3^- > Cl^- > ClO_4^- > SO_4^{2-}$ for aqueous MgX salt solutions. This order, except for Cl^-, agrees with those of Hofmeister. On the other hand, the value of electric potential and the separation ratio for MCl salts increase with increasing r_h in the order: $Cs^+ < Rb^+ <$
$NH_4^+ < K^+ < Na^+ < Li^+$. This order also agrees with those of Hofmeister [9].

Figure 6 illustrates the distribution patterns of the relative ion concentrations $\eta_i(\delta)$, charge density $\rho(\delta)$, surface electric charges σ_0, and an electric potential $\psi(\delta)$, respectively, at the gas-liquid interface around a small bubble. A detailed description of the ion concentration distributions is described in Fig. 6a. First, the $H^{\delta+}$ of polarized or oriented water molecules at the gas-liquid interface is considered to attract an anion (η_-) by an electric force while repeling those of the opposite sign (η_+). Selectively adsorbed negative ions bound to the $H^{\delta+}$ ions are partly or completely covered by positive ions. Some of the negative ions are removed from the surface by the thermal agitation of the bulk liquid to reach equilibrium. These, owing to

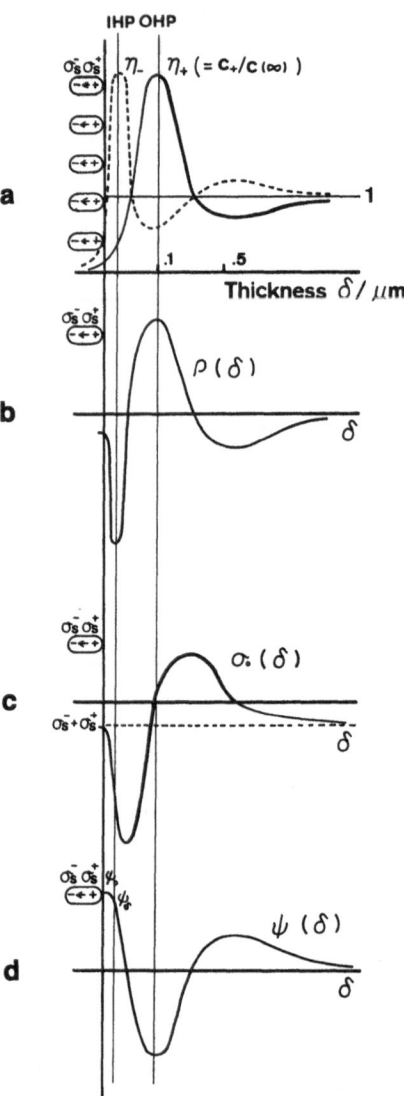

Fig. 6. Empirical distribution patterns of the ion concentrations (η_+, η_- (6a)), charge density $\rho(\delta)$ (6b), surface electric charges σ_0 (6c), and electric potential $\psi(\delta)$ (6d) in the electrical double layer at the gas-liquid interface around bubble. These patterns are clarified using the jet drop method

polarized water and the thermal agitation, make up a resultant field. Therefore, the surface negative electric charges acquired by a small bubble are due to the remaining exposed negative ions.

The evidence mentioned above is confirmed with ion or electric charge distributions obtained especially from the jet drop method as shown in Figs. 7 and 8. The so-called normal model of ion distribution in Fig. 6a have been well documented. An attempt is made in this paper to decide the ion distribution and the charge density $\rho(\delta)$, and a further object of the work

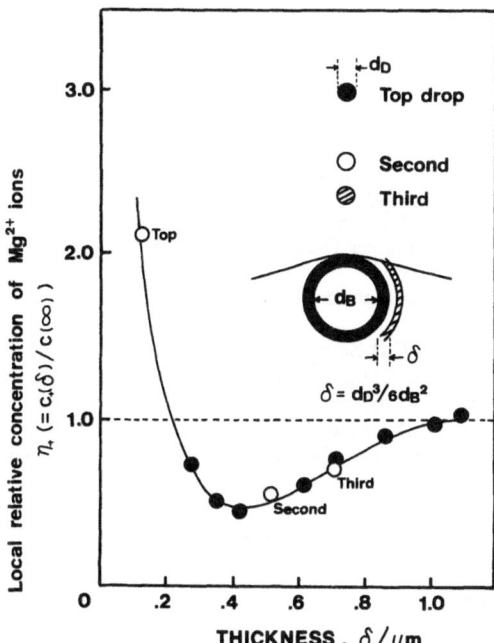

Fig. 7. Local relative concentration distribution of Mg^{2+} ion (η_+ vs thickness δ-plot) in 4.92 mM aqueous $MgCl_2$ $\cdot 6H_2O$ solution. Several jet drops (top, second, third) are produced from an onion-film shell around a small bubble

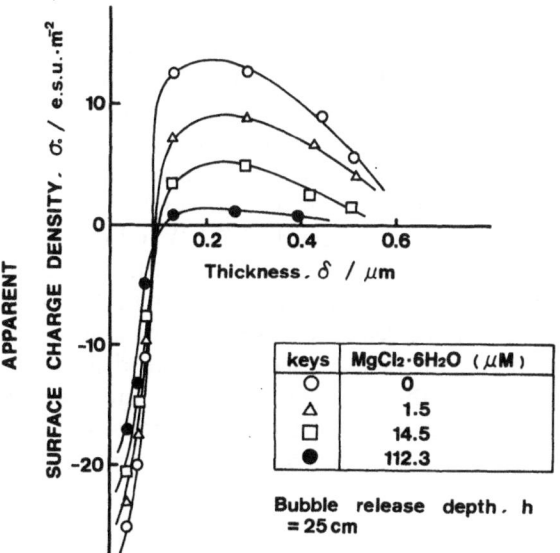

Fig. 8. Apparent surface electric charge distributions (σ_0 vs thickness δ-plot) as a function of aqueous concentrations of $MgCl_2 \cdot 6H_2O$ salt at the bubble release depth $h = 25$ cm

is to determine the apparent surface electric charges (σ_0) and an electric potential $\psi(\delta)$. Using Eqs. (10) and (11), the ion distributions in Fig. 6a lead to the charge density in Fig. 6b. An apparent surface charge in Fig. 6c leads to an electric potential, which is illustrated (Fig. 6d) using Eq. (6).

Figure 7 shows the relative concentration distributions of Mg^{2+} ions (η_+ vs thickness δ plot) in 4.92 mM aqueous $MgCl_2 \cdot 6H_2O$ solution by the analysis of Mg^{2+} ion concentration in various jet drops (top, second, third) [6] and only in the top drops of various sizes. The concentration of Mg^{2+} ion in bulk liquid and in the jet drops could be titrated with 0.01 or 0.0005 M EDTA using EBT as indicator. The number of the jet drops collected on a glass plate is namely 2000 which is the limit of analysis. An empirical distribution of Mg^{2+} ion concentration on or near the surface layer around a small nitrogen bubble by the jet drop method was essentially the same as that described by Fig. 6a.

Figure 8 shows an apparent surface charge distribution (σ_0 vs thickness δ-plot) from the measurement of the total electric charges only in the top jet drops, ejecting the thin film of thickness $\delta(= d_D^3/6d_B^2$; D: drop, B: Bubble) skimmed from the increased outlet of bubbles bursting into the air as the jet drops [6, 8]. On

this occasion, single nitrogen bubbles saturated with water vapor were released slowly at a controlled rate (more than $2.0\,s^{-1}$) from the tip of the capillary tube, which was vertically immersed to a depth of 25 cm beneath the liquid surface of 1.5, 14.5, and 112.3 μM aqueous $MgCl_2 \cdot 6H_2O$ salt solutions. An empirical surface electric charge (σ_0) measured in a manner similar to that of Blanchard [8], Iribarne and Mason [10] led to essentially the same conclusion as that described by Fig. 6c.

The experiments and methods used in Figs. 7 and 8 will be fully described elsewhere, and only a brief outline is given here.

5. Conclusions

By means of the rotating cylindrical cell technique and the jet drop method, the following conclusions are derived:
a) Negative surface electric charges ($-\sigma_0$) at the Stern layer decrease with increasing hydrated anionic radius r_h(Å) in the order: Br^- (3.30) > NO_3^- (3.35), Cl^- (3.32) > ClO_4^- (3.38) > SO_4^{2-} (3.79) for cation Mg^{2+} (3.74) in 4.92 μM aqueous MgX salt solutions. If the hydrated anionic radius of Cl^- ion is larger than that of ClO_4^- ion, this order agrees with that of Hofmeister. The surface potential differences (ΔV) [7] are in the same order.
b) Negative surface electric charges ($-\sigma_0$) at the Stern layer increase with increasing hydrated cat-

ionic radius r_h(Å) in the order: Li^+ (3.82)>Na^+ (3.58)>K^+ (3.31)>NH_3^+ (3.31)>Rb^+ (3.29)> Cs^+ (3.29) for the anion Cl^- (3.32) in 4.92 µM MCl salt solutions. This order agrees with both those of Hofmeister, and surface potential differences (ΔV) by Jarvis and Scheiman [7].

c) The interaction between Mg^{2+} (3.74) and Cl^- (3.32), and that between Li^+ (3.82) and Cl^- (3.32) in the Stern layer are smaller than those of other systems and it is confirmed that Stern layers are formed by the selective adsorption of anions.

d) Local ion or surface electric charge distribution patterns at the gas-liquid interface are clarified from an analysis of ion concentrations and total electric charges in the tops of jet drops.

Acknowledgements

The author wishes to thank Professor Emeritus Dr. R. Matuura and Dr. Y. Moroi of Kyushu Univ., and Professor Dr. K. Shirahama of Saga Univ., for valuable advice and stimulating discussions; he is also indebted to Professor Dr. S. Mori for his technical help with a part of the apparatus and for his encouragement, and to Mmes. A. Ohishi, K. Shigeta and K. Naganori, students in the Department of Chemical Engineering of Kyushu Sangyo University for their assistance in the experimental work.

References

1. Frumkin A (1924) Z phys Chem 109:34
2. McTaggart H (1922) Phil Mag 44:386
3. Currie BW, Alty T (1928) Proc Roy Soc A 122:622
4. Whybrew WE, Kuizer GD, Gunn R (1952) J Geophysical Research 57:459
5. McShea JA, Callaghan IC (1983) J Coll Polym Sci 261:757
6. MacIntyre F (1974) Scientific American 230 (51):62
7. Jarvis NL, Scheimann MA (1968) J Phys Chem 72:74
8. Blanchard DC (1963) Progess in Oceanography 1:71
9. Jarvis NL (1972) J Geophys Res 77 (27):5177
10. Iribarne JV, Mason BJ (1967) Trans Faraday Soc 63:2234

Received October 27, 1987;
accepted June 7, 1988

Author's address:

M. Sakai
Department of Industrial Chemistry, Faculty of Engineering
Kyushu Sangyo University
Fukuoka 813, Japan

Progress in Colloid & Polymer Science   Progr Colloid & Polymer Sci 77:143–145 (1988)

Thermodynamics of accumulation processes applied to colloidal systems

M. R. Mehandjiev

Scientific Department, "BIOTECH" Science & Engineering Co., Varna, Bulgaria

Abstract: A polyenergetic conjugation relationship existing between the intensity or capacity factors of state or both, has been developed in the thermodynamics of accumulation processes for processes and effects without mass transfer and with or without qualitative transformation in the affected system. The polyenergetically conjugated factors define different elementary energy kinds. The relationship is expressed by exponential equations. Results obtained by the use of derivative expressions of the polyenergetic conjugation relationship for investigation of the coagulation phenomena in $2.7 SiO_2 \cdot Na_2O - HCl - H_2O$ hydrosol systems show that those expressions can be used for the choice of the colloid system contents according to the coagulation rate needed, as well as in coagulation kinetic studies. The results confirmed the concentration exponent value in TAP (0.15) and indicate the possibility of application of time as capacity factor in the relationships with an exponent approximately equal to 0.02.

Key words: Polyenergetic conjugation, colloid systems, Thermodynamics of Accumulation Processes (TAP), coagulation kinetics

Introduction

At present, the "thermodynamics of accumulation processes" (TAP) and the "thermodynamics of irreversible processes" are the two branches of non-equilibrium thermodynamics, as shown in Fig. 1. TAP became an object of study in Bulgaria in 1967; it deals with phenomena associated with the introduction, transformation, and accumulation of mass or energy or both in the affected systems, i.e., in non-isolated, open, or closed systems. The phenomena's processes and effects, due to the work done upon the systems by positive exogenous impacts (influences) during a measurable impact time are called accumulation processes and effects. TAP always deals with processes which proceed with real rates and it is not opposed to thermodynamics of irreversible processes. On the contrary, TAP should be considered as a link between classic equilibrium thermodynamics and the thermodynamics of irreversible processes, since the beginning of every irreversible process is always preceded by the introduction and accumulation of mass, or energy, or both (i.e., by accumulation phenomena) in equilibrium, before the impact system. Proofs regarding the principle difference between the accumula-

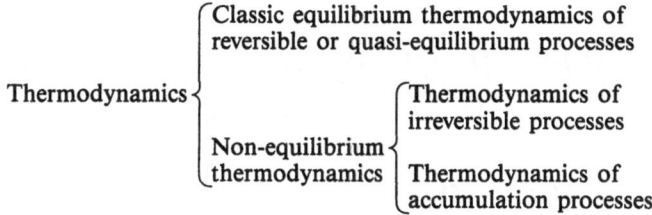

Fig. 1. A contemporary idea about the various branches of thermodynamic knowledge

tion and irreversible processes can be found in each hysteresis phenomenon.

All processes and effects are examined in TAP at the macroscopic level (which is characteristic for thermodynamics in general) using what we know about elementary energy (forms and types). The elementary energy kinds are defined by pairs of conjugated intensity (P) and capacity (X) factors of the system. The product of a given pair: P_r and X_r represent an extensive function E_r, which is considered as a quantity of energy of the elementary kind r. It is suggested that P_r and X_r of each E_r are measurable macroscopic factors. The conjugation existing between the intensity and capacity factors of one and the same elementary energy kind is called "monoenergetic conjugation".

According to TAP, each appearance of a new indispensable and observable elementary energy kind or each disappearance of an existing kind of energy is a qualitative transformation (QT) in the system. Most ordinary QT examples are the appearances and disappearances of phases or chemical substances in the systems phase and chemical QT.

Purposes of the study

A first application of TAP developed expressions in the investigation of the coagulation phenomena, as well as a verification of the possibility of participation of time t as a capacity factor in those expressions, are the objects of this work.

Derivative TAP expressions

A polyenergetic conjugation relationship existing between the intensity or capacity factors or both, which define different elementary energy kinds, has been developed in TAP for processes and effects without mass action and with or without QT in the affected systems [1]. The relationship between polyenergetic conjugated variables is expressed by exponential equations, e.g.,

$$\prod_{r=1}^{j} P_r^{\pm \theta_r} \prod_{i=j+l}^{n} X_i^{\pm \theta_i} = \text{constant} \;;$$ (1)

$$(r \neq i; \; r+i = n \geq 2)$$

$$T \cdot \prod_{r=2}^{j} P_r^{\pm \theta_r} \prod_{i=j+l}^{n} X_i^{\pm \theta_i} = \text{constant}$$ (2)

$$(r \neq i; \; r+i+l = n \geq 2) \;.$$

Assuming the exponent of the Kelvin temperature T to be equal to unity, as in Eq. (2), the relative exponents of all other P and X are approximately constant, $\theta_r \approx |0.3|$; $\theta_i \approx |0.4|$ in expressions for processes (effects) without QT, but vary for those attended by QT. The mean relative values of exponents for various elementary energy factors representing an impact upon a certain system experimentally obtained are shown in Table 1. In Eqs. (1), (2), n equals the number of the elementary energy kinds which are indispensable and sufficient for the characterization of an affected system, or of an impact in a degree corresponding to the purposes of the thermodynamic study. The significance of Eqs. (1), (2) has been proved in many impact cases [2, 3]. It has been also proved

Table 1. Mean relative values of some elementary energy factor exponents

Kind of impact in QT cases	Exponent value
Heat impact	1.00
Values experimentally obtained:	
Impact by increased surface area	0.04
Ultrasonic impact	0.04
Pressure impact (up to 5 MPa)	0.14
Chemical (concentration) impact	0.15
Maximum theoretical value for a possible impact with exception of heat impact	0.70

that the relative values of the exponents can be used independently of the T-participation in the expressions for QT or for non-QT cases.

From Eq. (1) written for two factors (intensity and capacity) only

$$P_r^{\pm \theta_r} \cdot X_i^{\pm \theta_i} = \text{constant} \quad (n = 2)$$ (3)

and assuming that the system concentration C of a reactant acting upon a hydrosol system will represent the intensity factor in Eq. (3) and that the coagulation time t of the hydrosol can be used instead of the capacity factor, we obtain

$$C^{0.15} t^{\theta_i} = \text{constant} \;.$$ (4)

According to TAP studies, both exponents in Eq. (4) have positive values, since the changes of both factors are antibatic ones; the value of C exponent is accepted in accordance with Table 1, since the coagulation of a hydrosol is an obvious QT. The value needed in Eq. (4) is that of the time exponent θ_i. It has been calculated by experimental results for the coagulation of a set of $2.7 SiO_2 \cdot Na_2O - HCl - H_2O$ hydrosol systems with various coagulation times. The calculation equation is easily derived from Eq. (4), written for two coagulation cases (indexed a and b). Thus, from

$$C_a^{0.15} t_a^{\theta_i} = C_b^{0.15} t_b^{\theta_i} = \text{constant} \;,$$ (5)

we obtain

$$0.15 \log \frac{C_a}{C_b} = \theta_i \log \frac{t_b}{t_a} \;.$$ (6)

The validity of $\theta_r = 0.15$ in colloid system cases has been previously confirmed in a set of experiments, calculating with another derivative expression asso-

Table 2. Examples of experimental results obtained by Eq. (6)

HCl concentration %	Test numbers and coagulation time t (min)				
$C_1 = 10$	$t_{11} = 1080$	$t_{12} = 1080$	$t_{13} = 1080$	$t_{14} = 1200$	$t_{15} = 1145$
$C_2 = 15$	$t_{21} = 178$	$t_{22} = 132$	$t_{23} = 116$	$t_{24} = 170$	$t_{25} = 184$
$C_3 = 20$	$t_{31} = 7.9$	$t_{32} = 5.9$	$t_{33} = 6$	$t_{34} = 6$	$t_{35} = 6$
Combined test results and O_i values calculated					
C_1, C_2 $\quad \theta_i =$	t_{11}, t_{21} 0.0332	t_{12}, t_{22} 0.0285	t_{13}, t_{23} 0.0268	t_{14}, t_{24} 0.0275	t_{15}, t_{25} 0.0332
C_1, C_3 $\quad \theta_i =$	t_{11}, t_{31} 0.0210	t_{12}, t_{32} 0.0204	t_{13}, t_{33} 0.0202	t_{14}, t_{34} 0.0191	t_{15}, t_{35} 0.0190
C_2, C_3 $\quad \theta_i =$	t_{21}, t_{31} 0.0141	t_{22}, t_{32} 0.0140	t_{23}, t_{33} 0.0141	t_{24}, t_{34} 0.0132	t_{25}, t_{35} 0.0133

Table 3. Examples of experimental results obtained by Eq. (8)

Coagulation time, (min)	Test pairs used in the equation		θ_r value calculated
1200	$C_0 = 2$ $C = 13.5$	$C_0 = 18$ $C = 1.5$	0.15
126 – 132	$C_0 = 11$ $C = 4.6$	$C_0 = 6$ $C = 10.5$	0.144
5.2 – 5.9	$C_0 = 14$ $C = 6$	$C_0 = 6.1$ $C = 13.9$	0.15

ciating the concentrations of $2.7\,SiO_2 \cdot Na_2O$ (symbol C_0) and of HCl (symbol C) in hydrosol systems with approximately equal coagulation time

$$C_0^{0.15} C^{\theta_r} = \text{constant} , \quad (n = 2) \tag{7}$$

or, for two systems, indexed a and b, we have

$$0.15 \log \frac{C_{0,a}}{C_{0,b}} = \theta_r \log \frac{C_b}{C_a} . \tag{8}$$

Experimental results

The experimental results for 70 coagulations of $2.7\,SiO_2 \cdot Na_2O - HCl - H_2O$ hydrosol systems are shown partly in Tables 2 and 3. Table 2 gives the coagulation time t in minutes, the combined test results in the calculations using Eq. (6), and the calculated time exponent values. Test results are numbered in accordance with the table matrix. The concentration ratios between $2.7\,SiO_2 \cdot Na_2O$ and HCl (in percentage) in hydrosol systems with approximately equal coagulation time and the θ_r values calculated by Eq. (8) using those ratios, are presented in Table 3.

We can see in Tables 2 and 3, that the value of the exponent of the time in polyenergetic conjugation expressions for QT cases is approximately equal to $0.01 - 0.03$ (mean value ≈ 0.02), the least value obtained in comparison with those in Table 1. That result indicates the possibility of interpreting time as a capacity factor in polyenergetic conjugation equations. The use of $\theta_r = 0.15$ for the concentration exponent in Eq. (6) is confirmed by the data of Table 3.

Conclusions

Results obtained in the use of polyenergetic conjugation relationship derivative equations for investigation of the coagulation phenomena in $2.7\,SiO_2 \cdot Na_2O - HCl - H_2O$ hydrosol systems show that the polyenergetic conjugation expressions can be used for the choice of the colloid system contents according to the coagulation rate needed, as well as in coagulation kinetics studies. The results confirmed the value of the concentration exponent in TAP (0.15) and indicate the possibility of applying time as capacity factor in the polyenergetic conjugation relationship with an exponent approximately equal to 0.02.

References

1. Mehandjiev M (1978) Abstr Vol of 8th Experim Thermodynamics Conference, Guildford, UK, pp 46–47
2. Mehandjiev M, Russev G, Dinev Z (1983) Environmental Protec Engng 1:95–100
3. Mehandjiev M (1984) Ecol Modelling 1–2:17–20

Received February 15, 1988; accepted June 7, 1988

Author's address:

Prof. M. R. Mehandjiev
279 Slivnitsa Blvd.
1202 Sofia, Bulgaria

Progress in Colloid & Polymer Science   Progr Colloid & Polymer Sci 77:146–151 (1988)

Evidence for secondary minimum coagulation in a silica hydrosol obtained by dynamic light scattering

P. Ludwig[1]) and G. Peschel[2])

[1]) Mütek GmbH (ETEC), Essen, FRG,
[2]) Institut für Physikalische und Theoretische Chemie der Universität Essen, FRG

Abstract: The coagulation kinetics of four different samples of a silica hydrosol differing in particle diameters (45, 101, 180, and 515 nm) were examined with the aid of photon correlation spectroscopy. The rate constants for rapid and slow coagulation were determined in the presence of LiCl and CsCl. The data analysis followed a method given by Versmold and Härtl. The stability ratio for a definite particle size at various electrolyte concentrations could be derived from the rate constants. The smallest particles were shown to be the most stable. This points to a coagulation in the secondary potential energy minimum, which could be proved by calculations of the stability ratio.

Key words: Colloid stability, coagulation kinetics, secondary minimum coagulation, silica hydrosol, photon correlation spectroscopy

Introduction

The stability of disperse systems [1, 2] has always been an important question in colloid chemistry. Colloid instability manifests itself as a coagulation process whose velocity is commonly regarded as a specific quantity of the state of the system. The profound effect is the change in the particle number and the increase in particle size.

A dispersion is considered to be stable if there exists a potential energy barrier between colliding particles that has a height larger than the average kinetic energy. This state is accounted for by the *Derjaguin-Landau-Verwey-Overbeek (DLVO)* theory. It is well known that electrolytes with increasing concentration are able to coagulate a sol by constricting the electrical double layers [2]. The kinetics of this process were considered in detail by Smoluchowski [3].

In order to get into contact, particles in a fluid medium, e.g., an electrolyte solution, usually have to overcome the electrostatic energy barrier resulting from the mutual overlap of electrical double layers and finally coagulate in the primary potential energy minimum [2]. Notwithstanding this ubiquitous mechanism there is much evidene [4–6] that in certain circumstances coagulation can also occur in the secondary minimum provided this is deep and the height of the barrier is large enough. This kind of coagula-

tion is considered to be strongly reversible [7] and is still poorly understood.

Another problem associated with coagulation phenomena is the dependence of the coagulation rate on particle size. Convincing evidence is presented by Ottewill and Shaw [8], Wiese and Healy [4], and Joseph-Petit et al. [9]. Whereas Ottewill and Shaw try to explain this effect by a variation of the Hamaker constant with particle size, the other authors believe in the coagulation in the secondary and also in the primary minimum in order to find their results in accord with the DLVO theory.

Theory

Rapid and slow coagulation

Coagulation kinetics can be rapid or slow according to the presence of a coagulation barrier. Rapid coagulation which is determined by the diffusional movement of colloid particles promptly occurs with every particle collision when no barrier exists [10], which is the case for sufficiently high electrolyte concentrations, e.g., above the critical coagulation concentration (CCC). A first treatment is due to Smoluchowski [3]. Details are explained in a preceding paper [11]. Smoluchowski in his calculations arrived at the half-life of coagulation

$$\tau = \frac{1}{8\pi D_1 a_1 n_0} \tag{1}$$

where D_1 is the diffusion coefficient of the single particles and a_1 their radius; the quantity n_0 is the initial number of primary particles. The rate constant of rapid coagulation was found to be

$$k_r = 8\pi D_1 a_1 . \tag{2}$$

The half-life of slow coagulation is characterized by the introduction of a factor α, which is a measure of the efficiency of particle encounter, so that Eq. (1) in this case reads

$$\tau = \frac{1}{8\pi D_1 a_1 n_0 \alpha} . \tag{3}$$

According to Fuchs [12] one can define a stability ratio

$$W = \frac{1}{\alpha} = \frac{k_r}{k_s} = 2a \int_{2a}^{\infty} \exp\left(V_{\text{tot}}(H_0)/k_b T\right) \frac{dH_0}{H_0^2} , \tag{4}$$

wherein a is the radius of the coagulating entity; k_s is the rate constant for slow coagulation, V_{tot} the total potential energy of interaction, and H_0 the shortest distance between two spherical particles.

Coagulation in the primary minimum

For primary minimum coagulation (PMC) a simple solution of Eq. (4) is given by [10]

$$W = \frac{1}{2\varkappa a} \exp\left(V_{\text{max}}/k_b T\right) . \tag{5}$$

V_{max} is the maximum value of the total potential energy, $\varkappa$ is the well known Debye-Hückel parameter.

Equation (4) can be modified in such a way that with respect to the approach of two particles, van der Waals and hydrodynamic forces can be taken into account [9, 13, 14]

$$W = \frac{\displaystyle\int_{2a}^{\infty} \beta(u) \frac{\exp\left(V_{\text{tot}}(H_0)/k_b T\right)}{H_0^2} dH_0}{\displaystyle\int_{2a}^{\infty} \beta(u) \frac{\exp\left(V_a(H_0) k_b T\right)}{H_0^2} dH_0} . \tag{6}$$

V_{tot} is according to

$$V_{\text{tot}} = V_D + V_R \tag{7}$$

composed of the long-range attractive dispersion energy V_D and the electrostatic repulsive energy V_R.

Honig et al. [14] introduced the relation

$$\beta(u) = \frac{6u^2 + 13u + 2}{6u^2 + 4u}$$

with

$$u = \frac{H_0 - 2a}{a} .$$

Plotting the logarithm of W by Eq. (3) for a chosen electrolyte against the logarithm of the electrolyte concentration yields the stability diagram which is distinguished by two straight lines of different slope, one for the range of slow, the other for that of rapid coagulation; these intersect at the CCC [15]. The stability diagram allows the calculation of the ψ_0 or ψ_δ potential. From these values the Hamaker constant A of the attractive particle-particle interaction can be derived [15, 16]

$$A = 9 \times 10^{-20} \frac{\gamma^2}{v^3 \sqrt{\text{CCC}}} /J \tag{8}$$

The CCC is inserted in mol dm^{-3} with

$$\gamma = \frac{e^{z/2} - 1}{e^{z/2} + 1} \quad \text{and} \quad z = \frac{ve\psi_0}{k_b T}$$

or

$$z = \frac{ve\psi_\delta}{k_b T} ;$$

e is the unit charge and v the ion valency.

The results obtained for the slow as well as the rapid coagulation can most often be discussed without difficulty in terms of the DLVO theory.

Coagulation in the secondary minimum

Coagulation is generally assumed to occur in the primary potential minimum; nevertheless, experimental evidence does exist for secondary minimum coagulation effects (SMC) [6, 9] which can in some way be treated by theoretical reasoning. The rate of SMC has not been a subject of intensive investigations, so many questions remain.

A simple formula is available [5] for evaluating W when SMC is considered and the depth of the energy minimum V_{min} is known

$$W = \frac{1}{1 - \exp\left(V_{\text{min}}/k_b T\right)} . \tag{9}$$

On the basis of the kinetic theory Marmur [6] has developed a coagulation model which incorporates both coagulation types in a "mixed" coagulation.

It is the purpose of this paper to examine the coagulation kinetics of a silica hydrosol and to scrutinize the results with respect to the question of which kind of coagulation has occurred.

Experimental

Method

The coagulation kinetics of a silica hydrosol have been determined by photon correlation spectroscopy which is a dynamic light scattering method [19–21] and an excellent tool for monitoring coagulation kinetics phenomena [22, 23]. The aim of the coagulation tests was to determine rate constants of rapid and slow coagulation. This, according to Eq. (4), allowed the evaluation of the stability ratio because it is a characteristic of the colloidal system.

Inspection of Eqs. (1), (2), and (3) shows that it is desirable to obtain the half-life of coagulation and the diffusion coefficient of the particle. It must be stressed that during the coagulation process particle aggregation proceeds so that on account of the slight polydispersity which arises, only a mean value of the particle diffusion coefficient can be observed by the dynamic light scattering experiment. Versmold and Härtl [22] have succeeded in presenting a fitting procedure when comparing theoretical with experimental results on the basis of an appropriately chosen value for τ, as is commented on in an earlier paper [11].

Apparatus and materials

Spectrometer: the dynamic light scattering or photon correlation spectroscopy (PCS) tests were carried out with a NICOMP Laser Particle Sizer Model 200. The PCS experiment was run in recycle-on-modus during the coagulation tests. The period for measurement lay in the range of 1–60 min according to coagulation velocity. A computer (Atari) connected with the PCS instrument was fed with data concerning the diffusion coefficient.

The silica particles for preparing the hydrosol were produced according to Stöber et al. [24] by controlled hydrolysis of tetraalkylsilicates.

Four different monodisperse samples were prepared containing spherical silica particles with diameters of 45, 101, 180, and 515 nm. The margin of error is about 1%. Further details are given in [11, 25].

Results

The rate constants of slow as well as rapid coagulation of the different hydrosol samples as described above were determined in the presence of dissolved LiCl or CsCl covering a broad concentration region $10^{-2}-2\,M$, i.e., far beyond the CCC.

In Figs. 1 and 2 the double logarithmic plot of the stability ratio against the concentration of LiCl and CsCl is depicted after the use of Eq. (4). It should be

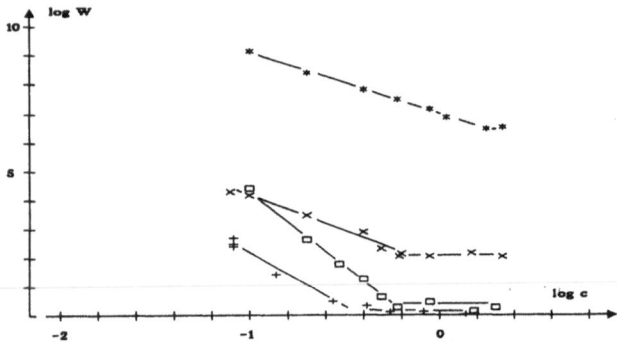

Fig. 1. Stability diagrams for silica particles of different diameters in the presence of LiCl. 45 nm ———*———, 101 nm ———×———, 180 nm ———□———, 515 nm ———+———

noted that the stability of the hydrosol samples grows higher with decreasing particle diameter, which appears unusual and deserves attention in the following.

For the regions of slow coagulation the slope [15]

$$\frac{d \log W}{d \log C} = m = -2.15\,\frac{a}{v^2} \tag{10}$$

can be derived from the stability diagrams. The value γ, as illustrated in the foregoing, incorporates the surface potential ψ_δ and allows the calculation of the Hamaker constant A by use of Eq. (8). Table 1 presents some values for ψ_δ when 515 nm particles are chosen and the kind of electrolyte is varied. Moreover, the value of the Hamaker constant is quoted.

In Table 2 the different values for the CCCs derived from Figs. 1 and 2 are listed. The largest values must be attributed to LiCl, which is a phenomenon well known in colloid chemistry. When regarding slow coagulation the increase in stability is rather dramatic with falling particle diameter.

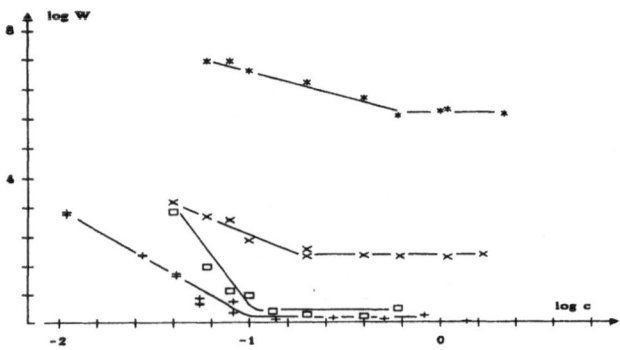

Fig. 2. Stability diagrams for silica particles of different diameters in the presence of CsCl. 45 nm ———*———, 101 nm ———×———, 180 nm ———□———, 515 nm ———+———, c in mol dm^{-3}; $T = 293$ K

Table 1. Comparison of the Hamaker constant A and the potential found in this work with the corresponding values given by Barringer [32]

	From this work ($H_0 = 515$ nm)			According to Barringer ($H_0 = 560$ nm) KCl
	LiCl	KCl	CsCl	KCl
At the CCC (in mV)	-6.2	-6.0	-5.6	-8.0
A (in J)	$1.1 - 2.9$	10^{-21}		$3.3 - 7.6$ 10^{-21}

In the present study special emphasis is given to the theoretical calculation of the stability diagrams as shown in Figs. 1 and 2, particularly to learn if these diagrams can be explained in terms of the DLVO theory. In order to evaluate the values for W one can refer to Eqs. (5) or (6) for the case of PMC, provided that the values for $V_{\max}$ and $V_{\min}$ are accessible. This is particularly important for Eq. (9). In principle $V_{\max}$ as well as $V_{\min}$ can be taken from the V_{tot}/H_0 plot for two interacting particles.

For performing this plot we refer to Eq. (7) and write

$$V_D = -\frac{a \cdot A}{12 H_0} \tag{11}$$

provided that $a \gg H_0$ [10]. A must be composed from individual components or can be taken from literature [16] ($A = 4 \cdot 10^{-21} J$). For the repulsive component it seemed preferable to use a formula given by Honig and Mul [26] which was derived under the assumption that the electrical surface charge remains constant during particle approach. This formula is highly accurate and is valid for $\psi_\delta < 58$ mV and $H_0 > 0.1 - 0.3$ nm according to the value of the Debye-Hückel-parameter. It reads

$$V_r = 0.064 \, N_A c \cdot k_b T S / \varkappa \tag{12}$$

Table 2. CCC in mol dm^{-3} for the particles under examination

Diameter (in nm)	CCC		
	Li$^+$	K^+	C_s^+
45	1.5	–	0.7
101	0.73	0.39	0.25
180	0.63	–	0.16
515	0.43	0.21	0.12

with

$$S = S_1 + S_2 + S_3$$

and

$$S_1 = \gamma^2 \exp(-\varkappa H_0)[1 + \tfrac{1}{2} \exp(-\varkappa H_0)]$$

$$S_2 = \frac{1}{2} \gamma^4 \exp(-2\varkappa H_0) \left[a - \frac{15}{(1+\gamma^2)} - 2\varkappa H_0 \right]$$

$$S_3 = -\frac{\gamma^2}{(1+45\gamma^2+900\gamma^4)} \cdot \left[\ln[1 - \exp(-\varkappa H_0)] + \frac{1}{2} \exp(-2\varkappa H_0) + \exp(-\varkappa H_0) \right] .$$

N_A is Avogadro's number and c is the electrolyte concentration in mol dm^{-3}.

Figure 3 shows a corresponding plot for the case of 515 nm particles and a 0.1 M aqueous electrolyte solution. The surface potential is chosen to be 25 mV. Additionally the particle distance is kept at a minimum value Δ during particle encounter. This minimum distance was introduced by Overbeek [27] to avoid the occurence of infinitely large attractive forces in the plot. The value Δ was estimated to be on the order of 0.3 nm.

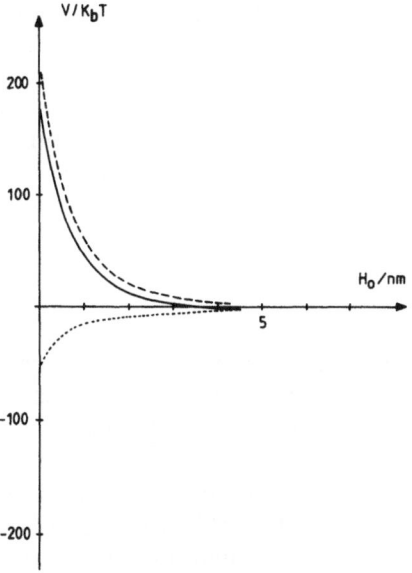

Fig. 3. Potential energy diagram of two interacting silica particles in 0.1 M aqueous electrolyte solution. $d = 515$ nm, $A = 4 \cdot 10^{-21} J$, $\psi_0 = 25$ mV, $\Delta = 0.3$ nm

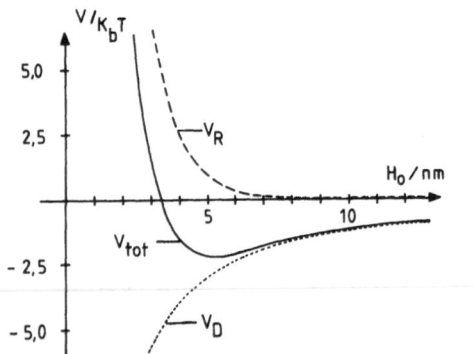

Fig. 4. Enlarged sector from the potential diagram in Fig. 3

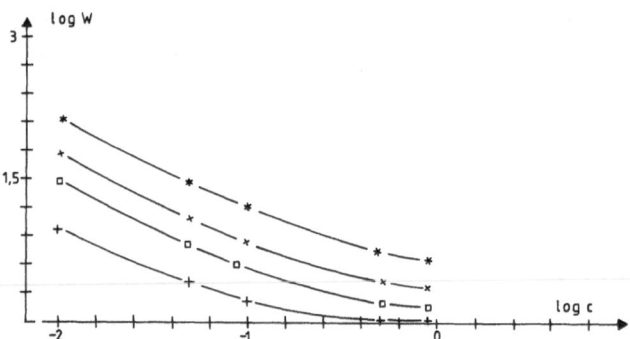

Fig. 5. Calculated stability diagrams for silica particles of different diameters in the aqueous solution of a univalent electrolyte. 45 nm ———*———, 101 nm ———×———, 180 nm ———□———, 515 nm ———+———, c in mol dm^{-3}; $T = 293$ K

In Fig. 3, because of the introduction of Δ, no primary minimum can be recognized. By omitting Δ one actually obtains a primary minimum, but it lies at a distance not larger than 0.1 nm which appears unrealistic. Figure 4 reflects the enlargement of a critical part of the diagram in Fig. 3. Without any doubt a secondary potential minimum appears at about 5 nm with a depth of $V_{min} \approx 2.1 - 2.2 \, k_b T$.

A further step was the calculation of the stability ratio W according to Eq. (6) which refers to the PMC by varying the particle size and the electrolyte concentration. Plotting log W against log C yielded stability diagrams which, since the largest particles exhibited the highest stability, were in excellent accord with the DLVO theory, but not with our experimental findings. From this fact it must be inferred that coagulation might take place in the secondary minimum as discernible in Fig. 4. Consequently calculations analogous to those performed for the PMC were done for the SMC which then led to the picture of Fig. 5, where just the reverse is found – the smallest particles prove to be the most stable. The theory cannot, of course, differentiate between the individual properties of Li$^+$ and Cs$^+$.

Discussion

Inspecting Fig. 5 one finds that for small-sized particles the values for log W are rather small when compared with those obtained by experiment (Figs. 1 and 2). These differences might originate in shortcomings of the theory. But what must be stressed is that in a rough manner the order of the stability diagrams for the individual particle sizes as found by applying Eq. (9) just covers that found by experiment. This inevitably should support the concept of SMC of the hydrosol under examination.

In the case of PMC the energy barrier displayed by the total potential energy determines the coagulation rate. The height of this energy maximum is proportional to the particle size which causes the larger particles to show enhanced stability. By reason of the strong van der Waals attractive forces at very small particle distances the decomposition of the particle aggregates is usually irreversible. Reverse conditions determine the SMC. The coagulation is, of course, controlled by the depth of the secondary potential energy minimum; but in this case aggregate decomposition is not made difficult by overcoming an energy barrier. The proportion of the energy depth of the minimum to the mean thermal energy of the particles is typically not large, that being the prerequisite for the extraordinary stability of the hydrosol.

The SMC is considered to be reversible, which we could support by desegregation tests after the coagulation process by short treatments with supersonic energy. Thus, without any difficulties primary particles displaying about the original size were obtained again.

The SMC is usually omitted from discussion about coagulation phenomena in literature. The reason is obviously that it is assumed that the SMC is of importance for large particles ($a > 1 \, \mu$m) only [28, 29] and only at rather large particle distances ($H_0 \geq 50$ nm) does a secondary potential energy minimum occur [30].

In the case of silica particles the Hamaker constant is rather small [10] and according to the electrolyte concentration these energy minima can be found at particle distances between 0.5 and 10 nm. The coagulation behavior will become complicated if the values for the Hamaker constants are of an order when PMC as well as SMC are both possible. Maximum stability will then be associated with a definite particle size and "mixed" coagulation effects [9].

It seems to the authors that since the Li^+ and Cs^+ ions exhibit different effects with respect to the sol stability both repulsive electrostatic forces and repulsive hydration forces may be effective. This is particularly evident when regarding the influence of Li^+ ions. That structural effects exerted by hydration water can have a serious influence on coagulation phenomena in the presence of inorganic salts was illustrated by Zimehl and Lagaly [30] when they tested latex dispersions. The change of water structure near silica particles and its affecting coagulation kinetics will be the subject of a forthcoming paper [31].

References

1. Derjaguin BV, Landau L (1941) Acta Physicochim USSR 14:633
2. Verwey EJW, Overbeek JThG (1948) Theory of the Stability of Lyophobic Colloids. Elsevier, New York
3. von Smoluchowski M (1916) Phys Z 17:557, 585; (1917) Phys Chem 92:129
4. Wiese GR, Healy TW (1970) Trans Farad Soc 66:490
5. Hogg R, Yang KG (1976) J Colloid Interf Sci 56:573
6. Marmur A (1979) J Colloid Interf Sci 72:41
7. Long JA, Osmond DWJ, Vincent B (1973) J Colloid Interf Sci 42:545
8. Ottewill RH, Shaw JN (1966) Discuss Farad Soc 42:154
9. Joseph-Petit A-M, Dumont F, Watillon A (1933) J Colloid Interf Sci 43:649
10. Sonntag H (1977) Lehrbuch der Kolloidwissenschaft, Berlin
11. Peschel G, Ludwig P (1988) Progr Colloid & Polymer Sci, in press
12. Fuchs N (1934) Z Phys 89:736
13. MacGown DNI, Parfitt GD (1967) J Phys Chem 71:449
14. Honig EP, Roebersen GJ (1971) J Colloid Interf Sci 36:97
15. Reerink H, Overbeek JThG (1954) Discuss Farad Soc 18:74
16. Sonntag H, Strenge K (1970) Koagulation und Stabilität disperser Systeme, Berlin
17. Bagchi P (1976) Colloid & Polymer Sci 254:890
18. Prieve DG, Ruckenstein E (1980) J Colloid Interf Sci 73:539
19. Berne BJ, Pecora R (1976) Dynamic Light Scattering, Academic, New York
20. Chu B (1974) Laser Light Scattering. Academic, New York
21. Pusey PN, Fijnaut H, Vrij A (1982) J Chem Phys 77:4270
22. Versmold H, Härtl W (1983) J Chem Phys 79:4006
23. Barringer EA, Novich BE, Ring TA (1984) J Colloid Interf Sci 100:584
24. Stöber W, Fink A, Bohn E (1968) J Colloid Interf Sci 26:62
25. Ludwig P (1987) Dissertation University of Essen
26. Honig EP, Mul PM (1971) J Colloid Interf Sci 36:258
27. Frens G, Overbeek JThG (1972) J Colloid Interf Sci 38:376
28. Kruyt HR (1952) Colloid Science, New York
29. Babenkov ED (1979) Kolloidn Zh 41:439
30. Zimehl R, Lagaly G (1986) Progr Colloid & Polymer Sci 72:28
31. Ludwig P, Peschel G (1988) being prepared
32. Barringer EA (1983) Dissertation, Massachusetts

Received February 29, 1988;
accepted June 7, 1988

Authors' addresses:

P. Ludwig
Mütek GmbH
Kruppstr. 82 (ETEC)
4300 Essen 1

G. Peschel
Institut für Physikalische und Theoretische Chemie
Universität Essen
Universitätsstr. 5–7
4300 Essen

Progress in Colloid & Polymer Science   Progr Colloid & Polymer Sci 77:152–157 (1988)

Sharp maxima in the flotation rate

M. Pitsch, K. Heckmann, H.-H. Kohler, and J. Strnad

Institute of Physical and Macromolecular Chemistry, University of Regensburg, Regensburg, West Germany

Abstract: The steady-state flotation rates of crystalline tin oxide, quartz, and glass, measured as a function of the reduced surfactant concentration (c/CMC), show sharp maxima at concentration values which are characteristic for a given mineral, and independent of the pH. A comparison of the flotation measurements using glass beads and ground glass particles shows that the flotation kinetics are strongly influenced by the surface edges. The sharp decline in the flotation rate at the high concentration side of the maxima is interpreted as being due to the formation of "rod-like (ad)micelles" along the surface edges. In the presence of bivalent counter ions a second bilayer forms and gives rise to a second maximum in the flotation rate.

Key words: Tin oxide, glass, flotation of quartz, pyridinium salts, effect of particle form, flotation rate maxima, phase transition

Introduction

The flotation of minerals is a complex process depending on a great number of known and unknown parameters. Many of these parameters cannot be varied independently, e.g., the concentration of the collector and the bubble size. The flotation process can be divided into three important steps: a) the movement of the mineral particle to the phase boundary liquid-gas; b) the formation of the three-phase contact; and c) the stabilization of the three-phase contact which prevents the particle from leaving the phase boundary again [1].

If the concentration of the mineral in the suspension, the size, and the number of the gas bubbles are kept constant, the steady-state flotation rate depends mainly on the formation of the three-phase contact, step b, which in this case is a function of the particle-surface hydrophobicity only.

In our previous paper we presented a first model for the adsorption of hexadecylpyridinium salts on quartz [5] on the basis of both adsorption and flotation measurements. Flotation measurements were made with a flotation apparatus that allowed the determination of the flotation rate under steady state conditions [6]. The flotation rates showed sharp maxima near the reduced surfactant concentration c/CMC = 0.5, independent of the total electrolyte concentration. In

the presence of bivalent counter ions, two flotation maxima appeared, the first one at a reduced surfactant concentration c/CMC = 0.5. To explain these findings we have postulated phase changes that occur at the surface at values of the reduced surfactant concentration and that are characteristic of a given mineral surface. The first flotation maximum was assumed to be reached when the first layer of surfactant (its hydrophobic groups oriented towards the bulk solution) is completed and a second layer – this time with the hydrophilic groups pointing towards the bulk solution – builds up. This phase change causes a steep decrease in the flotation rate. The second flotation maximum in the presence of bivalent counter ions was explained by the assumption of a multilayer adsorption with the counter ions acting as adsorption bridges between two bilayers.

A shortcoming of this first model is the fact that there is no discontinuity in the corresponding adsorption isotherm, which would clearly indicate such a sudden phase change in the surfactant layer [5]. This suggests that the phase changes in question occur in restricted regions of the particle surface. Actually, a mineral particle contacts a gas bubble nearly always via a corner or an edge. Therefore, phase changes occurring along corners and edges may play an important role in the flotation kinetics. This paper deals with that influence. For simplicity we will talk of edges only instead of corners and edges.

Experimental

Materials

The quartz, of p.a. purity, was from Merck, Darmstadt, FRG; the glass beads were from Dragon Werk, Bayreuth, FRG. The crystalline tin oxide was kindly provided by Dr. Helbig, University of Erlangen, FRG. The grain size of all samples used for the flotation tests was 0.14 – 0.16 mm. The hexadecylpyridinium salts were prepared by re-crystallization of hexadecylpyridinium chloride from the corresponding acids. The determination of CMC values and Krafft points of the surfactants are described in our previous papers [4, 7]., The abbreviation CP is used for the hexadecylpyridinium ion, and CPX for the corresponding salt with an anion X.

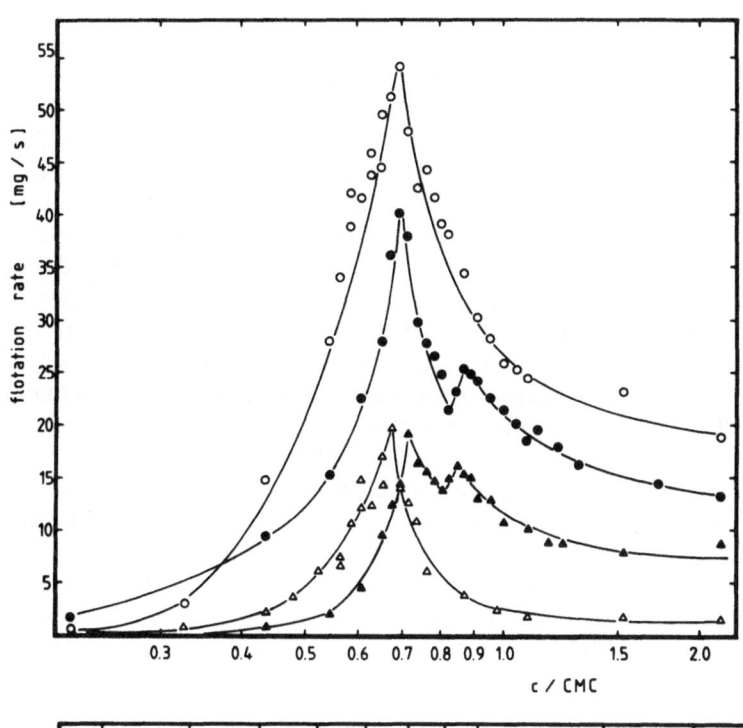

Fig. 1. Flotation rates of tin oxide with CPCl in 0.1 M KCl at pH 7 △ and pH 10 ○, and with CPHSO$_4$ in 0.1 M K$_2$SO$_4$ at pH 7 ▲ and pH 10 ●

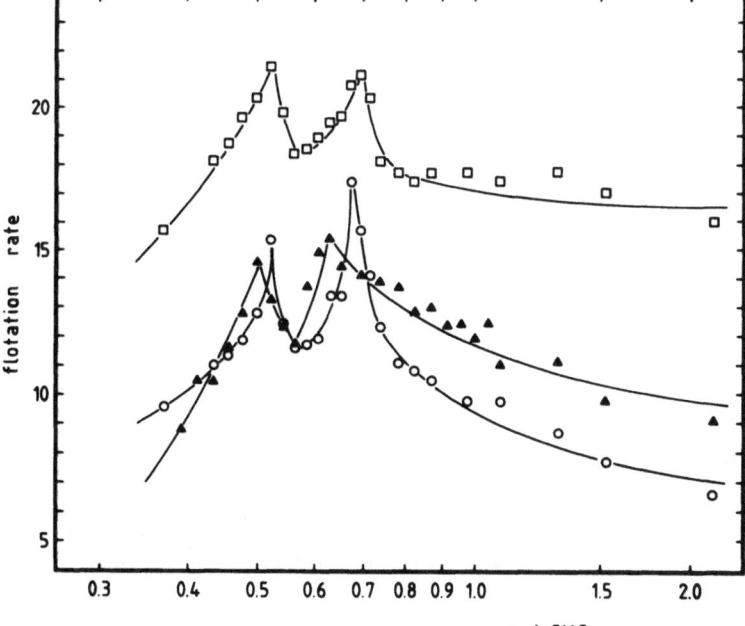

Fig. 2. Flotation rates of ground quartz ○, ground glass ▲ and ground quartz pre-treated by blowing □ with CPHSO$_4$ in 0.1 M K$_2$SO$_4$ at pH 7

Methods

All flotation measurements were carried out at 20 °C with a new type of flotation apparatus described in [5, 6]. This device allows steady-state flotation measurements. In order to obtain a high density of corners and edges the quartz particles and the glass beads were pre-treated by blowing them tangentially against a raw glass tube. The micrographs of the particles were made with a scanning electron microscope (Hitachi S-570).

Results and discussion

Flotation of tin oxide

Crystalline tin oxide was floated with CPCl as a collector in 0.1 M KCl and with $CPHSO_4$ in 0.1 M K_2SO_4 at pH 7 and pH 10. The results are presented in Fig. 1. The flotation rates in CPCl solution show a single, very sharp maximum at the reduced surfactant concentration $c/CMC = 0.67$, whereas two flotation maxima appear when tin oxide is floated with $CPHSO_4$. In this case the first flotation maximum appears also at $c/CMC = 67$, the second maximum at $c/CMC = 0.85$. The positions of the maxima on the c/CMC axis are independent of pH. In contrast, the absolute value of the flotation rate depends on pH, due to the variation of the mineral surface charge. As described earlier [5], the corresponding maxima in the flotation rate of SiO_2 are found at $c/CMC = 0.5$ and 0.7. Our recent findings support the hypothesis that, for a given mineral surface, the values of the reduced surfactant concentration leading to maximum surface hydrophobicity are independent of both pH and total electrolyte concentration.

Flotation of quartz and glass

Ground quartz and glass particles and quartz particles pre-treated by blowing were floated with $CPHSO_4$ in 0.1 M K_2SO_4 (Fig. 2). The flotation rates are in good agreement with our previous flotation tests [5]. The first and the second flotation maxima occur at $c/CMC = 0.5$ and 0.67, as found in [5].

According to our first model [5], the surfactant adsorbs as a bilayer in the presence of monovalent counter ions or a tetralayer in the presence of bivalent counter ions. The flotation rate maxima were assumed to be related to the phase transitions leading to the formation of these layers. However, there is no indication for such phase changes in the corresponding adsorption isotherms. On the other hand, as pointed out above, the edges of the mineral surface must play an important role in the collision of the particle with the gas bubble. We therefore suggest that the phase transitions leading to the peaks in the flotation rate actually occur at the edges of the mineral particle. The amount of surfactant adsorbed along the edges is certainly small compared with the total amount adsorbed on the surface and, thus, may be undetectable in the adsorption isotherm. For closer investigation, normal, nonporous glass beads (few edges) and glass particles pre-treated by blowing (many edges) were floated with $CPHSO_4$ in 0.1 M K_2SO_4 at pH 7. The results are

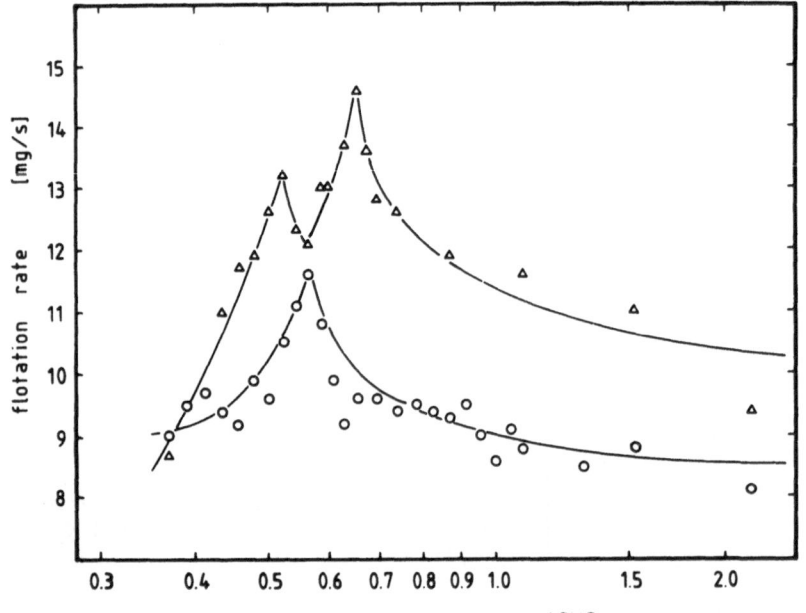

Fig. 3. Flotation rates of glass beads ○ and glass beads pre-treated by blowing △ with $CPHSO_4$ in 0.1 M K_2SO_4 at pH 7

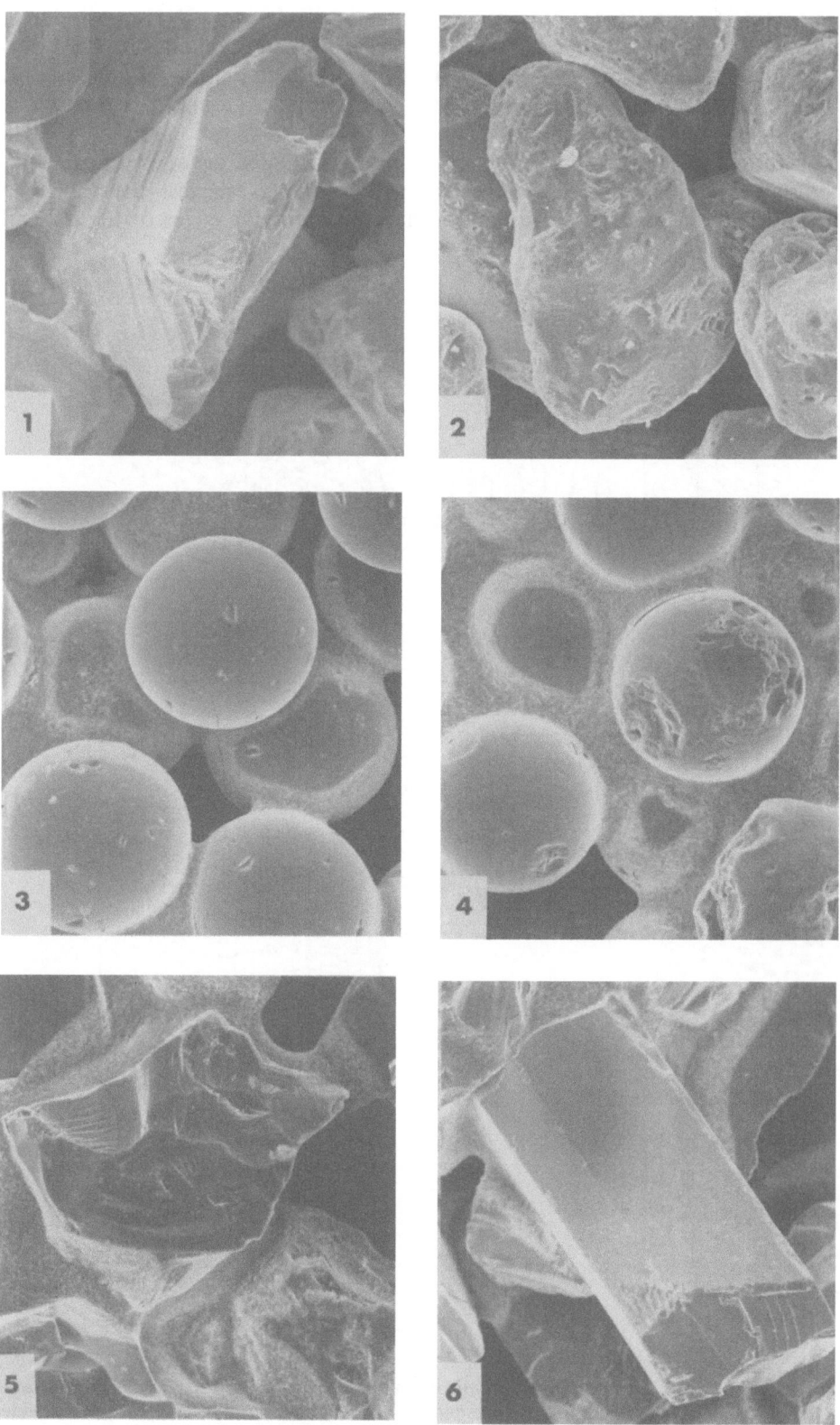

Fig. 4. ESM pictures of particles floated: 1) quartz particles ground; 2) quartz particles ground and pre-treated by blowing; 3) glass beads without pre-treatment; 4) glass beads pre-treated; 5) glass ground; and 6) tin oxide ground

presented in Fig. 3. The flotation rates of pre-treated glass beads show two flotation maxima, as in the case of tin oxide and quartz, whereas only one flotation maximum is seen with normal beads. In addition, this maximum is comparatively small. We assume that the density of edges on the surface of normal glass beads is too low to make a second flotation maximum visible, but this point needs further investigation. In any case, our experimental results stress the important role of corners and edges for the flotation rate. Only particles with a sufficiently high density of macroscopic edges, such as quartz particles ground *or* pretreated by blowing (Fig. 2), show higher flotation rates. The SEM pictures of all particles floated are shown in Fig. 4.

The experimental findings suggest the following flotation mechanism: along the edges of the mineral

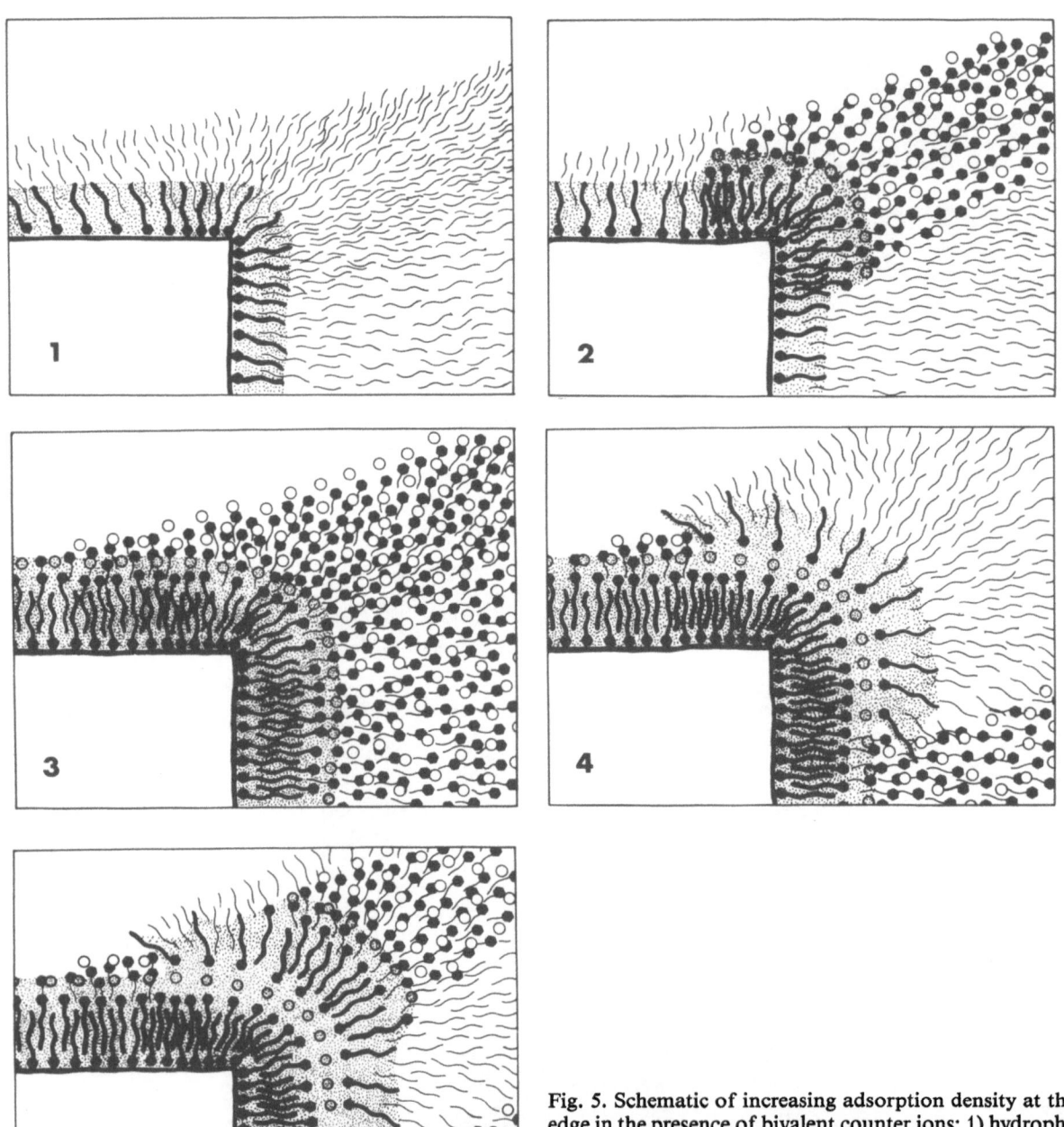

Fig. 5. Schematic of increasing adsorption density at the particle edge in the presence of bivalent counter ions: 1) hydrophobic edge – first flotation maximum; 2), 3) hydrophilic edge – decrease in the flotation rate; 4) hydrophobic edge with three monolayers – second flotation maximum; 5) hydrophilic edge – second decrease in the flotation rate

particles a first hydrophobic surfactant layer builds up at relatively low surfactant concentrations. With a further increase in the surfactant concentration, the adsorbed surfactant spreads from the edges into the surfaces. This results in an increase of the flotation rate. In the case of tin oxide, for example, the spreading is more pronounced at pH 10 (where the surface is more negatively charged) than at pH 7. Therefore, at pH 10 the flotation rate is higher. At a critical surfactant concentration a phase change occurs along the edges. "Rod-like micelles" are created, making the edges hydrophilic. This causes the steep decline in the flotation rate. The presence of bivalent counter ions leads to the formation of adsorption bridges between the second and a third surfactant layer so that the edges become more hydrophobic again. The second flotation maximum reflects a subsequent phase change leading to the adsorption of a fourth (hydrophilic) layer, very much like the formation of the second one. This concept is illustrated in Fig. 5. We would like to stress that the sharp maxima in the flotation rate are only observable when it is floated under steady-state conditions.

Conclusion

There is a certain reduced surfactant concentration where a maximum flotation rate is reached. This concentration depends on the surface properties of the mineral but is independent of the electrolyte concentration and the pH. With bivalent counter ions two such maximum values are usually observed.

Since corners and edges make the first contact between a mineral particle and a gas bubble, they play an important role in the flotation process. At corners and edges the adsorption density of surfactant is much higher than at plane surfaces. Maximum values of the hydrophobicity of corners and edges lead to maximum values of the flotation rate. At critical values of the reduced surfactant concentration the surfactant adsorbed to corners and edges undergoes a phase change. This makes corners and edges hydrophilic (with formation of "rod-like micelles") and causes the breakdown in the flotation rate.

References

1. Schulze HJ (1981) In: Physikalisch-chemische Elementarvorgänge des Flotationsprozesses. VEB Deutscher Verlag der Wissenschaften, Berlin
2. Frumkin A (1938) Z Fiz Chem 12:337
3. Wark JW (1933) J Phys Chem 37:623
4. Heckmann K, Schwarz R, Strnad J (1987) J Coll Interface Sci 120:114
5. Schwarz R, Heckmann K, Strnad J (1988) J Coll Interface Sci 124:50
6. Heckmann K, Schwarz R (1986) Chem Ing Tech 58:396
7. Schwarz R, Strnad J (1987) Tenside Surfactants Detergents 24:143

Received February 16, 1988;
accepted June 7, 1988

Authors' address:

J Strnad
Institute of Physical and Macromolecular Chemistry
University of Regensburg
8400 Regensburg

Progress in Colloid & Polymer Science

Progr Colloid & Polymer Sci 77:158–164 (1988)

Pigment dispersion in organic solvents

M. Liphard and W. von Rybinski

Henkel KGaA, Düsseldorf

Abstract: The interactions between pigments, solvents, and additives play a decisive part in the formulation of paints and coatings. These interactions were examined by means of calorimetric measurements of the enthalpies of displacement and by determining the adsorbed additive amounts of pigments from non-polar solvents. The combination of the two method proved particularly favorable for examining the type of interaction between additive and pigment.

In most cases the polar groups of both components were decisive for the adsorption of additives on the pigment surfaces from non-polar solvents. The differential enthalpy of displacement gives hints in regard to the type of bond between the additive and the pigment. The behavior in multicomponent systems can be appraised in this way. The adsorption and stability tests on the pigment dispersions are shown to correlate.

Key words: pigments, dispersion adsorption from solution, calorimetry

1. Introduction

Many solvent-based paints and coatings with high pigment concentrations include surface-active substances that are used to improve the dispersion of the pigment. This dispersion is of utmost importance for the properties of paints and coatings, for example the hiding power, color intensity, and uniform coloring without changes in shade. There are also advantages with regard to the storage because the sedimentation behavior and the redispersibility of sediments can be improved. In the manufacture of paints and coatings, pigment disagglomeration can be enhanced by adding such additives as grinding auxiliaries. The properties mentioned are all closely connected with the particle size of the pigments in the paint or coating. That is to say, with the degree of their agglomeration or flocculation [1–6]. The additives should therefore serve to maintain pigment particles in a deflocculated state or to cause only a weak, and reversible flocculation in order to guarantee optimal properties of paints and coatings.

The preparation of stable, concentrated pigment dispersions in organic solvents requires a steric stabilization, i.e., the individual pigment particles must be surrounded by an adsorbed layer of surfactants or polymers [7–9]. As there are only repulsive forces between two particles when the adsorbed layers start to overlap, the range of steric interactions is relatively small. This is why there are certain differences in the stabilization by polymers or by surfactants. The latter, due to their low molecular weight, are limited as to the maximum thickness of the adsorbed layer that can be reached. However, two essential preconditions apply for both types of additives: the molecules must be thoroughly anchored on the surface and they must be compatible with the solvent. Surfactants in most cases comply with this requirement due to the different polarities of the head an terminal groups. In the case of polymers [10] *A-B* block copolymers are preferred, where *A* and *B* are segments of different polarity.

This paper reports on two surfactant-type additives. They are particularly suitable for basic investigations due to their relatively simple molecular structures. They are a dicarboxylic acid and an aliphatic amine. The investigation deals both with the anchoring of the functional groups (carboxylate and amino groups) on the surface of titanium dioxide and carbon black and with the influence these additives exert on the behavior of pigments in concentrated dispersions.

2. Experimental

2.1 Materials

The pigments used were a carbon black (Gasruß FW1) made by Degussa, Frankfurt, FRG, specific surface area 334 m²/g (BET), and a titanium dioxide made by Kronos Titan GmbH, Leverkusen, FRG, specific surface area 16 m²/g (BET). Both pigments were used without further purification. The solvent was analytical reagent grade toluene. The model additives used were *n*-dodecyl amine (98%, Aldrich Chemie, Steinheim, FRG), a technical alkane dicarboxylic acid (dimeric fatty acid, molar mass approx. 570 g, Henkel KGaA, Düsseldorf, FRG) as well as a mixture of amine and dimeric fatty acid in a molar ratio of 2:1.

2.2 Experimental parameters

First, the adsorbed amounts m^a (mg/m²) were determined as a function of the equilibrium concentration c of the solution by the depletion method. Then suitable calorimetric measurements were carried out. The so-called enthalpy of displacement was determined, i.e., the enthalpy change during the exchange of solvent molecules on the pigment surfaces against the corresponding quantity of additive molecules from the solution.

With identical surface areas occupied by the solvent and the dissolved substance, this displacement reaction can be written as

$$(1)^a + (2)^1 \rightleftharpoons (1)^1 + (2)^a$$

where component (1) represents the solvent and component (2) represents the additive. The indexes a and 1 refer to adsorbed molecules and molecules in the solution. The corresponding enthalpy of reaction for an exchange of one mole of additive, is the differential molar enthalpy of displacement $\Delta_d H_2$. It may depend on the degree of surface coverage as well as on the composition of the liquid phase. With the calorimeter used, the enthalpy of displacement per unit area, $\Delta_d H$, is measured as a function of the initial solution concentration. The differential molar enthalpy of displacement $\Delta_d H_2$, results from the slope of the function $\Delta_d H = f(n^a)$. In this case, two measured variables, the adsorbed amount $n^a = m^a/M$ (mol/m²) and $\Delta_d H$ (J/m²) refer to the same equilibrium concentration of the solution. To determine $\Delta_d H_2$ it is therefore useful to keep the ratio of the amounts of all components constant for the calorimetry and adsorption measurements, so that the equilibrium concentrations are also known for the measurements of $\Delta_d H$. The differential molar enthalpy of displacement should allow conclusions with regard to the type and strength of the bond between the adsorbed molecules and the surface to be drawn.

In addition, the influence of the additives on the dispersion stability was investigated qualitatively. For this purpose the sediment volumes of concentrated pigment dispersions were measured in the presence of different concentrations of additive. Because the degree of surface coverage was known from the adsorption measurements, it was possible to establish the relationship between the dispersion stability and the build-up of the adsorbed layer. The sediment volumes of pigments often allow qualitative conclusions to be drawn with regard to the size of the pigment agglomerates and, consequently, to the stability of the dispersion.

Generally speaking, small, well-dispersed pigment particles settle to much more compact sediments than large-sized pigment agglomerates, and thus form smaller sediment volumes.

2.3 Methods

2.3.1 Determination of adsorbed amounts

For the adsorption measurement, 6 g TiO₂ or 1 g carbon black and in the case of the pigment mixture, 3 g TiO₂ + 0.5 g carbon black, each were mixed with 22 g additive solution and shaken for 1.5 h. The initial concentrations were in the range of 0–2% wt. The pigment was then centrifuged off and the supernatant solution analyzed for its additive concentration. The concentrations of acid and amine in toluene were determined via the density of the solutions. This was done with the aid of a vibrating tube densitometer (Heraeus/Paar, model DMA 601/60).

2.3.2 Calorimetric measurements

The enthalpies of displacement were determined in an isothermal calorimeter (Tronac Inc., Orem, Utah, USA) equipped with a top-mounted titration unit (model 1250). 12 g TiO₂ or 2 g carbon black each dispersed in 44 g toluene were introduced into the reaction cell. A concentrated additive solution was added by titration in steps, so that the enthalpy isotherm could be measured cumulatively. Blank measurements without pigment were made to determine the enthalpy of mixing produced by diluting the solution, and corresponding corrections for the enthalpies of displacement were taken into account. The concentration and initial weights were selected in accordance with the adsorption measurements so that the equilibrium concentrations of the solution were known after the adsorption had taken place in the calorimeter.

2.3 Determination of sediment volumes

The amounts 2.727 g TiO₂ or 0.455 g carbon black, or in the case of the pigment mixture 1.363 g TiO₂ + 0.228 carbon black were mixed with 10 g additive solution each and shaken for 1.5 h in 10 ml glas jars. To determine the equilibrium sediment volumes, the pigment heights were observed for several days until they reached a constant value. As in the adsorption tests, the equilibrium concentrations as well as the adsorbed amounts were known by selecting identical ratios of pigment, solvent, and additive, as in the adsorption measurements.

3. Results and discussion

3.1 Titanium dioxide

Figure 1 shows the adsorption isotherms of the dimeric fatty acid, of the dodecyl amine, and of the mixture of acid : amine = 1 : 2 from toluene on TiO₂. The adsorbed amounts are given as mass per unit area, the equilibrium concentrations as percentages by weight of the additive dissolved in each case. Based on the assumption that the polar head groups of these molecules are anchored at the surface and that the hy-

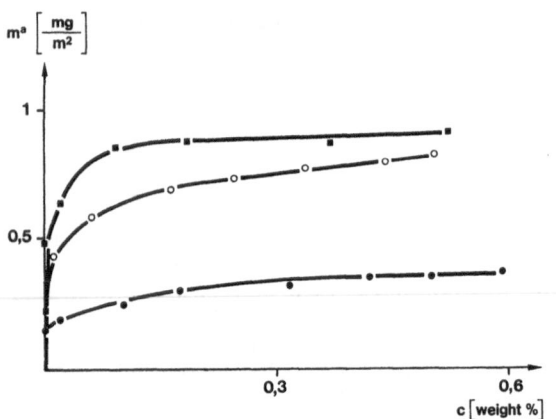

Fig. 1. Adsorption isotherms of dodecyl amine ●, dimeric fatty acid ○ and a 1:2 mixture of acid and amine ■ on TiO₂ from toluene

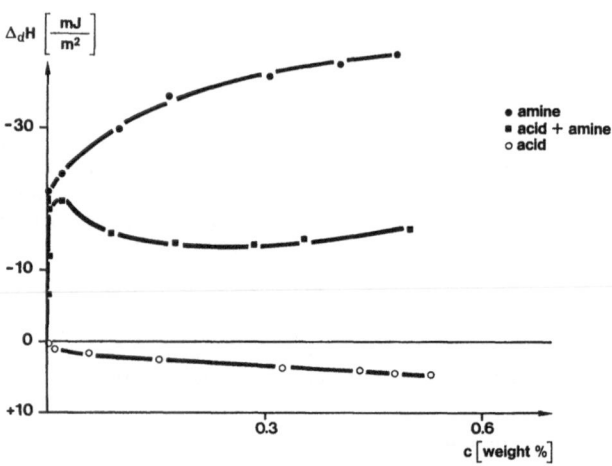

Fig. 2. Enthalpy of displacement isotherms for dodecyl amine ●, dimeric fatty acid ○ and a 1:2 mixture of acid and amine ■ on TiO₂ from toluene

drophobic chains become arranged in the direction of the surrounding solvent, the differences found for the plateau values of the isotherms of the individual components can be anticipated because of the different molecular weights and the different number of polar groups of the acid and the amine. The corresponding molar adsorbed amounts are $2.6 \cdot 10^{-6}$ mol/m² for the amine and $1.7 \cdot 10^{-6}$ mol/m² for the acid. This corresponds to a space requirement of 0.64 nm² or 1.0 nm² per molecule on the TiO₂ surface.

The different are occupied per molecule can be attributed to the different number of functional groups of the acid and the amine. However, no maximum packing density is reached. Very closely packed, vertically arranged alkyl chains occupy an area of only 0.20 nm². For this reason, the adsorbed amounts of the individual components can possibly be explained by the independent adsorption of acid and amine at different active sites on the surface. Also conceivable would be an increased adsorption due to the interaction of the functional groups of the mixture. The calormetric measurements were meant to clarify this point.

Because the calorimetric data obtained correspond to the enthalpy changes accompanying exchange reactions, it is necessary to know what species are being exchanged. Both acids and amines can form associates in non-plar, aprotic solvents via hydrogen bonds [12–14]. The NH...H bond is weaker than the OH...H bond [14], so that in dilute solutions the amines are associated to a lesser extent than the acids. Simple carboxylic acids associate to form dimers [14] whereas intramolecular hydrogen bonds can be formed by the dimeric fatty acid used. Therefore we

must assume that these bonds are still closed even at very low concentration. If the solution contains mixtures, acid-base associates can also be formed. Here hydrogen bonds between the molecules or between the corresponding pairs of ions are possible, depending on the strength of the acid or base; even pure ion bonds between the acid and the base are possible in part.

Figure 2 shows the enthalpies of displacement of amine, acid, and the mixture of toluene on TiO₂ as functions of the equilibrium concentrations. These enthalpy values have been assigned to the respective adsorbed (exchanged) amounts in Fig. 3. From the slope, the respective enthalpy of displacement per gram (or per mole) of exchanged molecule can be calculated.

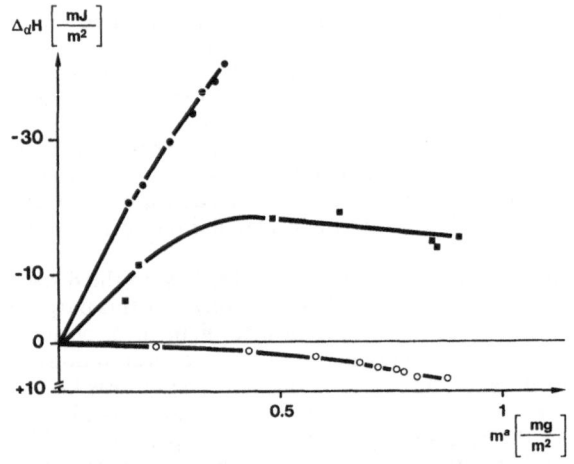

Fig. 3. Enthalpy of displacement as a function of the adsorbed amounts for the systems dodecyl amine ●, dimeric fatty acid ○ and 1:2 mixture ■/toluene/TiO₂

For the individual components this differential enthalpy of displacement does not depend on the degree of surface coverage by the respective substance, as the almost constant slope of the function $\Delta_d H = f(m^a)$ shows. This means the exchange process taking place is always the same, irrespective of the extent to which the TiO_2 surface is already covered with acid or amine. Because of the energetics, the amine exchange is highly favored. Every exchange probably produces a new bond between the amino group of the dodecyl amine and certain active centers of the surface. The differential molar enthalpy of displacement amounts to -23 kJ/mole, i.e., this should be the value of a hydrogen bond. The corresponding value for the dimeric fatty acid is considerably smaller and positive, i.e., over all no additional bonds are formed in this case. Because a dimeric fatty acid most probably forms intramolecular hydrogen bonds in solution, it must therefore be assumed that these bonds have to be broken up before they can be closed again with active sites on the surface.

In contrast with the individual components, the differential enthalpy of displacement of the mixture depends essentially on the surface coverage. This shows that different exchange processes take place. The differential enthalpy of displacement is still negative for lower surface coverage. This allows the conclusion that the initially preferred amine adsorption takes place. For higher degrees of coverage, it then becomes slightly positive which indicates a preferential adsorption of acid. As results from the steep increase of the adsorption isotherm and the plateau value which is already higher for very low equilibrium concentrations (Fig. 1), however, the adsorption of the amine does not constitute a precondition for the adsorption of the acid. It must therefore be assumed that amine and acid are adsorbed independently at different active sites on the TiO_2 surface.

The different adsorption behavior is also reflected by the behavior of the dispersions. In general, the sediment volumes decrease as the surface coverage increases; the comparison between the adsorbed amounts and sediment volumes at the same equilibrium concentrations in the solution illustrates this (Fig. 4). As can be anticipated in view of the steric stabilization of the TiO_2 pigment particles by the adsorbed layer, the higher molecular weight acid is slightly more effective than the lower molecular weight amine. The combination of both is particularly effective. The simultaneous adsorption of both components leads to the formation of a densely packed adsorbed layer even at very low equilibrium concentrations and thus causes a corresponding steric stabilization in this

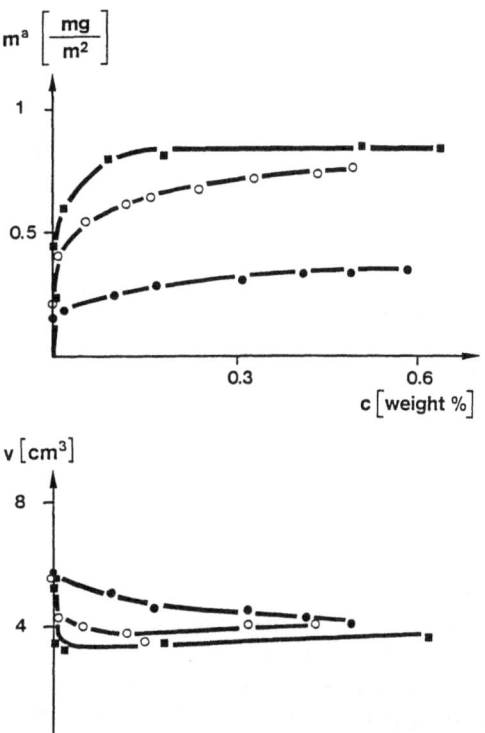

Fig. 4. Sediment volumes of TiO_2 (2.727 g) in the presence of different equilibrium concentrations of dodecyl amine ●, dimeric fatty acid ○ and 1:2 mixture ■ in comparison with the corresponding adsorbed amounts

range. With a view to applications as an additive for paints and coatings, this means that an optimum stabilization by the mixture is already possible at the minimal residual concentration and, consequently, at the initial concentration of the solution.

3.2 Carbon black

Figure 5 shows the adsorption isotherms of the individual components as well as of the 1:2 mixture of carbon black. They differ from the corresponding systems with TiO_2 in some essential points. The adsorption of the acid is weaker than that of the amine, as can be seen from the smaller initial slope of the isotherm. The comparison of plateau values also shows that the number of acid molecules adsorbed is smaller than that of amine molecules ($3.3 \cdot 10^{-7}$ mole/m² acid, $8.2 \cdot 10^{-7}$ mole/m² amine), even when considering the different numbers of functional groups.

This is also in agreement with the highly varied differential molar enthalpies of displacement for acid and amine (Fig. 7). They range from -11 kJ/mole for

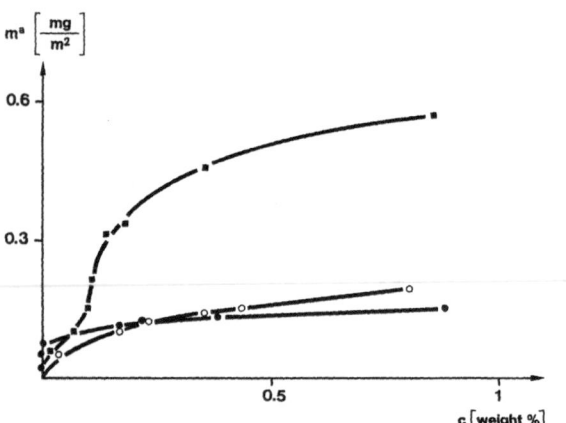

Fig. 5. Adsorption isotherms of dodecyl amine ●, dimeric fatty acid ○ and a 1 : 2 mixture of acid and amine ■ on carbon black from toluene

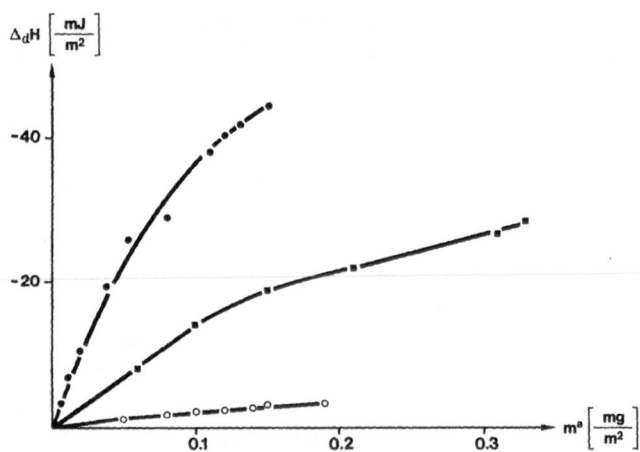

Fig. 7. Enthalpy of displacement as a function of the adsorbed amounts for the systems dodecyl amine ●, dimeric fatty acid ○ and 1 : 2 mixture ■/toluene/carbon black

the acid to -82 kJ/mole for dodecyl amine. For the amine this applies only in the case of low surface coverages. If the adsorbed amounts are higher, the slope of the function $\Delta_d H = f(m^a)$, i.e., the differential enthalpy of displacement, is slightly smaller. This may be due to the fact that the solution already contains slightly more amine associates in the case of the corresponding higher equilibrium concentration.

The relatively low negative value of the differential molar enthalpy of displacement and the initial slope of the isotherm indicate an unspecific adsorption in the case of the dimeric fatty acid. It can be assumed that acid molecules with intramolecular hydrogen bonds as already present in the solution are adsorbed. Such a hydrophobic interaction with the surface is easily conceivable in the case of the carbon black

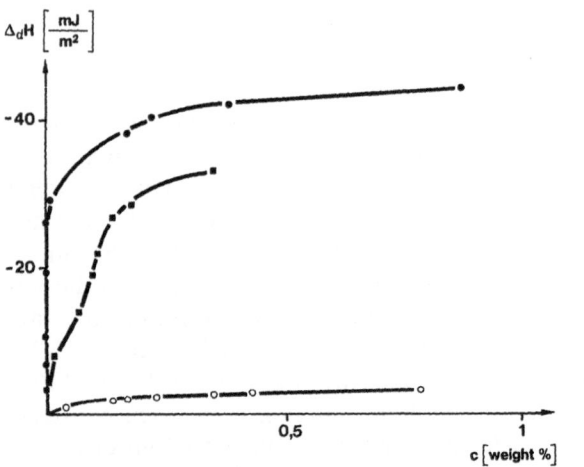

Fig. 6. Enthalpy of displacement isotherms for dodecyl amine ●, dimeric fatty acid ○ and a 1 : 2 mixture of acid and amine ■ on carbon black

because the carbon blacks surface is partly hydrophobic. In the case of the adsorption of fatty acid dimers from non-polar solvents onto graphitized carbon black comparable values of differential enthalpy of displacement were found [15].

The very high negative value of the differential molar enthalpy of displacement in the case of the amine indicates stronger bonds with active sites at the carbon black surface. Simple hydrogen bonds as in the case of the TiO_2 cannot be assumed because of the value of -82 kJ/mole[2]. The formation of hydrogen bonded into pairs, similar to those that can also be formed with acid-base mixtures in non-polar, aprotic solvents are conceivable.

In contrast with the TiO_2, the adsorption of the 1 : 2 mixture on carbon black is low at the beginning and then steeply rises as the equilibrium concentration increases (Fig. 5). The corresponding isotherm of the enthalpy of displacement (Fig. 6) shows a similar curve. Such and S shape can be due to cooperative effects in the adsorbed layer [15–18] or to multi-layer adsorption [19].

Information concerning these adsorption processes, in turn, can be obtained from the combination of calorimetry and adsorption measurements. Figure 7 shows that the differential enthalpy of displacement of the mixture clearly depends on the degree of surface coverage. The quantity $\Delta_d H$ is strongly negative at low surface coverage, i.e., up to approx. 0.15 mg/m². This adsorbed amount corresponds to the maximum surface coverage with amine, so that a pure amine adsorption can be assumed in this range. In fact, this was to be expected in view of the essentially weaker adsorption of the acid. The fact that the differential enthalpy of displacement is smaller than in the case of the pure amine adsorption may be due to the different association in the solution. While the pure amine contains sufficient monomers at small concentrations, a greater share of acid-base associates must be anticipated for the mixture. This is also apparent from the measurements of the enthalpy of mixing. They are much more endothermic in the case of the acid-base mixtures than in the case of pure amine.

At higher surface coverage, the differential enthalpy of displacement is considerably smaller. Due to the large amounts adsorbed in this range, that exceed the sum of the individual components, it must be assumed that additional acid-base associates are adsorbed, possibly in a second layer. Such a phenomenon can be explained by the preceding strong hydrophobization of the carbon black by the amine adsorption. In contrast with TiO_2, the adsorption of the amine in this case is a precondition for the further adsorption of the other component.

3.3 Pigment mixtures

Figure 8 shows the adsorption isotherms of the acid-amine 1:2 mixture on TiO_2, carbon black and on a

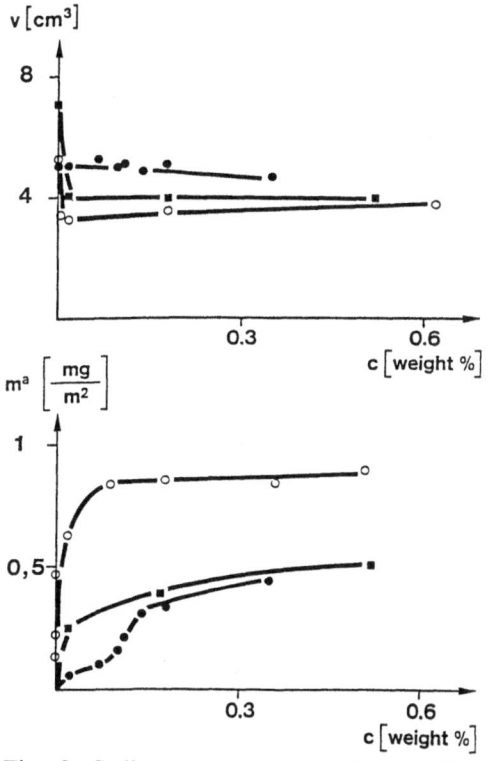

Fig. 8. Sediment volumes of TiO_2 (2.727 g) ○, carbon black (0.455 g) ● as well as of a mixture of TiO_2 (1.363 g) and carbon black (0.228 g) ■ in the presence of different equilibrium concentrations of the acid-base mixture in comparison with the corresponding adsorbed amounts

TiO_2-carbon black mixture (mass ratio TiO_2 : carbon black = 6 : 1, surface ratio 1 : 3.5), and the corresponding sediment volumes at different equilibrium concentrations. The sediment volume of the mixture in the absence of additives (7.0 cm^3) is markedly higher than expected (5.2 cm^3). This indicates a relatively strong co-flocculation between the titanium dioxide and carbon black particles unless an additive is present. However, the flocculation considerably decreases when the pigment surface is covered with additive, as the comparison of the adsorption isotherms with the corresponding sediment volumes shows (Fig. 8).

In this context, it can be clearly seen that the adsorption isotherm of the mixture corresponds to the sum of individual isotherms in the plateau range, but that the adsorbed amounts are higher than expected with small concentrations. This means that the TiO_2-carbon black mixture influences the adsorption behavior of the pigments in the case of low surface coverage. The acid-base mixture leads to a sufficient layer formation at very low equilibrium concentrations that prevents the strong co-flocculation of the pigments even in this range. For this reason, this additive consisting of two components should be considered particularly important for use in paints and coatings with pigment mixtures.

4. Conclusion

The combination of adsorption and calorimetric measurements has shown that the polar groups of the two additives studied plays an important part in the adsorption mechanisms. Both the amino group and the carboxylate group can produce hydrogen bonds to a pigment surface if the latter contains corresponding active sites. The amine bonds are energetically more favorable than the acid bonds due to the smaller degree of association in the solution.

As the corresponding differential enthalpies of displacement show, hydrogen bonds are possible both between un-ionized groups (TiO_2) and between groups corresponding to an ion pair (carbon black). The strong association of an acid in solution may lead to circumstances where only a hydrophobic interaction with the surface takes place (carbon black) instead of the hydrogen bonds (TiO_2). In the adsorption of the acid-amine mixture, both a simultaneous, independent adsorption (TiO_2), and a mutually influenced, increased adsorption (carbon black) may occur. In the case of the pigment mixture, an increased adsorption already occurs at an unexpectedly low concentration. These adsorption phenomena essentially lead to a stabilization of the pigments in the dispersion already at the small residual concentrations in the solution. In practical operation, this allows very low initial concentrations of the additive mixture. The positive effects become apparent, particularly in the case of titanium dioxide and with the TiO_2-carbon black mixtures.

Acknowledgement

We would like to thank Mr. A. Arnold and Mr. A. Soenges for their care in performing the experiments.

References

1. Report/Diskussionstagung der Deutschen Bunsengesellschaft f Phys Chem am 20./21. 10. 1966, Frankfurt a. M., on: Physikalische Grundlagen der anwendungstechnischen Eigenschaften von Pigmenten (1967) Ber Bunsenges Phys Chem 71:239
2. Brockes A (1964) Optik 21:550
3. Crowl VT (1967) J Oil Col Chem Assoc 50:1023
4. Carr W (1982) J Oil Col Chem Assoc 65:373
5. Völz HG (1972) XIe Congrès Fatipec, Florenz 11.–16. 6. 1972. Edizione Ariminum, Mailand
6. Tadros TF (1986) Colloids Surf 18:427
7. Tadros TF (1986) Colloids Surf 18:137
8. Vincent B (1983) Org Coat 5:169
9. Marcovic J, Ottewil RH (1987) Colloids Surf 24:69
10. Jakubauskas HL (1986) J Coatings Technology 58:71
11. Lange H (1966) Kolloid Z Zeitschr Polymere 211:106
12. Rowlinson JS, Swinton FL (1982) Liquids and Liquid Mixtures, 3rd ed. Butterworth, London, S 166/65
13. Popovych O, Tomkins RPT (1981) Nonaqueous Solution Chemistry. John Wiley & Sons, New York
14. MacLean Davis M (1968) Acid-Base Behavior in Aprotic Organic Solvents. National Bureau of Standards Monograph 105
15. Liphard M, Glanz P, Pilarski G, Findenegg GH (1980) Prog Colloid Interface Sci 67:131
16. Findenegg GH, Liphard M (1987) Carbon 1:119
17. Bien-Vogelsang U, Findenegg GH (1986) Colloids and Surfaces 21:469
18. Findenegg GH, Koch C, Liphard M (1983) Adsorption from Solution. Ottewill RH (ed). Academic Press, London, p 87
19. Wolf F, Wurster S (1970) Tenside 7:140

Received November 20, 1987;
accepted June 3, 1988

Authors' address:

M. Liphard
Henkel KGaA
P.O. Box 1100
4000 Düsseldorf

Progress in Colloid & Polymer Science   Progr Colloid & Polymer Sci 77:165–170 (1988)

On the sediment volume of colloidal aggregates
1. A fractal approach to the problem

M. Zrinyi*), M. Kabai-Faix, and F. Horkay

Department of Colloid Science, Loránd Eötvös University, Budapest, Hungary, *)Abteilung für Experimentelle Physik, Universität Ulm, Ulm, F.R.G.

Abstract: A new approach based on the fractal geometry of random aggregates is used to describe the concentration of the sediment formed by the settling of colloidal aggregates. It is predicted that the sediment concentration depends not only on the adhesion forces, but also on the size of the settling particles. Experimental results performed on iron(III) hydroxide sols qualitatively support our ideas.

Key words: Brownian aggregation, fractal dimension, sediment volume, coagulation, iron(III) hydroxide sol

Introduction

Early studies on the sediment volume of coarse suspensions have shown that the concentration or the volume of sediments basically depend on both the particle shape and the particle interactions. Anisometric particles settle to a looser packing than do spheres under the same conditions. Another important factor is the strength of adhesion between the settling particles. The stronger the attractive forces are, the less dense the sediment is.

Although a tremendous amount of work has been done in order to understand the structure and properties of sediments, our knowledge is relatively poor in comparison with our understanding of the basic colloid physics or other topics in colloid science. Even the above mentioned qualitative rules cannot be generalized for every kind of sediment.

In this paper we set out our basic concepts of the sediment volume of colloidal aggregates and evaluate their usefulness as applied to some new experimental results. Our approach is restricted only to colloidal aggregates which form by coagulation or flocculation of primary colloidal particles.

Models of Brownian aggregations

One of the characteristics of colloidal particles is their Brownian motion. When coagulation occurs as a

result of several factors, e.g., when the sum of screened coulombic repulsion and the van der Waals attraction at the barrier becomes smaller than the average kinetic energy of the particles due to extra electrolyte added to the system, the particles meet on Brownian trajectories, and the Brownian motion can be supposed to make its influence felt on the structure of aggregates.

The first numerical approach to the problem of random aggregation was proposed by Vold in 1963 [1]. She developed a method in which a computer was used to simulate the formation of aggregates. Since this pioneering work there has been considerable progress in computational techniques and scientific approaches. As a result of these, a wide variety of models are now available. The structure of random aggregates formed by irreversible coagulation of colloidal particles can be studied by means of these models.

Among these the most important for us now are the diffusion limited aggregation (DLA) and the clustering of clusters (CC) models.

Diffusion limited aggregation

Diffusion limited aggregation [2] is a kinetic growth process in which many particles assemble around an initial seed by diffusing from far away. The lattice version of DLA can be given as follows: consider a lattice and place a single particle at the origin of the lattice. The second particle is chosen at random at a large distance from the first (seed) particle. This particle undergoes a random walk until it visits a site adjacent to the seed. It then stops and sticks to form an ag-

---

*) On leave from Department of Colloid Science, Loránd Eötvös University, Budapest, Hungary

gregate of two particles. Then another particle is released, walks and joins the cluster when it touches it. This process continues until an aggregate of the required size is formed. The DLA model can be used to describe the irreversible aggregation of Brownian particles in infinitely dilute sols.

Clustering of clusters

In the DLA model only one particle is present in the vicinity of the aggregate at any time. For this reason DLA is somewhat deficient as a proper model of colloidal aggregation in real systems.

The CC model [3, 4] includes the effects of finite particle concentrations and multiple stationary growth sites. A cluster consists of a group of occupied lattice sites connected via nearest neighbor occupancy. In this model single particles as well as clusters of particles are allowed to diffuse together. The Brownian motion is represented by the displacements of the aggregates (clusters) by one unit on the lattice. Clusters are selected, one at time, at random, with a probability proportional to their diffusion coefficient. The dependence of the diffusion coefficient on mass is also taken into account. If the cluster contacts other clusters via nearest neighbor occupancy, the contacting aggregates are merged and the diffusion coefficient of the new cluster is computed. In this way small clusters have a mobility greater than the large ones. During the random walks two clusters are assumed to stick rigidly at their first contact. The simulation process continues until only one cluster or a network of particles remains. This model is expected to describe the structure of aggregates that form by irreversible, rapid coagulation.

In addition to the above mentioned models, others have been developed, taking into account e.g., reversible coagulation, fragmentation, disaggregation and chemical effects [5, 6].

Common features of Brownian aggregates

Although several models were introduced in order to understand the structure of aggregates they all show similar and characteristic properties. Among them the most important geometrical property is the self-similarity. This means that the geometrical properties of aggregates are indistinguishable as a function of length or resolution. Examining the structure of these aggregates with much higher, or with much lower resolution, would not result in any discernable difference in geometry. Mandelbrot was the first to point out that many artificial and natural structures exhibit

self-similar geometry [7, 8]. In general, objects with such properties are called *fractal* objects. The concept of fractal structure has potential utility because it is possible that structures which appear as a result of random events can be described within a geometric framework. For example in case of fractal-like aggregates, the total number of particles, N in aggregates is found to scale as the average radius R with the law

$$N \sim R^D \tag{1}$$

where D is the fractal dimension which, in general, is less than the spatial dimension. Analogously, the number of particles inside a sphere of radius R for a homogeneous (non-fractal) structure scales as $N \sim R^d$, where the exponent d is the corresponding Euclidean dimension (d is 3 in our usual world).

Aggregates which grow by accretion according to the rules described previously form large ramified structures with $D < d$. That is, they tend to exhibit inhomogeneous arrangements of particles, with large amounts of open spaces and irregular structures.

The fractal dimension of some simulated aggregates, determined in $d = 2$ and $d = 3$ Euclidean space are summarized in Table 1, where it can be seen that D is considerably smaller than d. This means that there is no constant relation between mass and volume as the length scale is changed. The amount of mass inside a sphere of radius R for a homogeneous structure scales as $M|R| = AR^d$, where A is a numerical constant. Analogously the amount of mass inside a sphere of radius R for fractal structure scales as $M|R| = BR^D$, where B is another constant. Since the volume of a sphere with a radius R varies as $V|R| = CR^d$, where $C = 4\pi/3$, one obtains the average density of the fractal structure as a function of its size R

$$\rho|R| = \frac{M|R|}{V|R|} = \frac{BR^D}{CR^d} = KR^{D-d} \tag{2}$$

where $K = B/C$.

Table 1. Fractal dimension of different Brownian aggregates grown in $d = 2$ and $d = 3$ dimensional Euclidean space

Model	$D|d=2|$	$D|d=3|$
Aggregation with linear trajectories	–	2.78
Diffusion limited aggregation	1.75	2.5
Diffusion limited aggregation with disaggregation	1.54	–
Clustering of clusters	1.5	1.78
Reversible diffusion limited cluster aggregation	1.57	2.03
Chemically limited cluster aggregation	1.55	2.00

Now it is obvious, that the density of the aggregates falls with increasing length, implying that the fractal object of large size would be extraordinary light. Equation (2) says that the density of aggregates depends on their size in contrast with the density of homogeneous structures, where $D = d$. If one considers aggregates of different sizes, $\rho|R|$ goes to zero for $R \to \infty$. This is called scaling behavior, which means if R is increased by the factor b than ρ decreases by a factor $g|b|$ where $g|b|$ is independent of R. At this point two remarks must be made. At short lengths, $r \leqslant$ size of primary colloidal particles, the local geometry is not fractal, while at very long lengths, $r \gtrsim R$, the structure may appear to be homogeneous, implying a lower and an upper cutoff for r when applying Eq. (2). Below the lower cutoff as well as beyond the upper cutoff ρ no longer depends on the size.

Concentration of sediment formed by settling of colloidal aggregates

In the previous section we dealt with the random aggregation process. Only the Brownian motion of particles was taken into account. As the size of aggregates continues to increase, the Brownian motion gradually loses its importance and the aggregates settle down and form a sediment due to the strengthening of hydrodynamical forces. Thus, not only the Brownian motion of primary colloidal particles, but also the settling of aggregates must be taken into consideration. As an approximation we suppose that as the size of aggregates increases the anisotropic Brownian motion due to hydrodynamic forces does not change the fractal character of the aggregates. As the effect of Brownian motion is overwhelmed by the sedimenting field, the aggregates settle down and form a sediment. This sediment has fractal geometry at lengths shorter than the radius of gyration of aggregates and has homogeneous (non-fractal) geometry at greater lengths.

The concentration of sediment basically depends on the average concentration of primary colloidal particles in the aggregates, as well as on the packing of the aggregates.

Let us consider a sediment with a final volume of V. The number of aggregates is denoted by n_R, and R represents the radius of gyration of the monodisperse aggregates. An aggregate is composed of N primary colloidal particles. The sediment volume can be expressed

$$V = \frac{4\pi}{3} p|a|n_R R^3 \qquad (3)$$

where $p|a|$ takes into account the effect of adhesion on the packing of aggregates.

The concentration of the sediment – given by the volume fraction, φ_s, of colloidal particles in the sediment – can be given

$$\varphi_s = \frac{(4\pi/3)n_R N r^3}{V} = \frac{1}{p|a|} \frac{N r^3}{R^3} \qquad (4)$$

where r represents the radius of primary colloidal particles.

In case of fractal aggregates N scales with R^D (see Eq. (1)) and Eq. (4) can be written as

$$\varphi_s = \frac{\text{const.}}{p|a|} R^{D-3} . \qquad (5)$$

Equation (5) says that not only the average concentration of colloidal particles in aggregates (see Eq. (2)), but also the concentration of the sediment depends on the size of aggregates. The smaller the fractal dimension D, the stronger this dependence.

We must mention that Eq. (5) does not result in the same size dependence in case of non-fractal aggregates, where $D = 3$. In this case φ_s is independent of R. This is not strictly valid, because at very small particle sizes the volume of solvate layer may affect the volume or the concentration of the sediment. The concentration of the sediment increases with increasing size of the particles, however this dependence is pronounced only when the thickness of the solvate

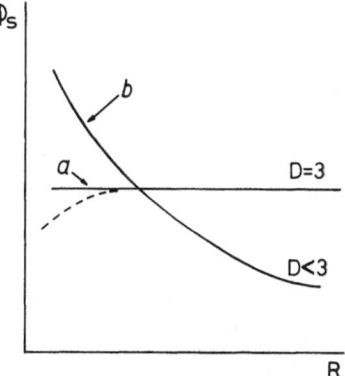

Fig. 1. Schematic representation of the dependence of sediment concentration on the radius of settling particles: a) non-fractal particles; b) fractal aggregates. The dotted line represents the effect of solvate layer in the case of coarse suspensions [10]

layer is comparable with the radius of particles. In the case of aggregates this is not true.

Figure 1 shows how the sediment concentration depends on the size of aggregates when the adhesion forces are unaltered ($p|a|$ = constant). At this stage our purpose is to make sure if our hypothesis given by Eq. (5) can or cannot be accepted. This is reported in the experimental part.

Experimental

We made several experiments with different sols ($Fe(OH)_3$, As_2S_3, PS latex, and TiO_2 suspension). For the coagulation, electrolytes of different types (KNO_3, K_2SO_4, NaOH, NaCl, Na_3PO_4) were used. Because all of the studied systems gave qualitatively the same results, we focus only on the results obtained with iron(III) hydroxide sol. The iron(III) hydroxide sol was prepared by Graham's [9] method. The particles are positively charged with a surface charge density of $86 \mu C\,cm^{-2}$. Average particle radius is 2.8 nm. The electrolytes mentioned were used for the coagulation of the sol.

Because the sediment concentration is expected to depend on both the adhesion forces and the size of the settling aggregates, our intention was to vary only the size of aggregates, while keeping the adhesion forces constant. This can be done by coagulating the sols at different degrees of dilution. At each dilution the concentration of the electrolyte used for the coagulation was the same. It was assumed that the sizes of the aggregates formed in concentrated sols were smaller than

Table 2. Coagulation of iron(III) hydroxide sol by KNO_3 at different dilution. The concentration of sol was 1.07%

The quantity of			Concentration of	
Iron(III) hydroxide sol cm³	Equilibrium medium cm³	1 M KNO_3 solution cm³	sol/%	electrolyte/M
3	0	3	0.534	0.5
3	47	50	3.2×10^{-2}	0.5
3	97	100	1.6×10^{-2}	0.5
3	197	200	8.0×10^{-3}	0.5
3	397	400	4.0×10^{-3}	0.5
3	997	1000	1.6×10^{-3}	0.5

in the case of the diluted sols. That is, if coagulation is performed at very high dilution, one obtains fewer aggregates with bigger sizes than in the case of less diluted systems. The sols were diluted with their equilibrium media prior to coagulation.

A typical series of experiments is given in Table 2. The same type of experiment was repeated with more concentrated sols ($c = 9.29\%$) and with different electrolyte solutions. The volume of sediment was determined by letting the coagulated samples settle in sealed glass tubes (see Fig. 2). It usually took three weeks to attain the final volume of sediment. In order to characterize the size of the aggregates, viscosity and settling velocity measurements were made. A Ubbelohde type viscometer was used for the viscosity measurements and sealed glass tubes were used for the settling kinetic measurements. All experiments were performed at 25 °C.

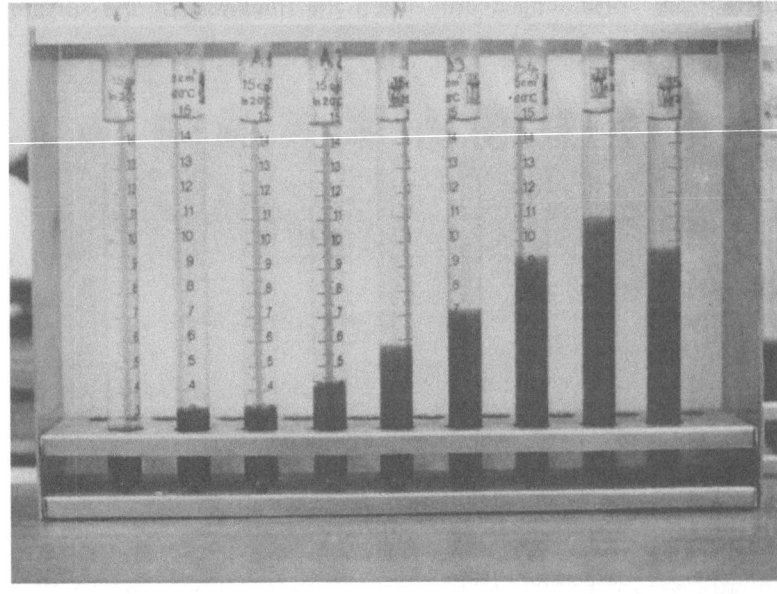

Fig. 2. The sediment volume of coagulated iron(III) hydroxide sols. All the sealed tubes contain 6.4×10^{-2} g iron(III) hydroxide particles. The percentage of concentrations of sols at which the coagulation was performed are from left to right: 4.645, 1.006, 0.503, 0.534, 0.107, 0.064, 0.032, 0.016 and 0.0006%. The concentration of KNO_3 is 0.5 M for each sample. This photo was taken one month after the coagulation had been performed

Results

One of our most important experimental observations is that the sediment volume of the same amount of sol particles is strongly affected by the sol concentration at which the coagulation was performed. This is illustrated in Fig. 2.

It can be seen that as the sol concentration at coagulation decreases the sediment volume drastically increases. The same dependence was observed in the case of other electrolytes as well as of other sols.

In order to be sure that this strong effect on the sediment volume is not the consequence of the changed adhesion forces due to the formation of a new adsorption layer as a result of dilution, other experiments were performed. At fixed sol concentration we systematically varied the pH and the concentration of electrolyte. The pH of the system was varied in the range of 2–13 using HCl and NaOH solutions. The effect of pH on the sediment volume was less than 5%. The strongest effect (5%) was observed at pH 13.

Similar results were obtained when the concentration of electrolyte was varied. The effect of KNO_3 concentration in the range 0.3–1.3 M was negligible. On the basis of these experimental results one can conclude that the increase of sediment volume (or the decrease of sediment concentration) is not the consequence of the possible modification of the adsorption layer due to dilution prior to coagulation.

According to our approach (see Eqs. (4) and (5)) the other important factor which makes its influence on the sediment concentration is the size of the settling colloidal aggregates.

Larger aggregates form looser sediments than smaller ones. We have supposed that the coagulation at very high dilution favors the formation of large aggregates, which should form a loose sediment with large sediment volume. To check the validity of this assumption, viscosity and settling velocity measurements were made.

For the viscosity measurements the sediments of the colloidal aggregates were suspended in their equilibrium media and the viscosity number as a function of concentration was determined. Figure 3 shows this dependence. It must be mentioned that these viscosity data are only partly indicative because it was found that the shear stress developed in the capillary can break the structure of aggregates. However these data qualitatively support our idea, according to which looser sediments are composed of larger aggregates.

The same conclusion was obtained on the basis of the settling velocity measurements. Figure 4 shows how the height of the sediment depends on time. The

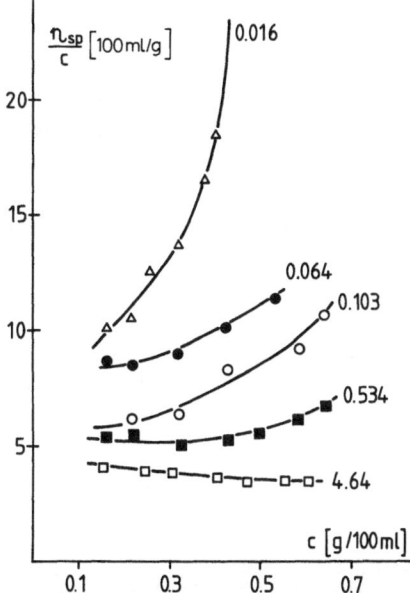

Fig. 3. Dependence of the viscosity number on the concentration of the suspension formed from the sediments. All the sediments contain 6.4×10^{-2} g iron(III) hydroxide and the concentration of KNO_3 is 0.5 M. Concentration of colloidal particles at coagulation: [%]/volume of sediment [cm³], △ 0.016/10.7, ■ 0.534/4.3, ● 0.064/7.00, □ 4.645/3.2, ○ 0.103/5.75

smallest sediment velocity belongs to the highest equilibrium sediment volume. Because the sediment velocity is related to the size of the settling particles, one can conclude that different sediments contain aggregates of different sizes, and the larger the sediment volume (or the smaller the sediment concentration), the bigger the sizes of the settling particles.

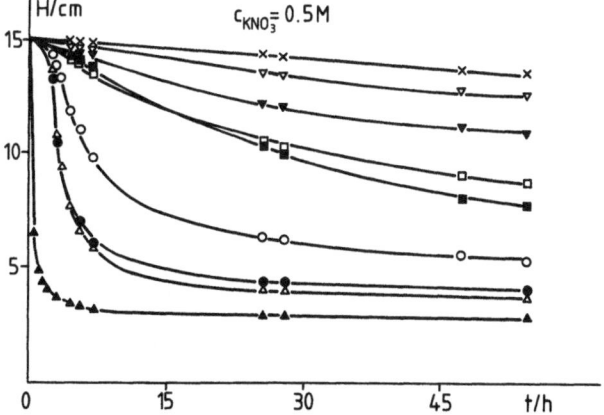

Fig. 4. Dependence of the height of sediment on time. All the sediments contain 6.4×10^{-2} g iron(III) hydroxide sol particles. The concentration of KNO_3 is 0.5 M. Concentration of colloidal particles at coagulation [%]/final volume of sediment [cm³]: × 6.4×10^{-3}/9.6, ▼ 3.2×10^{-2}/9.25, ■ 1.0×10^{-1}/5.75, ● 0.5/3.3, ▲ 4.6/2.6, ▽ 1.6×10^{-2}/10.7, □ 6.4×10^{-2}/7.0, ○ 0.53/4.3, △ 1.0/3.2

Progress in Colloid & Polymer Science, Vol. 77 (1988)

Conclusion

We have discussed the concentration of sediments formed from colloidal aggregates. A new approach based on the fractal geometry of random aggregates is proposed by means of which the dependence of sediment volume on the size of settling particles can be predicted. Experimental results performed on iron(III) hydroxide sols qualitatively support our ideas.

Acknowledgements

The support of the Hungarian Academy of Sciences (OTKA No.: 1204/86) is gratefully acknowledged. One of the authors (M. Z.) thanks the Alexander von Humboldt Stiftung for financial support.

References

1. Vold MJ (1963) J Coll Sci 18:684
2. Witten TA, Sander LM (1981) Phys Rev Lett 47:1400
3. Kolb M, Jullien R, Botet R (1983) Phys Rev Lett 51:1119
4. Meakin P (1983) Phys Rev Lett 51:1123
5. Stanley HE, Ostrowsky N (1985) On growth and form. Martinus Nijhoff Publishers, Dordrecht
6. Herrmann NJ (1986) Physics Reports 136:153
7. Mandelbrot BB (1977) Fractals: form chance and dimension. W. H. Freemen and Co. San Francisco
8. Mandelbrot BB (1982) The fractal geometry of nature. W. H. Freeman and Co. San Francisco
9. Kabai-Faix M, Rohrsetzer S, Wolfram E (1969) Ann Un Sc Bud Rol Eötvös Nom S Chim 11:89
10. Patzkó A, Vàrkonyi B, Szàntó F (1971) Acta Physica et Chimica 17:91

Received October 14, 1987;
accepted June 9, 1988

Authors' address:

M. Zrinyi
Department of Colloid Science
L. Eötvös University
H-1088 Budapest, Puskin n. 11–13

Progress in Colloid & Polymer Science

Progr Colloid & Polymer Sci 77:171–175 (1988)

Application of bentonite suspensions as supporting media in foundation engineering an other underground workings

F. Hilbert, H. Schweiger and I. Huber

Institute of Chemical Technology of Inorganic Substances, Technical University Graz, Austria

Abstract: Investigations of the physicochemical properties of suspensions of active (sodium-) bentonite in water and of the interactions of such suspensions with loose soil show that the following phenomena are responsible for the applicability of bentonite suspensions as supporting media for the walls of underground excavations (slurry wall technique):

i) The long-known formation of a watertight film or penetration zone at or in the loose soil which makes possible the transmission of the full hydrostatic pressure of the suspension to the walls.

Besides that, there are further important phenomena that have not been considered in technical literature concerning the slurry wall technique up till now:

ii) The development of an osmotic pressure between groundwater and bentonite suspension, which is caused by the formation of a semipermeable structure of montmorillonite particles at the phase boundary.

iii) The development of an underpressure in the penetration zone in the soil, which is caused by a very slow volume contraction of the solidified suspension in the soil pores.

iv) A considerable increase of the soil cohesion in the penetration zone of the suspension, caused by the formation of a supporting structure of montmorillonite platelets linking the soil grains together.

The additional effects of ii) and iii) cause a formal increase of the hydrostatic pressure of the suspension in the excavation and prevent or restrict, together with the cohesion increase iv), the detachment of single grains from the walls, which is the initiating step of wall collapse in cohesionless soils.

Key words: Application of bentonite underground excavations, slurry wall technique, (sodium) montmorillonite, application of montmorillonite, (active) bentonite

Bentonite and suspensions of bentonite have found various technical applications. Of increasing importance is the utilization of suspensions of active bentonite as supporting media for walls of underground excavations (slurry wall method). If an excavation is held constantly filled during the digging with a proper suspension of active bentonite, the bentonite suspension stabilizes the walls and renders mechanical support unnecessary. In this way it is possible to excavate trenches and pits of great depth even in completely cohesionless soils and below the groundwater table. After the excavation has reached the desired depth, it may be filled up with concrete, which on account of its higher density displaces the bentonite suspension upwards.

This slurry wall technique and analogous techniques are more economical and faster than traditional techniques for the construction of foundations, building trench enclosures, underground sealing walls, etc.; the technique offers some additional advantages [1].

The practical realization of the technique is well elaborated for all these underground workings, but the calculation methods and approved working procedures are mostly based upon purely empirical observations.

Macroscopically it is easy to see that bentonite suspensions form a watertight film on the walls of the excavations and that the suspensions also penetrate into the soil and solidify thixotropically in the soil pores.

Due to this sealing of the soil, the full hydrostatic pressure of the suspension is transmitted to the walls of the excavation.

However, practical experience shows that in many cases the hydrostatic pressure of the suspension is not sufficiently high to explain the stabilization of the walls, especially if the excavation gets below the water table [2]. For stabilization, only the difference between the hydrostatic pressure of the suspension and that of the groundwater is effective. The densities of the practically used suspensions are only somewhat higher than the density of water (about 1.01 to 1.04). Therefore the resulting pressure differences are small and in many cases insufficient to compensate the calculated soil pressure. It is also remarkable that suspensions of ordinary clay are ineffective, even if they cause the same sealing of the soil as bentonite suspensions [1].

Fig. 1. Transmission electron micrograph of a carbon replica of a thixotropically solidified suspension of active bentonite (4%) (see text). (Magnification 23 100×)

These observations lead to the conclusion that besides the sealing action of bentonite, there have to be other phenomena which take part in the wall stabilization. It is evident that the stabilization cannot be explained by the frictional force transmitted to the soil during the inflow of the bentonite suspension, as it is assumed in several technical papers [3]. Experimentally it is easily shown [1, 4] that the suspension solidifies quickly in the soil pores and that after this solidification the walls remain stable.

Starting point of our investigations was the observation that the slurry wall technique is only feasible with bentonite suspensions. Therefore, the additional phenomena mentioned above have to be effects of the particular properties of active bentonite.

The properties of active bentonite are determined by the properties of the main constituent, the clay mineral sodium montmorillonite. Bentonite is the natural degradation product of volcanic ash rich in glassy constituents. Besides montmorillonite, it contains in minor quantities other minerals (in most cases illite, quartz, mica, feldspar, lime, pyrite), quantity and type of the accompanying minerals depending upon the particular source. Most of the mines deliver inactive bentonite, which has to be activated by ion exchange in a treatment with sodium carbonate [5].

Montmorillonite is a three-layer clay mineral composed of clearly distinguishable components [6]:

i) the montmorillonite mineral layers themselves, normally stacked parallel, carrying negative electrical charges.

ii) exchangable cations, which compensate the negative charge of the mineral layers, occurring between the layers.

iii) neutral water molecules associated with the cations and the mineral layers.

The negative charge of the mineral layers is due to isomorphous substitution of lower valence cations and crystal defects [6]. The water is bonded in the space between the layers in nonstoichiometric amounts. With sodium montmorillonite, only a part of this water may be given off and taken up again reversibly.

Suspensions of active bentonite in water are markedly thixotropic. At high enough concentration the suspensions solidify after a short time to a gel, which is fluidized again by mechanical stirring.

This solidification is brought about by the particular structure of montmorillonite, which consists of anions of colloidal size (crystalline silicate layers) and charge compensating cations between the layers. In aqueous suspensions of sodium montmorillonite the result of ionic dissociation is a solution of colloidal anions and sodium cations. The silicate layers are negatively charged as a whole, but the charge distribution is not uniform. The electrical charge on edges and corners is under most conditions different from that on the faces, the sign and the extent of the difference depending upon pH and salt additions to the suspension [6]. This inhomogenous charge distribution is basically the cause of the formation of solidification structures resembling a house of cards or ribbons [6].

Figure 1 shows a transmission electron micrograph of a carbon replica of such a "house of cards" structure. For preparation a thin layer of a solidified suspension of active bentonite was shock-frozen in liquid nitrogen (to prevent the formation of crystalline ice), then the sample was vacuum-etched (part of the ice was sublimated into vacuum) and a carbon replica of the surface made. In the picture the narrow, darker colored areas are the montmorillonite platelets, sticking mostly nearly perpendicular to the picture plane in the remaining ice; really perpendicular platelets are

narrow black bands, platelets deviating from this orientation are lighter colored and bigger with increasing deviation. The large, light gray areas in the picture are the surface of the remaining amorphous ice. The edge-to-face bonding of the platelets forming the "house of cards" structure is clearly discernible.

Suspensions of bentonite show the colligative properties of solutions. A known phenomenon is the osmotic pressure between such suspensions and water. It is easily shown with a Pfeffer cell separating the two phases by a semipermeable membrane (e.g., parchment). Table 1 gives some of our results.

Even without separation by an artificial membrane an osmotic pressure may develop between bentonite suspensions and water. To effect this, the water phase has to be mechanically stabilized, e.g., by using water saturated sand or grit as second phase. An experimental setup like this resembles closely the conditions of the practical slurry wall technique. Figure 2 shows the experimental setup, Table 2 some results of these experiments.

From the existence of an osmotic pressure the existence of a semipermeable membrane at the phase boundary between bentonite suspensions and water saturated soil has to be concluded. In our case there

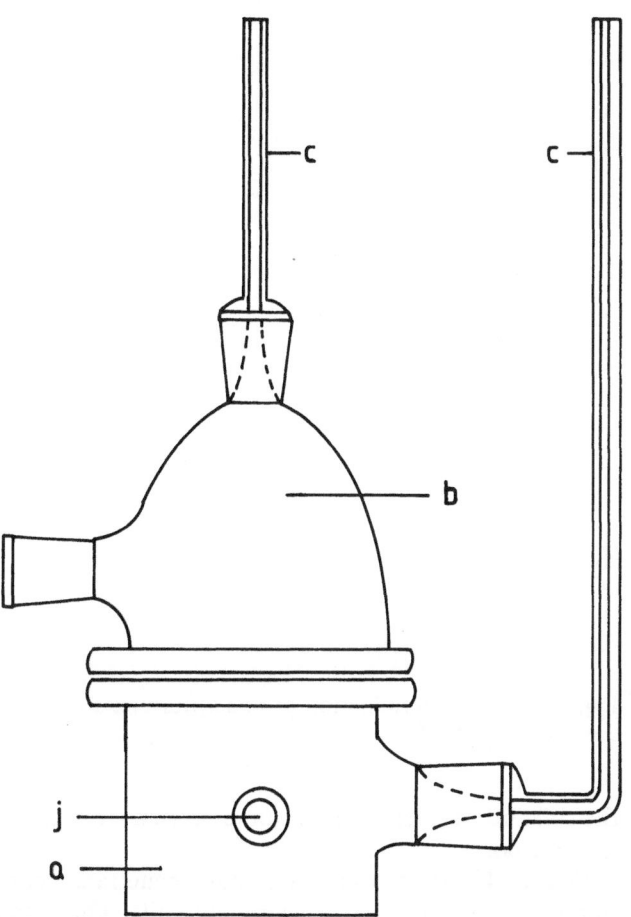

Fig. 2. Cell for determination of osmotic pressure; *b* bentonite suspension, *j* junction for levelling device, *c* capillaries, *a* water saturated sand

Table 1. Osmotic equilibrium pressures of bentonite suspensions against water. Diaphragm: cellophane

Suspension weight − %	Pressure N/m^2
4.0	883
3.0	697
2.0	481
1.5	351
0.5	128

Table 2. Osmotic equilibrium pressures of suspensions of bentonite against water saturated sand; no artificial diaphragm

Sand	Bentonite	Pressure (N/m^2)
seasand	4% Na-bentonite	398
seasand	4% raw bentonite	395
calcareous sand	4% Na-bentonite	437
seasand + 5% clay	4% Na-bentonite	772
seasand + 3% clay	4% Na-bentonite	760
seasand + 1.5% clay	4% Na-bentonite	583
calcareous sand + 1.5% clay	4% Na-bentonite	768
seasand + both phases 0.002 M NaCl	4% Na-bentonite	461

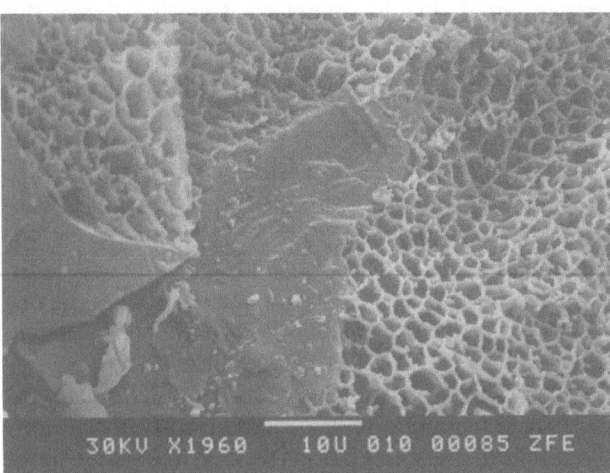

Fig. 3. Scanning electron micrograph of the structure of solidified bentonite between sand grains. (1960×)

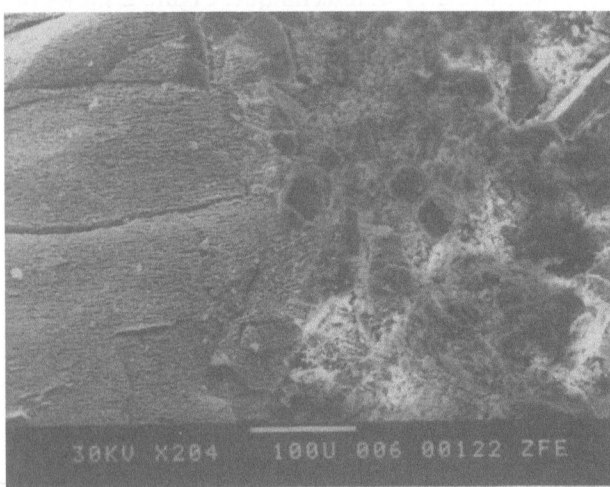

Fig. 4. Scanning electron micrograph of the structure of solidified bentonite in the border zone of suspension and water saturated sand. (204×)

is only the possibility that this membrane is formed by the bentonite itself. This proves that an osmotic pressure will develop in the practical application of the slurry wall technique too and may be one of the additional factors contributing to wall stabilization.

To elucidate the semipermeable structure at the phase boundary, scanning electron investigations were carried out. Figures 3 and 4 demonstrate the structure at the phase boundary after thixotropic solidification of the active bentonite. The solidification in the soil pores produces a closed structure, which must have the effect of a semipermeable membrane.

Besides that, the scanning electron micrographs reveal the penetration of the bentonite suspension into

Fig. 5. Apparatus for determination of volume contraction

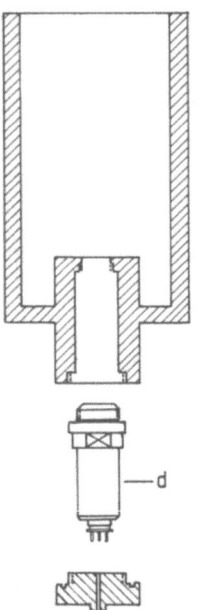

Fig. 6. Cell for determination of underpressure; d differential pressure sensor

the soil and the formation of an intergrain structure in the soil pores.

Figure 4 shows that bentonite penetrates considerably even into very finegrained soils. The mean grain size of the sand in Fig. 4 was below 0.1 mm and the overpressure only 3 mm water height; nevertheless, the penetration depth is still some 0.5 mm. With sand

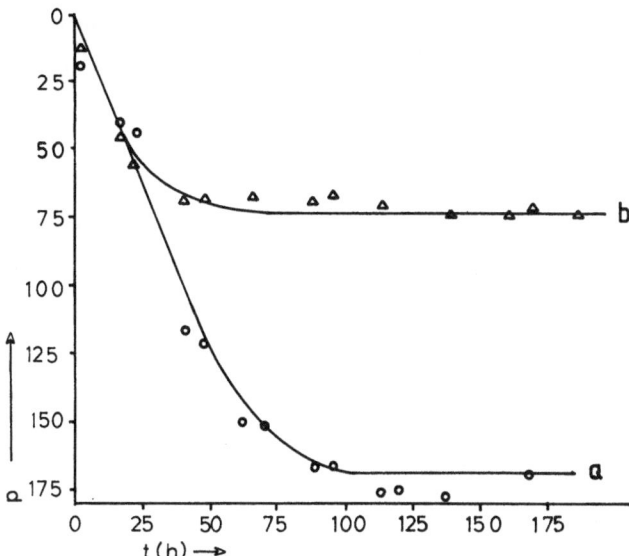

Fig. 7. Time dependence of underpressure. Bentonite suspension 7.3 weight % (a) 3.9 weight % (b); pressure *p* in cm water column

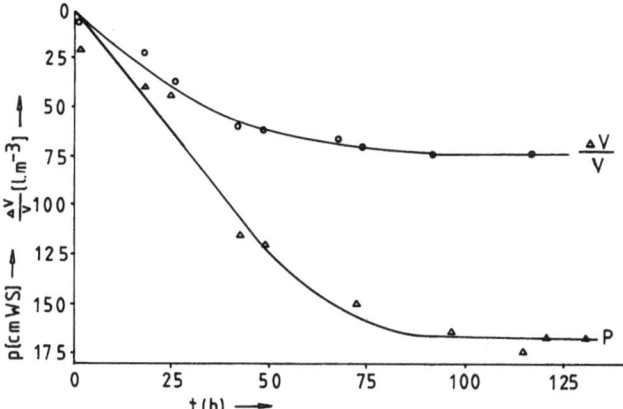

Fig. 8. Time dependence of underpressure and volume contraction. Bentonite suspension 7.3 weight %, pressure *p* in cm water column

or grit of coarser grain penetration depths of 20 to 50 cm or more may be reached [7].

There was clear evidence that the first step of the thixotropic solidification is the bonding of the montmorillonite platelets to the surface of the sand grains. That shows that the bonding between the grains and the platelets is stronger than that between the platelets themselves. This means that the three-dimensional network of the solidification structure forms a supporting framework fo the sand grains.

The "glueing together" of the soil grains in the penetration zone prevents or at least restricts the detachment of single grains from the wall, which in cohesionless soil is always the first step of wall collapse [8]. Together with the development of an underpressure in the solidified suspension (see below), the glueing effect seems to be the most important additional factor for wall stabilization. A quantitative determination of the cohesion increase in the penetration zone is under way, model experiments have already given proof of the increase [9].

Also very important seems to be another widely unknown property of bentonite suspensions. During determinations of specific densities it was found that the densities of the suspensions increase slowly with time. Clearly this increase has to be caused by a corresponding volume contraction.

The volume contraction of suspensions of active bentonite was determined in the experimental cell of Fig. 5. It was found that the contraction is rather slow and continues for several days. With a suspension of 7.3 weight %, the final value is reached after 155 h.

If the bentonite suspension is enclosed in space, the result of the volume contraction is a corresponding underpressure. Such an underpressure could also be an important factor for wall stabilization, because the solidification in the penetration zone forms an enclosed space in which an underpressure may develop.

The development of underpressure in enclosed space was determined in the experimental cell shown in Fig. 6. An electronic sensor measures the pressure and makes possible the recording of the time dependence. Fig. 7 shows the pressure change for suspensions of 7.3 and 3.9 weight %, Fig. 8 gives volume contraction and pressure change for a suspension of 7.3 weight %.

Finally we would like to remark on the reversibility of water loss and uptake. Mostly it is taken for granted that the giving off and taking up of water by sodium montmorillonite is fully reversible. According to our results this is not true. The drying of active bentonite is only reversible if not too much water is given off. Drying with the aid of desiccating agents or at elevated temperatures deteriorate or eliminate the characteristic properties. Even mild drying over silica gel at room temperature deteriorates active sodium bentonite so much that a 4% suspension will not solidify.

Acknowledgement

The authors express their gratitude to the Fonds zur Förderung der wissenschaftlichen Forschung in Österreich for substantial financial support and to the Zentrum für Elektronenmikroskopie Graz, especially Doz. Dr. W. Geymayer, for carrying out the SEM and TEM investigations.

References

1. Hilbert, F, Veder Ch, Foidl G, Huber I (1984) ÖIAZ 3:85
2. Veder Ch (1963) Der Bauing 38:378
   Lorenz H (1950) Die Bautechnik 27:313
   Niemann H-J (1960) ibid 37:310
3. Weiß F (1967) Die Standfestigkeit flüssigkeitsgestützter Erdwände. Bauing-Praxis, Heft 70, Ernst W und Sohn, Berlin
   DIN 4126 (1986) Ortbeton – Schlitzwände – Konstruktion und Ausführung
4. Veder Ch, Hilbert F (1983) Proc 8th European Conf Soil Mechanics and Foundation Engng, Helsinki, Sess 9, p 175
5. Hofmann U (1956) Angew Chemie 68:53
6. Brindley G, Brown G (1980) Crystal Structures of Clay Minerals and their x-ray Identification. Min Soc Monograph No 5, London

    Lagaly G (1986) In: Kulicke W (ed) Fließverhalten von Stoffen und Stoffgemischen 147–186. Hüthig, Basel
7. Möbius H, Günther T (1977) Die Bautechnik 8:267
8. Morgenstern N, Tamassheb E (1965) Geotechnique 15:387
   Lorenz H (1967) Baumasch u Bautechnik 220
9. Fuchsberger M, Schönstein K (unpublished results)

Received January 12, 1988;
accepted June 22, 1988

Author's address:

Prof. Dr. F. Hilbert
Institute of Chemical Technology of Inorganic Substances
Technical University Graz
A-8020 Graz, Austria

Progress in Colloid & Polymer Science   Progr Colloid & Polymer Sci 77:177–179 (1988)

Effective volume and immobilization concentration of dispersed particles

J. Pfragner

Institut für Physikalische Chemie der Universität Graz, Austria

Abstract: We report flow curves of china clay slurries measured over a wide range of concentration and shear rate, therewith calculating the hydrodynamic volume and the maximum packing fraction of particles under shear, following Mooney. Results are consistent with limiting viscosity number measurements, and show the decrease of effective volume like shear thinning behaviour. To the contrary, the concentration of immobilization decreases after a maximum towards a high shear value. This makes it conceivable that shear forces make the dispersion change from one state to another, represented by two limiting flow curves. The transition between both produces a viscosity maximum.

Key words: Dispersion, viscosity, effective volume, maximum packing fraction, flow curve analysis, china clay

Introduction

The flow behaviour of inorganic dispersions depends on concentration, zetapotential, temperature and shear forces. At constant temperature, under present conditions of pH value and additives determining the particle surface charge and electroviscous effects, the relative dispersion viscosity should be a function of solid content (weight fraction c or volume fraction ϕ) and shear rate $\dot{\gamma}$. Although of outstanding practical interest, there exists no satisfying functional description, especially for china clay slurries with some rheological properties not easy to understand. In the dilute limit, viscosity tends to obey Einstein's equation

$$\eta_r = 1 + k\phi \tag{1}$$

where k is identical to the limiting viscosity number $[\eta]$. For solid spheres, k equals 2.5. Otherwise it mirrors anisotropy and depends on particle size and shear rate, as is the case with china clay. For medium concentration a great variety of empirical functions has been reported, as reviewed in [1]; these are often extensions of Einstein's equation. But increasing ϕ lets the viscosity grow much steeper, and approaching immobilization shows the necessity for introducing a maximum packing fraction ϕ_i or c_i. Several examples are listed in the cited references. An analogue result is produced by Mooney's self crowding factor s [2] and Eq. (2).

$$\ln \eta_r = \frac{k\phi}{1 - s\phi} \tag{2}$$

The denominator figures the free part of dispersion medium not occupied with particle contact. Obviously s is the reciprocal of ϕ_i and k resembles $[\eta]$. A good data fit with spherical [3] and nonspherical [4] dispersions has been reported, despite the claim that Mooney's equation had occasionally caused incredibly high values of ϕ_i for spherical particles [5]. Advantageously, it is easy to linearize in order to separate the effects of hydrodynamic volume and immobilization concentration.

With increasing shear rate, there is the common difficulty of functional description, even in the case of monotonously decreasing η_r from η_{r0} to $\eta_{r\infty}$. Moreover, china clay dispersions have a viscosity maximum. Thus we take the shear rate as a parameter and a concentration dependence is computed for constant $\dot{\gamma}$; the results are values of $[\eta]$ and ϕ_i indicating the particle state under shear.

Experimental

We prepared slurries of china clay (Dinkie A) in water with a full concentration range up to $c = 0.7$ (70% w/w). The pigment's density was found to be $2652 \, \text{kg m}^{-3}$. The dispersion obtained polysalt S, above $c = 0.3$ in a constant weight relation of $3 : 1000$ polysalt : solid. The zetapotential was $-28 - -29 \, \text{mV}$, evaluated by means of Repap and a

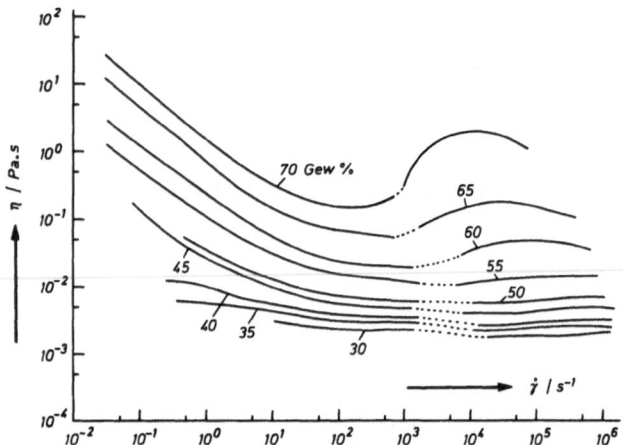

Fig. 1. China clay (Dinkie A) slurry in water, containing 3 g Polysalt S : 1000 g solid, at 20 °C. Slurry viscosity versus shear rate with different solid contents as weight percentage. Between rotational and capillary flow measuring range interpolation is shown as dotted line

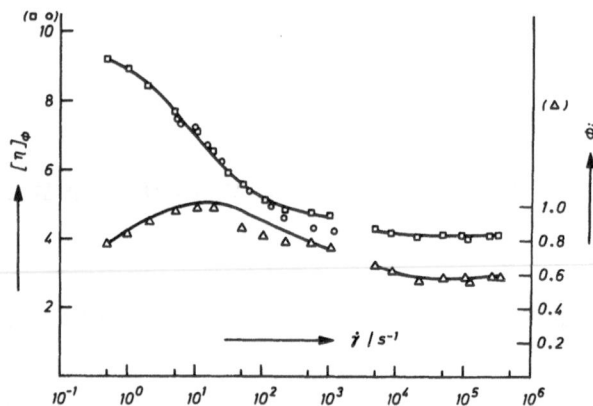

Fig. 2. China clay dispersion. Shear dependence of limiting viscosity number $[\eta]$ and immobilization concentration ϕ_i, both dimensionless volume fractions. The measured values of $[\eta]$ (circles) agree with the calculated ones (squares). Also using Eq. (3), ϕ_i has been calculated (triangles) according to the scale at the right

Mass Transport Analyser device. The pH was 7.2–7.3. All measurements were done at 20 °C. We obtained flow curves at shear rates $10^{-2}-10^{-6}$ s^{-1} by means of rotational viscosimeters (Rheomat 30 and Low Shear, Contraves, Zürich; RM 7200, Rheometrics, New Jersey) and capillary flow (HV, Paar, Graz). In calculating the viscosity values for the higher concentrations shown in Fig. 1, we eliminated the kinetic energy loss and yield stress influence [6] and the considerable apparent wall slip [7]. Shear heating and Reynold's numbers were kept in check. We made an effort to limit viscosity number measurements, due to sedimentation effects. We used a set of Ubbelohde viscosimeters with efflux times from 40 seconds to three minutes; glass capillaries of 1.5 m length, and 0.15 to 0.9 mm in diameter; horizontal, vertical upstream and downstream flow was controlled with microburettes and a piezoresistive differential pressure transducer; rotational flow was controlled in the Low Shear device. But so doing we covered a shear range of 7–2000 s^{-1}. The upper limit resulted from turbulent onset, the lower one from exceeding sedimentation effects when measuring times necessarily increase. As for dispersions beyond $c = 0.3$, dilution was done with 0.1 % polysalt S in water, in order to keep the zetapotential in level due to desorption equilibrium at particle surface. The specific viscosity was related to the particle volume fraction ϕ, yielding $[\eta]$.

Results and discussion

Figure 2 contains the directly accessible limiting viscosity numbers, marked as circles. With increasing shear rate, they slope down from 7.8 to 4.2. This is common with structural viscosity, but we cannot draw a distinction between anisotropic particle orientation in the flow field or flock disruption under stress, of which the latter may release immobilized solvent. We

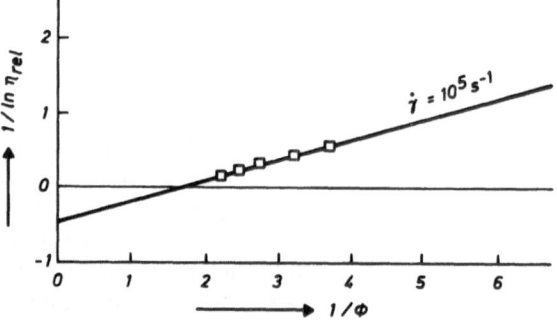

Fig. 3. Graphic representation of Eq. (3), entering relative viscosity with different volume fractions ϕ of china clay, as seen in Fig. 1. The slope of the linear fit is $1/[\eta]$; the intercept at $1/\phi = 0$ is $-1/[\eta]\phi_i$

use the values to prove the following flow curve analysis.

At constant shear rate we take (Fig. 1) viscosity values at different concentrations, convert them to η_r and ϕ and enter Eq. (2), which is rewritten in our terms

$$\frac{1}{\ln \eta_r} = -\frac{1}{[\eta]} \cdot \frac{1}{\phi_i} + \frac{1}{[\eta]} \cdot \frac{1}{\phi} . \qquad (3)$$

$1/\ln \eta_r$ turns out to be a fairly linear function of $1/\phi$, as depicted in Fig. 3. We obtain $[\eta]$ from the slope and ϕ_i from the intercept at $1/\phi = 0$. These computed values of the effective volume are in good agreement with the measured ones. They seem to approach a high

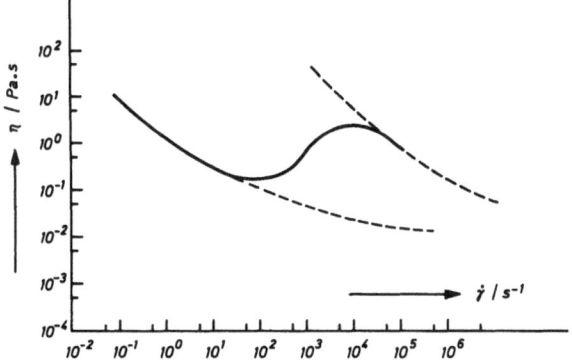

Fig. 4. China clay dispersion 75% w/w; A model of two limiting flow curves which shows general shear thinning behaviour. Shear forces cause flock disrupture and increase of solvent layer, so that the actual flow curve turns from one state to another

shear limit at about 4.1. As shown in Fig. 2, the immobilization concentration ϕ_i starts increasing with shear rate, runs through a maximum of about 0.95 at $\dot{\gamma} = 10 \, \text{s}^{-1}$ and tends to a high shear limit of 0.58. An increasing ϕ_i fits well to the decreasing $[\eta]$ with particle orientation, but when both slope down the orientation effect has been exhausted or at least overwhelmed by repulsive particle interaction. This brings us to the potential curve which has an outer or secondary minimum whenever a nonhydrophilic dispersant yields stabilizing surface charge. A secondary minimum at a fixed particle distance must lead to a constant immobilization concentration, unless orientation effects may lend interstitial liquid under shear, thus increasing ϕ_i. But in fact ϕ_i decreases and indicates the elongation of particle distance at minimum. That means shear forces have altered the flock structure.

We complete this with some information that is hidden in $[\eta]$, due to rotational diffusion constant D_r. If we may take $[\eta] = 8$ as the low shear limit and we make use of Perrin's formula [8], we receive for the axis ratio p the absolute values of the half axes in nanometers: in the low shear region, $D_r = 3.6$; $p = 6.5 = 650/100$ nm/nm in the case of prolate ellipsoids; $p = 0.1 = 47/470$ nm/nm if oblate.

In the high shear region the concentrations give $D_r = 50$; $p = 6.5 = 270/42$ nm/nm, if prolate; $p = 0.1 = 21/196$ nm/nm, if oblate. Obviously a considerable flock disrupture takes place. The radius of gyration, 282 nm, resulting from laserlight measurements, mat-

ches the particle size at low shear conditions. We regard this as a fundamentally different state of dispersion, compared with high shear conditions. If we draw the flow behaviour of both the states as two limiting flow curves, for example with $c = 0.75$ w/w in Fig. 4, we limit the actual flow curve that exhibits the transition from one to the other. In this point of view we agree with Schurz [9].

Conclusion

We wish to separate two parallel effects produced by shear forces: on one hand the flocks suffer disaggregation, as can be seen from $[\eta]$ with the help of Perrin's formula; on the other hand there is more liquid trapped in particle contact, presumably for shielding the double layer distortion, as can be seen from ϕ_i. There is an intermediate range of shear rate where decreasing ϕ_i makes shear-thickening against all shear-thinning effects. A viscosity maximum results. The low shear conditions rule extensive flocks, high shear conditions produce a disaggregated state and both are of qualitatively different flow behaviour. These two states may be represented by two different limiting flow curves. The transition between them produces a viscosity maximum with higher concentrations.

References

1. Rutgers IR (1962) Rheol Acta 2:305
2. Mooney M (1951) J Colloid Sci 6:162
3. Simha RJ (1952) J Appl Phys 23:1020
4. Brodnyan JG (1959) Trans Soc Rheol 3:61
5. Wildemuth CR, Williams MC (1984) 23:627
6. Neubauer JG (1983) PhD Thesis, Karl-Franzens-Univ. Graz
7. Pfragner J (1985) Radiusdependent Capillary Flow of Dispersions. Presented at the Annual Meeting of the German Rheological Society, Berlin
8. Schurz J (1974) Struktur − Rheologie. Berliner Union Kohlhammer, Stuttgart
9. Schurz J (1984) Wochenbl f Papierf 8:275

Received February 26, 1988; accepted June 9, 1988

Author's address:

J. Pfragner
Institut für Physikalische Chemie
Universität Graz
A-8010 Graz, Österreich

Progress in Colloid & Polymer Science   Progr Colloid & Polymer Sci 77:180–194 (1988)

Correlation between latex stability data determined by practical and colloid chemistry-based methods *)

J. P. Fischer and E. Nölken

Forschung und Entwicklung GB G/H, Hoechst AG, F.R.G.

Abstract: In order to test the applicability of colloid chemistry-based methods for the evaluation of latex stability, poly(vinyl acetate) latices were modified by copolymerization and emulsifier addition. The copolymerization with ethylene sulfonic acid sodium salt (ESASS) proved to be an effective procedure to enhance the latex stability. The addition of a nonylphenol-poly(ethylene oxide) sulfate ammonium salt (Hoe S 2928) was less effective.

By these methods one series each was prepared with the amount of comonomer or emulsifier increased gradually. A different latex stability was obtained upon shearing and storage at 70 °C, depending on concentration.

The different stability could easily be measured by practical methods, e.g., solids content and particle size distribution determined by Laser Aerosol Spectroscopy (LAS) and Photon Correlation Spectroscopy (PCS).

The data of Critical Coagulation Concentration (CCC) with $CaCl_2$ specified by turbidity measurements correlate with the stability determined by practical methods, as well as the determination of surface charges by polyelectrolyte titration with a Streaming Current Detector (SCD).

On the contrary, Zeta Potential determined by Mass-Transport Electrophoresis and Microelectrophoresis of the original, dialyzed or thoroughly cleaned latices did not correlate with latex stability. A good correlation, however, could be found between Zeta Potential and the other criteria of the latex stability during a coagulating titration with a cationic polymer.

Key words: Latex stability tests, zeta potential, critical coagulation concentration, streaming current detector, equilibria of surface charges

1. Introduction

The improvement of latex stability is a permanent object of industrial research and colloid science as well. The conditions for latex stability under practical requirements during production, upon storage and application, are often different from the conditions of methods to evaluate latex stability by colloid chemistry-based measurements. For example, contradictions have often been found between Zeta Potential and stability in practical application.

It was therefore the aim of this work to prepare two series of latices with progressively varied stability in order to correlate results for latex stability by practical and by colloid chemistry-based methods. For that reason a simplified latex-recipe was chosen knowing that in practice, latex stability does not only concern production and handling of the latex itself, but also the application in combination with salts, pigments or flocculating additives, e.g., in paints or paper coatings.

It is known that latex stability can be improved by copolymerization or addition of ionic-charged additives neglecting the stabilization by protective colloids or non-ionic tensides. Usual latices are prepared by emulsion polymerization with persulfate initiator, preferably ammonium persulfate (APS). In addition, an anionic emulsifier ®Hostapal BV = Alkyl-aryl-poly(ethylene oxide) sulfate sodium salt, was used in order to obtain a constant particle size and particle size distribution for all latices investigated. Due to the fact that sulfate groups were unavoidable in our simplified practical latex system, additional stabilization was obtained by sulfate and sulfonate groups; this method limited the stabilizing groups to these two types.

---

*) Dedicated to Professor Dr. Heinz Harnisch on the occasion of his 60th birthday.

Latices of this type were modified by two methods to improve stability. The most effective procedure to improve latex stability was the copolymerization with ethylene sulfonic acid sodium salt (ESASS) as comonomer. In a second series the addition of an emulsifier with sulfate groups, Hoe S 2928 = nonylphenol-poly(ethylene oxide) sulfate ammonium salt with roughly 23 units of ethylene oxide, was less effective, but both series were able to demonstrate the different effect of copolymerization and emulsifier addition for latex stability.

2. Experimental Part

2.1 Preparation of the latices

The latices were prepared in a 2 l three-necked flask starting with 544 g H_2O heated to 45 °C and with the addition of 0.75 g ammonium persulfate (APS) in 13.3 g H_2O. At 50 °C, 2.5% of the total monomer emulsion was added — this was previously prepared by emulsifying 1 000 g vinyl acetate in a solution of 3 g $Na_4P_2O_7$ and 60 g ®Hostapal BV (50%) in 278 g H_2O adding 0.75 g APS in 13 g H_2O.

After the aqueous phase reached 70 °C the residual 97.5% of the monomer emulsion was added continuously over a period of four hours. At the end of the selfheating polymerization, 0.5 g APS in 26.7 g H_2O were added. The polymerization was completed within 30 minutes and allowed to cool to room temperature.

In the case of a copolymerization with 0.1, up to 2% ethylene sulfonic acid sodium salt (ESASS, Series A), the comonomer, was dissolved in the aqueous phase before preparation of the monomer emulsion. The addition of 0.5 with up to 4% of the emulsifier Hoe S 2928 = a nonylphenol-poly(ethylene oxide) sulfate ammonium salt with roughly 23 units of ethylene oxide (series B) was performed with an emulsifier solution of 32% solid content after the polymerization of the base latex without comonomer. The percentage of various additions is related to a total of 100% vinyl acetate.

2.2 Determination of particle size and particle size distribution

The particle size and particle size distribution were measured by two methods:

a) Laser Aerosol Spectroscopy (LAS)

This method was developed in cooperation with a research group [1] which uses a similar apparatus as a particle counter and sizer of airborne particles. The utilization of Laser Aerosol Spectroscopy (LAS) for determination of particle size distribution of latices in the range of 80 nm up to 600 nm with a very good resolution has been already described [2, 3]. The method was improved in range up to 1 200 nm upon addition of a He/Ne-laser measuring at small scattering angle (4.6−11.3°) and it offered easy handling by automatic control (Fig. 1), [2, 3].

All of the 13 latices investigated in this work show nearly the same particle size and particle size distribution as the example of original latex No. 40 in Fig. 2.

Table 1 summarizes the data of the latices investigated, characterized by the three commonly used particle size averages:

d_n = number average,
d_w = weight average,
d_z = Z-average

b) Photon Correlation Spectroscopy (PCS)

These measurements were performed with a commercial system Malvern 4700. A diagram of this spectrometer is shown in Fig. 3. The determination of particle size distribution was usually carried out at an angle of 90° by using the software Auto-Sizer II, supplied by Malvern Instr. Ltd.

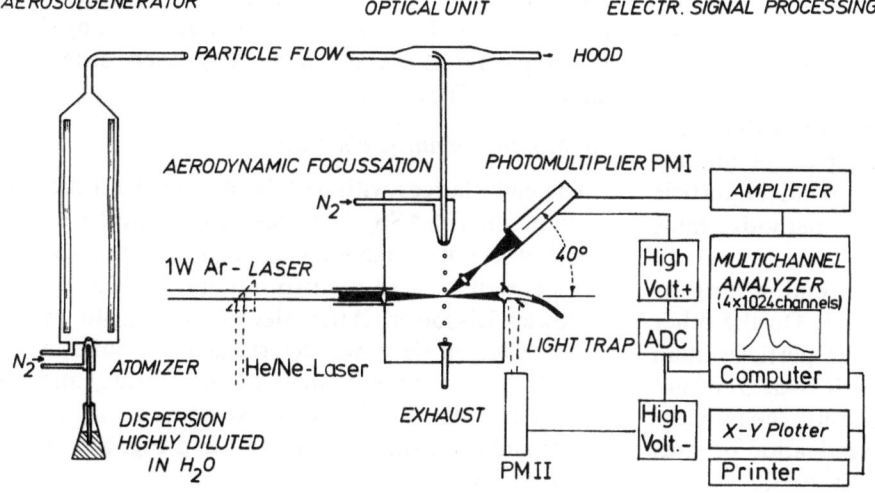

Fig. 1. Scheme of the Laser Aerosol Spectrometer (LAS) as a laser particle counter for measuring particle size distribution of latices with high resolution (for details see [1−3])

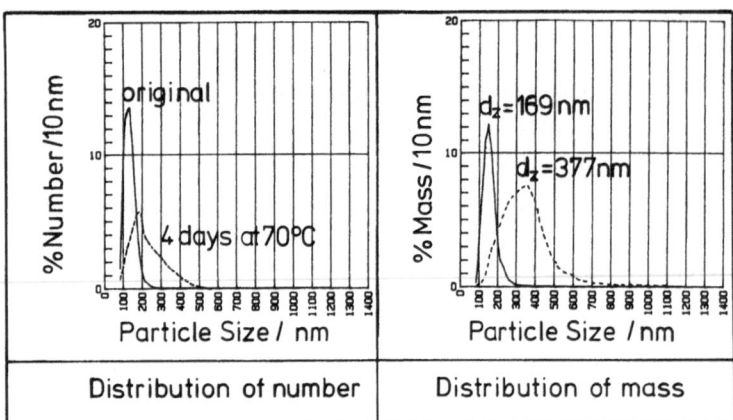

Fig. 2. Particle size distribution (left: distribution of number; right: distribution of mass) for poly(vinyl acetate) latex No. 40 determined by Laser Aerosol Spectroscopy (LAS) before (solid line, $dz = 169$ nm) and after storage at 70 °C for four days (dotted line, $dz = 377$ nm)

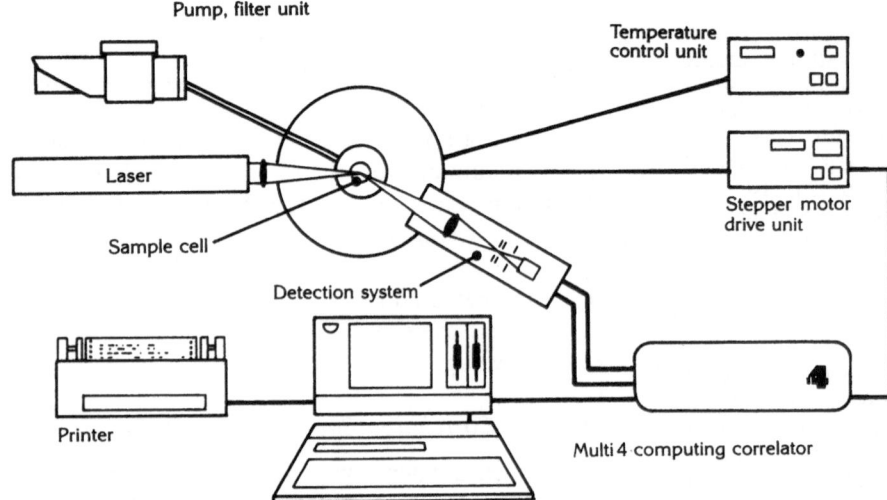

Fig. 3. Scheme of the spectrometer Malvern 4700 used for Photon Correlation Spectroscopy (PCS) to measure particle size dz and particle size distribution

The latices in Fig. 2 well characterized by Laser Aerosol Spectroscopy (LAS) (Poly(vinyl acetate) latex No. 40, before and after storage for four days at 70 °C), have a somewhat different particle size distribution (than that shown in Fig. 4) if Photon Correlation Spectroscopy (PCS) is used. Especially, it is the large particle size distribution after storage at 70 °C which is very well characterized by LAS; Figure 4 shows a bimodal particle size distribution with unlikely parts of small particle size and a gap at 500 nm due to Mie-light scattering. Nevertheless, identical particle size Z-average d_z is obtained by both methods (Table 1, Figs. 2, 4).

In some cases the particle size distribution measured with PCS is broader than with LAS. On the other hand, PCS is able to measure larger particles than LAS because the upper limit of PCS is at 3000 nm higher than 1200 nm for LAS. Therefore, the application of PCS is well suited to our investigation of latex stability after shearing and storage at 70 °C.

Aggregates formed during shearing or storage can be detected more sensibly by Z-average d_z *determined by PCS. Having proved the identity of d_z for both* methods (LAS and PCS) for some latices (as with latex No. 40 in Figs. 2, 4), stability against shearing and storage at 70 °C can well be characterized by simply comparing Z-average d_z among each other (see Tables 1, 2, Figs. 5, 6).

2.3 Stability against shearing

The eight latices with 0–2% ethylene sulfonic acid sodium salt (ESASS) in series A and the five additional latices with 0.5–4% emulsifier Hoe S2928 in series B were treated for two minutes at 5000 rpm in a Cowles-Dissolver. After sieving with a 40 micrometer ® Perlon-sieve we determined the amount of coagulate, the solids content of the filtrate, and the particle size distribution with averages, d_n, d_w and d_z before and after shearing. Table 1 and Fig. 5 show the effect of shearing.

Table 1. Characterization of Poly(vinyl acetate) latices investigated with respect to solids content and particle size distribution before and after shear (2 min 5000 rpm in a Cowles Dissolver)

Series	Latex No.	Additive [%]		Solids content [%]		Residue on sieve [%]	Particle size distribution/averages in [nm]							
		ESASS	Hoe S2928	Orig.	After shear		Original latices before shear						After shear	
							Aerosol Spectroscopy			Photon Correlation Spectroscopy				
							dn	dw	dz	dn	dw	dz	dz	Modal.
A	40	0	0	51.6	43.1	16.5	128	143	151	133	152	167	277	bi
	41	0.1	0	51.6	47.7	7.5	144	156	165	129	152	171	267	bi
	42	0.2	0	51.4	50.1	2.5	127	144	158	128	140	151	266	bi
	43	0.3	0	51.7	51.6	0.14	125	141	152	120	142	167	257	bi
	44	0.5	0	52.0	52.0	0.02	131	148	159	123	144	161	228	bi
	45	1	0	51.8	51.8	0.03	131	145	155	110	158	210	218	mono
	46	1.5	0	51.4	51.4	0.12	129	143	155	160	203	223	206	mono
	47	2	0	51.9	51.9	0.6	134	149	162	120	155	187	275	bi
B	40	0	0	51.6	43.1	16.5	128	143	151	133	152	167	277	bi
	40/0.5	0	0.5	51.4	50.7	1.1	128	143	151	133	152	167	181	mono
	40/1	0	1	51.3	49.9	2.2	128	143	151	133	152	167	188	mono
	40/2	0	2	51.0	47.9	3.1	128	143	151	133	152	167	244	bi
	40/3	0	3	50.7	47.0	5.5	128	143	151	133	152	167	240	bi
	40/4	0	4	50.4	45.0	8.1	128	143	151	133	152	167	238	bi

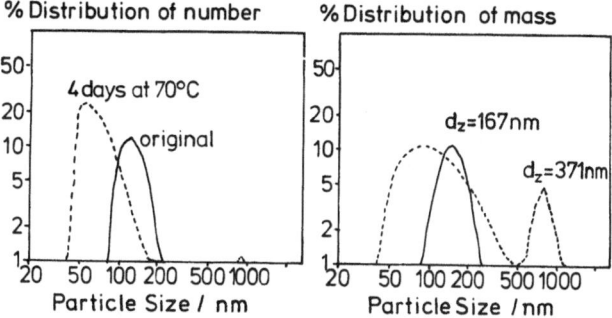

Fig. 4. Particle size distribution (left: distribution of number; right: distribution of mass) for poly(vinyl acetate) latex No. 40 determined by Photon Correlation Spectroscopy (PCS) before (solid line, $dz = 167$ nm) and after storage at 70 °C for four days (dotted line, $dz = 371$ nm)

For series A solids content and residue on sieve indicate only a minor stability against shear stress for samples with low (0–0.2%) or highest (2%) content of the ESASS-comonomer. The minimum residue on sieve for 0.5–1% ESASS is less sensitive for the indication of coagulation as particle size distribution analysis by LAS or PCS. Monomodal particle size distributions by PCS and roughly unchanged Z-averages dz are only observed for 1 or 1.5% content of the ESASS-monomer, thus characterizing the region of maximum stability for series A.

In series B the particle size measurements show the maximum stability against shear upon the addition of 0.5–1% emulsifier Hoe S2928; the same result is drawn as the results of solids content and residue on

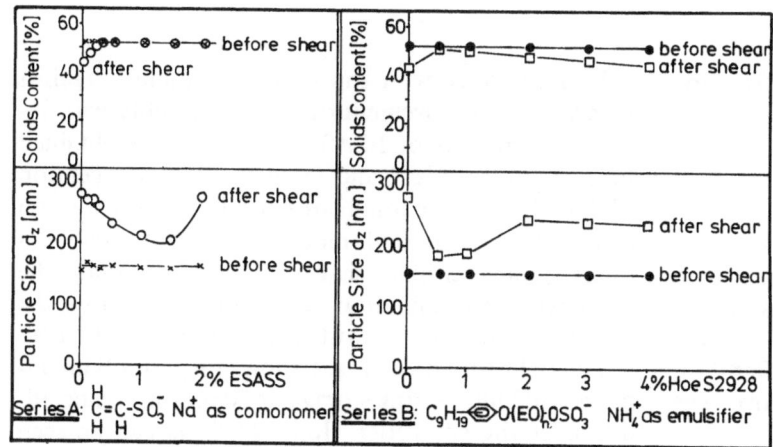

Fig. 5. Change of particle size dz and solids content of two series of poly(vinyl acetate) latices by shear (2 min, 5000 rpm in a Cowles Dissolver)

Table 2. Measurements of latex stability with respect to solids content and particle size distribution before and during storage at 70 °C for at least 21 days (bi = bimodal particle size distribution)

Series	Latex No.	Additive [%]		Solids content [%] after sieving (40 µm)						Particle size Z-average d_z via PCS					
		ESASS	Hoe S 2928		After storage of ... days						After storage of ... days				
				Orig.	2	4	7	14	21	Orig.	2	4	7	14	21
A	40	0	0	51.6	51.0	45.2	6.1	just koag.		167	273	371 bi	440 bi	just koag.	
	41	0.1	0	51.6	51.5	51.5	51.5	50.9	50.6	171	219	249	352 bi	420 bi	492 bi
	42	0.2	0	51.4	51.4	–	51.1	50.5	50.1	151	173	178	197	273 bi	399 bi
	43	0.3	0	51.7	51.6	–	51.6	51.6	51.6	167	172	172	171	170	225
	44	0.5	0	52.0	52.0	–	52.0	52.0	52.0	161	182	185	189	186	219
	45	1	0	51.8	51.8	–	–	–	51.8	210	203	200	209	207	205
	46	1.5	0	51.4	51.4	–	–	–	51.4	223	193	200	210	207	221
	47	2	0	51.9	51.8	–	–	–	51.7	187	206	210	216	239	353 bi
B	40	0	0	51.6	51.0	45.2	6.1	just koag.		167	273	371 bi	440 bi	just koag.	
	40/0.5	0	0.5	51.4	51.1	50.9	17.6	5.0	koag.	167	174	206	216	434 bi	koag.
	40/1	0	1	51.3	51.0	50.7	24.6	5.3	koag.	167	182	180	225	414 bi	koag.
	40/2	0	2	51.0	50.5	49.5	32.9	6.6	koag.	167	167	172	179	315 bi	koag.
	40/3	0	3	50.7	50.5	49.4	35.6	10.5	koag.	167	167	170	175	268	koag.
	40/4	0	4	50.4	49.9	49.4	40.7	17.8	koag.	167	168	170	169	250	koag.

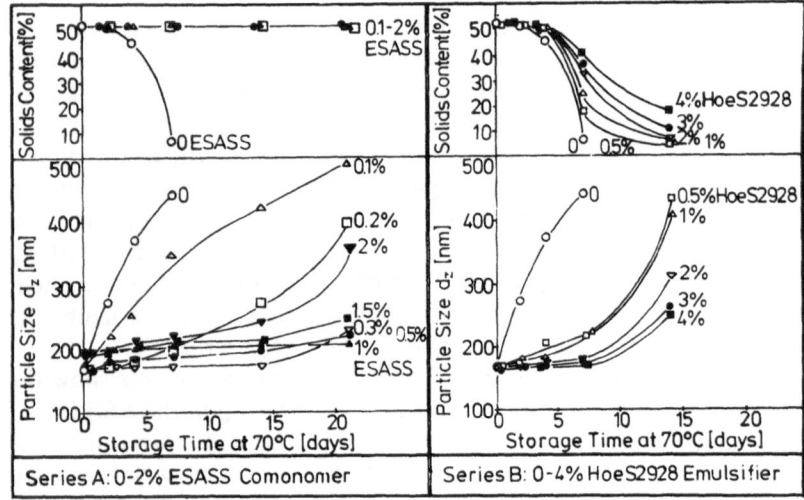

Fig. 6. Change of particle size dz and solids content of two series of poly(vinyl acetate) latices (left: series A with 0–2% ESASS-comonomer; right: series B with 0–4% emulsifier Hoe S 2928) during storage at 70 °C

sieve in Table 1. The addition of the emusifier Hoe S 2928 is less effective vs shearing than the use of ESASS as comonomer in series A.

2.4 Stability during storage at 70 °C

In order to characterize latex stability during storage at room or higher temperature, latex samples have been stored at 70 °C. The stability was measured by solid content and particle size distribution for the filtrate of the 40 micrometer ® Perlon-sieve after 2, 4, 7, 14 or 21 days respectively, as described above. The results are given in Table 2 and Fig. 6.

In accordance with the rating of stability against shearing an improvement of storage stability with increasing amounts of ESASS-comonomer is obtained for series A (see Fig. 6 and Table 2), with the exception of 2% ESASS. The results of particle size distribution measurements (d_z and modality by PCS) reveal a much more sensible scale for stability than the gravimetric determination of the solid content on the relatively stable latices with the ESASS-comonomer.

The right column of Fig. 6 shows for series B an improved stability during storage at 70 °C with rising addition of the emulsifier Hoe S 2928, as can be seen from a strong decay of solids content and a rise in particle size with simulated bimodal particle size distributions by PCS. This emulsifier, Hoe S 2928, only gives stabilization for a short period. Higher amounts of Hoe S 2928 improve storage stability but are detrimental to stability against shearing, as discussed in section 2.3.

Therefore, the simple addition of emulsifier Hoe S 2928 is not sufficient for stabilization from a practical point of view.

2.5 Polyelectrolyte titration with a Streaming Current Detector (SCD)

In a previous work it has been shown that the distribution of charges between latex particle surface and serum can be measured by polyelectrolyte titration with a Streaming Current Detector (SCD) [3]. The same method was applied to latices of series A with the ESASS-comonomer by titrating with a 0.001 n solution of Poly-Di-Allyl-Di-Methyl-Ammonium-

Chloride (Poly-DADMAC Hoe S 3529). The emulsifier Hoe S 2928 could not quantitatively be detected by the SCD-system because of its low molecular weight.

The application of a Streaming Current Detector for polyelectrolyte titration was developed by Schempp and others [4, 5]. With the self-constructed apparatus shown in Fig. 7 (now commercially available in a slightly modified system [6]) it is possible to titrate charges on polymers or on surfaces of colloids. Different amounts of Poly-ESASS homopolymer can be stoichiometrically titrated by Poly-DADMAC (see Fig. 8).

The results of titrations of the original latices (dilution 1 : 80) are given in Fig. 9, the total negative charge for 1 g solid is dependent on pH and the content of the ESASS-comonomer. The small rise of charges between pH 3 and pH 10 is caused by carboxylic groups (roughly 10 micro-equ/g solids) which belong to water soluble material. This can be seen in Table 3 from the low carboxylic content after centrifugation.

For each of the eight latices the distributions of the total sulfate+sulfonate groups have been determined according to Fig. 10. The individual results are summarized in Table 3. The latices were dialyzed by ultrafiltration to give a first ultrafiltrate serum as a measure of the components soluble in water. The part adsorbed or bound on surface before dialysis is ob-

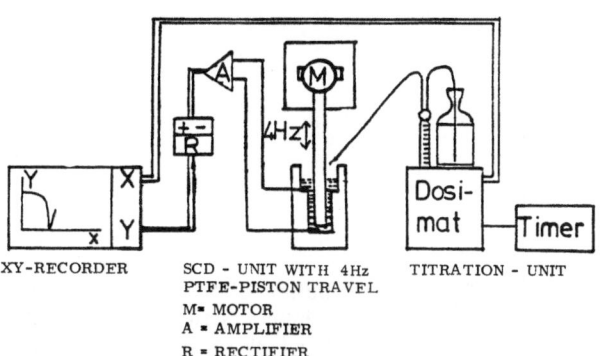

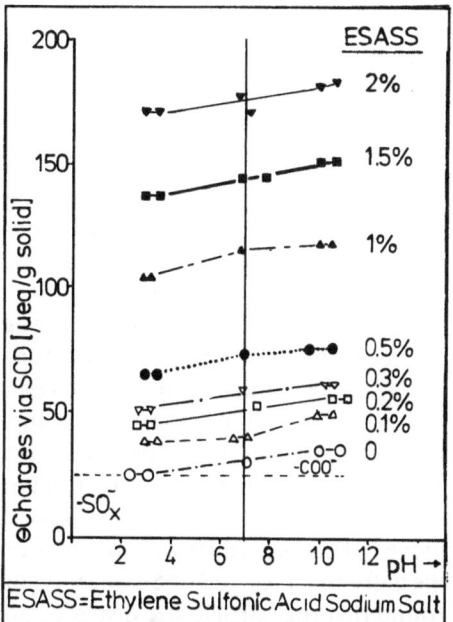

Fig. 7. Scheme of the Streaming Current Detector (SCD) for determination of charges bound on surfaces or polymers by polyelectrolyte titration

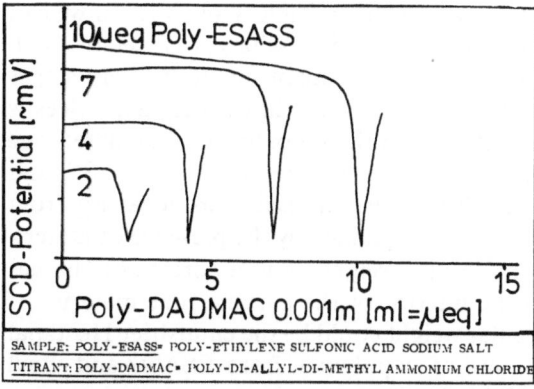

Fig. 8. Examples of titration curves for various amounts of Poly-ESASS (ethylene sulfonic acid sodium salt) with Poly-DADMAC (Di-allyl-di-methyl ammonium chloride) by means of a Streaming Current Detector (SCD)

Fig. 9. Dependence of total charge determined by SCD-Titration on different pH for the poly(vinyl acetate) latices of series A with 0–2% ESASS-comonomer

Table 3. Comparison of theoretical and experimental data for the distribution of charges determined by polyelectrolyte titration by means of a Streaming Current Detector (SCD) for latices and sera at various stages of cleaning (APS = Ammonium Persulfate, BV = Hostapal BV emulsifier)

Latex No.	Additive ESASS [%]	Total theor. charge $\left[\dfrac{\mu \text{ equ}}{\text{g solid}}\right]$	Experimental charge content determined by SCD [μ equ/g solids or g serum]									
			Sulfate and sulfonate groups at pH 3					Carboxylate groups at pH 10				
			Latices			Sera		Latices			Sera	
			Orig.	Dial.	Centrif.	Ultraf.	Centrif.	Orig.	Dial.	Centrif.	Ultraf.	Centrif.
40	0	APS: 19 +BV: 43 = $\overline{62}$	25	16	15	2	10	10	9	1	9	9
41	0.1	72	38	20	18	5	20	10	9	1	9	9
42	0.2	81	45	20	20	6	25	11	9	2	9	9
43	0.3	91	51	21	21	10	30	10	9	1	9	9
44	0.5	101	65	30	21	13	44	10	9	1	9	9
45	1	139	106	45	22	20	84	13	9	2	9	11
46	1.5	177	140	78	25	33	115	14	13	2	9	12
47	2	216	172	108	32	41	140	11	12	2	9	12

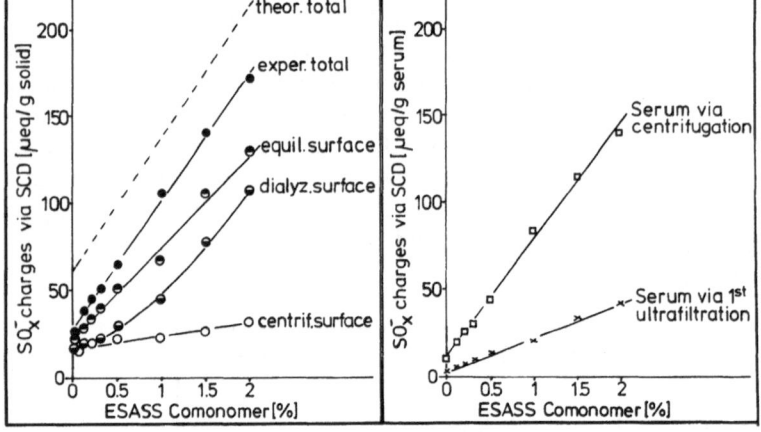

Fig. 10. Distribution of SO_x^--charges of poly(vinyl acetate) latices (series A) with 0–2% ESASS-comonomer at various stages of surface cleaning

tained by subtracting this part from the total charge content of the original latex. If we continue repeating ultrafiltration with new water the status of the dialyzed latex is reached at a minimum conductivity of 10 μS/cm.

The second procedure to clean latex surface was centrifugation to receive a centrifugation serum containing those parts of charges and polymers which can be easily desorbed by a strong diminution of the surface during centrifugation. Then the centrifuged latex was cleaned by threefold redispersing and centrifuging in order to obtain the status where all adsorbed charges were removed. This procedure causes a loss of solid content especially for a low ESASS-comonomer content. It was found that the particle size distribution of the resulting latices was always the same before and after filtration with a ® Perlon-sieve, with the exception of the latex without ESASS. This latex showed an

enhancement of d_w from 143 nm to 161 nm with a broader particle size distribution by LAS.

Comparing all results in Table 3 and Fig. 10 we can see that the amount of SO_x^--charges correlates with the ESASS-comonomer content in all different stages of cleaning. The total charge in the original latices is somewhat lower than the theoretically calculated one on the basis of total persulfate APS, the ESASS, and the Hostapal BV emulsifier used. This departure from the theory may be explained by the presence of material of low molecular weight, or to a portion of charges bound in the interior of the particle. The material of low molecular weight could be the emulsifier Hostapal BV or a part of copolymer from ESASS and vinyl acetate soluble in the water phase.

The serum after centrifugation contains about 90% of the SO_x^--charges bound in a water soluble copolymer of ESASS and vinyl acetate. For the latex with 2%

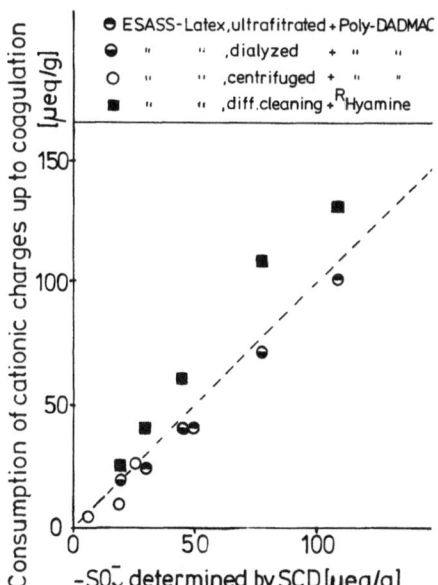

Fig. 11. Correlation between consumption of cationic charges up to coagulation of latices at various stages of cleaning by addition of Poly-DADMAC (◒◒○) or ®Hyamine (■) with SO_x^--surface charge determined by SCD titrations

of ESASS-comonomer, 9.7% solid content were found in the centrifugation serum with one-third of ESASS and two-thirds of vinyl acetate in the copolymer.

According to this proportion of SO_x^--charges being desorbed during centrifugation the latices show only a very low amount of residual charges after the threefold centrifugation and redispergation. From Table 3 and Fig. 10 we conclude that only 10% of the total ESASS-SO_3^--charges are strongly bound on the latex surface. A big proportion of the tightly bound charges comes from persulfate initiation with ammonium persulfate.

On the other hand we observed a higher amount of SO_x^--charges for dialyzed latices, which is only adsorbed due to the high surface of the dispersed system. This part will also be active if the latices are simply diluted.

These different parts of charges either bound or only adsorbed on the particle surface are responsible for the stability. In addition to the rating of stability against shearing and storage at 70°C via measurements of solids content or particle size distribution we will demonstrate in the next section that this is analogous to results of the Critical Coagulation Concentration (CCC). Plotting the CCC against the measured SO_x^--charge content, Fig. 17 shows that concerning original, dialyzed or centrifuged latices, all CCC from different stages of cleaning (shown in Fig. 16)

come together for one common dependence if we take the charges determined via SCD as the x-axis. Thus, the charge determined by the SCD method is a good criterion for latex stability against electrolytes, as well.

The principles of the SCD-method also show that SO_x^--charge content is a good parameter for the latex stability upon addition of opposite charges. During the titration with the cationic Poly-DADMAC, coagulation occurs shortly before the stoichiometric point of zero charge. In Fig. 11 the consumption of Poly-DADMAC up to the fast coagulation point correlates well with the SO_x^--charge measured by SCD, or for a cationic tenside ®Hyamine*). This material of lower molecular weight is less effective than the Poly-DADMAC, but the consumption of ®Hyamine correlates well to surface charges determined by SCD, too.

The sequence and shape of the SCD-titration will be discussed in Section 2.7 in connection with the results for Zeta Potential.

2.6 Stability to electrolyte addition

Latex stability, upon addition of electrolyte, is very important for many applications in combination with pigments, e.g., paints and paper coatings. Measurement of this kind of stability is also a very interesting feature for characterizing the electrostatic stability from a theoretical point of view.

Ottewill and Shaw [7] have developed a very good method to get a Critical Coagulation Concentration (CCC) by measuring the rates of coagulation at different levels of salt concentration. CCC characterizes the concentration of salt where coagulation changes to fast coagulation according to the theory of Smoluchowski [8]. This method was applied in some publications [9–12] in order ot characterize latex stability upon salt addition and to correlate those results with the DLVO-theory.

In this work we determined CCC especially for $CaCl_2$ since Ca^{2+}-ions are often present in industrial applications of latices. Bivalent cations like Zn^{2+} and Mg^{2+} showed similar effects as Ca^{2+}. As will be discussed at the end of this chapter, the range for the determination of CCC is more convenient for bivalent ions than for either mono- or three-valent ions.

Coagulation has been detected by means of turbidimetric measurements with a commercial photometer (Model 616 Fa. Metrohm) equipped with glassfiber optics and using slow magnetic stirring. Two different procedures have been proved:

---

*)   ®Hyamine = (N-benzyl-N,N-dimethyl-N-(4-(1,1,3,3-tetramethyl-butyl)phenoxyethoxyethyl) ammonium chloride.

a) The Rate Method

This procedure is in accordance with the method of the mentioned authors [7, 9–12]. At constant conditions of 25 °C, pH 7.0, 0.06% solids content and comparable particle size distribution coagulation was observed at different levels of $CaCl_2$ concentration as can be seen in Fig. 12. The slope of the decay for the transmission vs time (dT/dt) at the beginning is taken as the relative rate constant k to calculate the Stability Factor W in Eq. (1) by dividing the rate of fast coagulation (dT/dt)$_{fast}$ at high salt concentration through the actual rate of coagulation (dT/dt)

$$W = \frac{(dT/dt), \text{fast}}{(dT/dt)} = \frac{k, \text{fast}}{k} . \qquad (1)$$

The stability factors W calculated from the initial slopes according to Fig. 12 were extrapolated to the value $W = 1$ as presented in Fig. 13 in the form of log W against log C; C was taken as the molar concentration of $CaCl_2$ present. The intersections at the C-axis are the CCC-values (Critical Coagulation Concentration) which increase along with the increased amount of ESASS-comonomer during latex preparation. The latices shown in Fig. 13 were previously dialyzed by thorough ultrafiltration (serum-replacement with water) through a membrane that excludes molecular weights above 100000, until a conductance of 10 μS/cm was reached.

For the latex with 2% of the ESASS-comonomer, a minimum can be seen in Fig. 13, indicating a higher stability for high electrolyte concentrations. Such a behavior was sometimes found for latices with high stability against electrolyte, without reasonable explanation.

b) The Addition Method

A more convenient method to measure Critical Coagulation Concentration (CCC) was developed which uses continuous addition of $CaCl_2$-solution of suitable concentration to the latex, and proper addition rate. In Fig. 14 some of the turbidimetric titrations with the same dialyzed latices as in Fig. 13 are shown resulting in two possible methods of specification. In the upper part of Fig. 14 light transmission shows typical coagulation curves. First an aggregation is observed as a point of inflection for the electrolyte concentration at the maximum rate of coagulation. Then, transmission passes through a minimum followed by flocculation of large particles. The curves in the lower part of Fig. 14 are the derivative forms obtained by measuring with a potentiograph model E 536 of Metrohm with the dT/dt function. By this method the

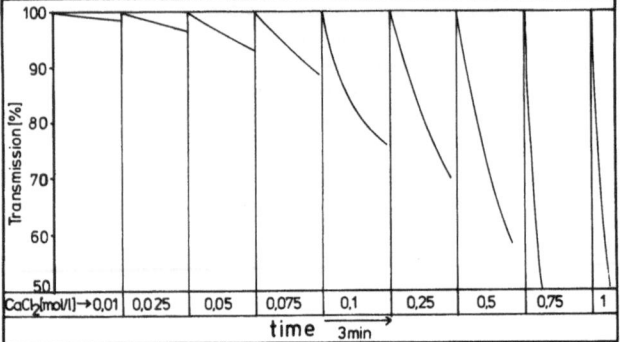

Fig. 12. Determination of initial coagulation rate by turbidimetry at various levels of $CaCl_2$ concentration

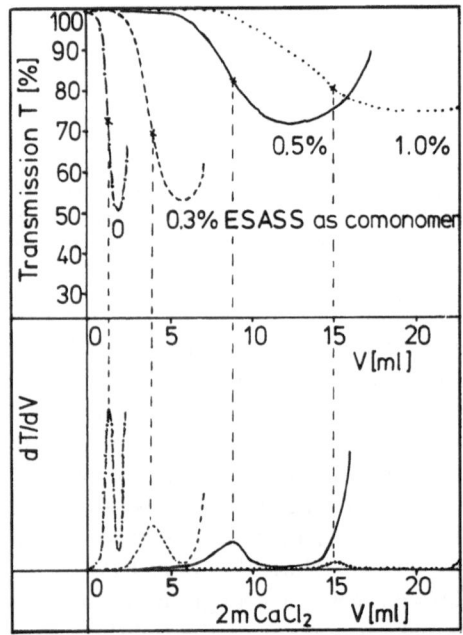

Fig. 13. Determination of Critical Coagulation Concentration (CCC) by plotting stability factors W versus concentration of the $CaCl_2$ solution for dialyzed poly(vinyl acetate) latices (series A with 0–2% ESASS-comonomer)

first maximum can be used easily in order to calculate CCC.

Figure 15 shows a good correlation of both Rate Method and Addition Method. Within a certain range both methods can be used to measure CCC, but the Addition Method is the most convenient if CCC is not too high and the concentration of the salt solution can be taken in such a range that the addition does not change the latex concentration for more than 20%.

The results for both series of latices are summarized in Table 4 and plotted in Fig. 16. For series A, Critical Coagulation Concentration (CCC) increases with

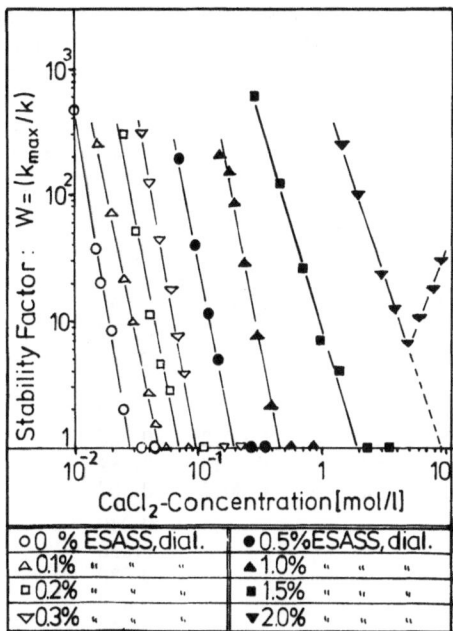

Fig. 14. Determination of Critical Coagulation Concentration (CCC) during continuous addition of 2 m $CaCl_2$ solution for poly(vinyl acetate) latices (series A with 0, 0.3, 0.5 or 1% ESASS-comonomer)

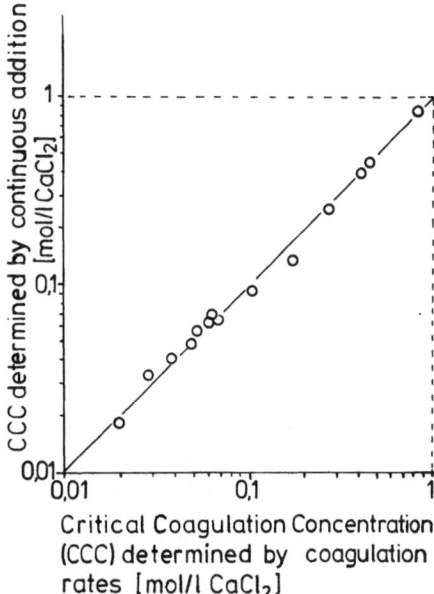

Fig. 15. Comparison between Critical Coagulation Concentration (CCC) from rate method and from continuous addition method

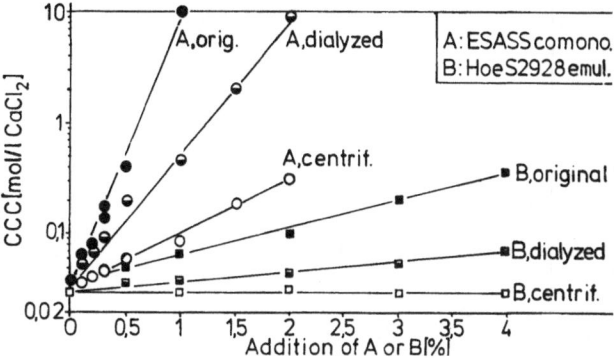

Fig. 16. Critical Coagulation Concentration (CCC) for poly (vinyl acetate) latices (series A with ESASS-comonomer; series B with Hoe S 2928 emulsifier) in $CaCl_2$ solution at various levels of cleaning

higher ESASS-comonomer content for more than two orders of magnitude for the original, and also at a lower level for the dialyzed latices. The same latices threefold centrifuged and redispersed in water show a much lower stability against $CaCl_2$. The tenfold higher stability with 2% of the ESASS-comonomer is much more effective than the addition of 4% of the emulsifier Hoe S 2928 in the original latices. After dialysis, and especially after centrifugation and redispergation, there is no further stabilizing activity of the previously added emulsifier. It applies to both series A and B as well that dialysis and centrifugation lower the stability, a fact which can be explained by diminuation of the charges adsorbed on the surface as determined in section 2.5 by SCD.

As has been shortly mentioned in section 2.5, the data for series A in Fig. 16 coincide to one common dependence in Fig. 17 if we do not take the amount of added ESASS-comonomer, but rather the effective SO_x^--charge content, determined by SCD as the x-axis. The good correlation between electrolyte stability determined by CCC, surface charge, stability against shearing, stability upon storage at 70 °C, and addition of cationic additives, show the wanted good correlation between practical and colloid chemistry-based methods. The ambiguity of Zeta Potential will be discussed in the next section.

In Table 4 some results are reported using KCl and $AlCl_3$ or $Al_2(SO_4)_3$, respectively, in order to prove the Schulze-Hardy-Rule, which approximates the effect of sensitivity from monovalent to divalent and trivalent ions $1 : 80 : 640$. More recently refinements [11] have altered this slightly to the 6th power rule (2)

cation activity for coagulation:
$$K^{1+} : Ca^{2+} : Al^{3+} = 1^6 : 2^6 : 3^6 = 1 : 64 : 729 . \qquad (2)$$

The Critical Coagulation Concentrations (CCC) of dialyzed latices in Table 4 for use of KCl are only twice as high as for $CaCl_2$ on the base of equivalents. For

Table 4. Characterization of latex stability against salt addition by Critical Coagulation Concentration (CCC) for Poly (vinyl acetate) latices at various stages of cleaning (RM = Rate Method, AM = Addition Method, n.m. not measurable)

Series	Latex No.	Additive [%]		Critical Coagulation Concentration (CCC) [mol/l]							Centrif. latices
		ESASS	Hoe S2928	Original latices		Dialyzed latices					
				CaCl$_2$		CaCl$_2$		KCl	Al$_2$(SO$_4$)$_3$	AlCl$_3$	CaCl$_2$
				RM	AM	RM	AM	RM	AM	AM	AM
A	40	0	0	0.037	0.040	0.028	0.033	0.147	0.013	0.001	0.030
	41	0.1	0	0.068	0.065	0.048	0.056	0.264	0.020	0.00105	0.037
	42	0.2	0	0.080	0.080	0.071	0.068	0.306	0.028	0.0013	0.040
	43	0.3	0	0.175	0.135	0.096	0.088	0.371	0.043	0.0011	0.047
	44	0.5	0	0.40	0.40	0.20	0.198	0.736	0.10	0.0011	0.060
	45	1.0	0	10	n.m.	0.45	0.45	1.85	0.21	0.0012	0.086
	46	1.5	0	n.m.	n.m.	n.m.	2.0	6.4	0.95	0.0013	0.184
	47	2	0	n.m.	n.m.	n.m.	9	n.m.	n.m.	0.0012	0.307
B	40	0	0	0.037	0.040	0.028	0.033	–	–	–	0.030
	40/0.5	0	0.5	0.050	0.050	0.039	0.039	–	–	–	0.030
	40/1	0	1	0.063	0.068	0.042	0.040	–	–	–	0.030
	40/2	0	2	0.105	0.093	0.047	0.042	–	–	–	0.032
	40/3	0	3	0.22	0.192	0.050	0.055	–	–	–	0.029
	40/4	0	4	0.35	0.35	0.065	0.072	–	–	–	0.030

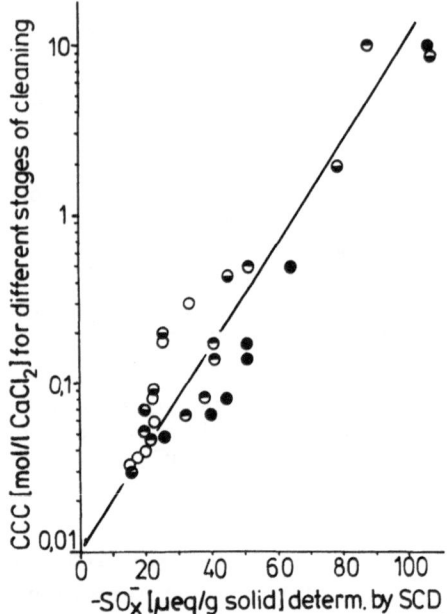

Fig. 17. Dependence of Critical Coagulation Concentration (CCC) with CaCl$_2$ on effective SO$_x^-$ surface charge determined by SCD-Titration for poly(vinyl acetate) latices (series A with various ESASS-comonomer content) at various levels of cleaning: (●) original; (◐) ultrafiltrated surface; (◑) dialyzed surface; (○) threefold centrifuged surface

Al$_2$(SO$_4$)$_3$, the coagulation activity is nearly the same as for CaCl$_2$ on the base of equivalents, but for AlCl$_3$ the activity is much higher and does not depend on the ESASS-comonomer. The two Al^{3+}-salts gave a shift of pH to the acidic range. The deviations from the

Schulze-Hardy-Rule which we observed should be investigated in the future, but it is of no consequence in the rating of stability against electrolytes such as CaCl$_2$, which could be enhanced progressively by the ESASS-comonomer or the Hoe S2928 emulsifier in Series A or B.

2.7 Zeta-Potential measurements

Zeta Potential is often assumed to be a criterion for colloid stability from the theoretical point of view [13]. It was the aim of this work to prove its relevance for the latices investigated that showed very different stabilities against shearing, storage, salt addition or additives with opposite charge in the preceding sections 2.3 – 2.6.

Two different methods have been applied:

a) Mass-Transport Electrophoresis (MTE)

The method for measuring Zeta Potential at practical solid content of 50% for latices has been reported in a preceeding paper [3].

In a special chamber, Electrophoretic Mobility (EM) is measured by the Electrophoretic Mass Transport into a plug formed near the anode. Electrophoretic Mobility is simply transformed with the Smoluchowski Eq. (3) in order to get the Zeta Potential in units of (mV):

Zeta Potential (mV)

$$= \frac{4\pi\eta}{\varepsilon}\,\text{EM} \approx 13\,000\,\text{EM}\ ((\text{cm/s})/(\text{V/cm})) \qquad (3)$$

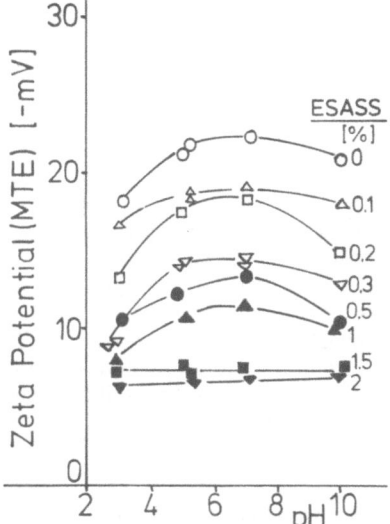

Fig. 18. Dependence of Zeta Potential determined by Mass Transport Electrophoresis at 51% solids content on pH for poly(vinyl acetate) latices series A with various ESASS-comonomer content

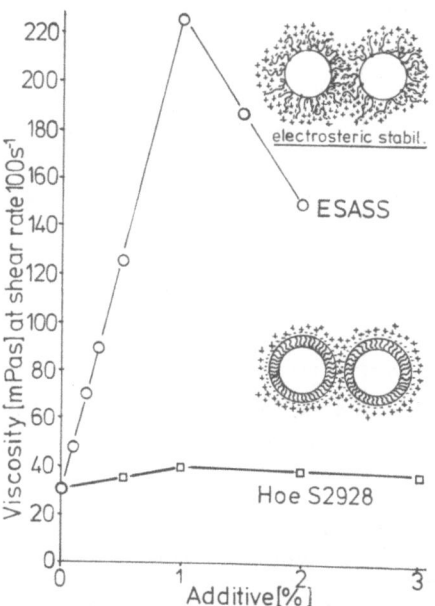

Fig. 19. Latex viscosity for poly(vinyl acetate) latices at 51% solids content at various levels of ESASS-comonomer (series A) or emulsifier Hoe S2928 (series B) addition, measured at shear rate $100 \, sec^{-1}$

The validity of Eq. (3) for latices with 50% solids is not clear, but the actual measured EM is only transformed by the factor 13000, to the more convenient Zeta Potential.

Figure 18 shows that the Zeta Potential of latices from series A depends on the pH values. Unexpected latices with no or a low content of the ESASS-comonomer and a low stability, discussed in preceeding sections, show the highest values for Zeta Potential. On the contrary, most stable latices with a higher content of the ESASS-comonomer show a low value of Zeta Potential without any dependence on pH. The low dependence on pH was expected since particle charge is only formed by sulfate or sulfonate groups which are disassociated at any pH investigated. According to [3], the generally low values of Zeta Potential at high solid content were also expected.

The reason for the reciprocal dependence of Zeta Potential from stability, measured by the other methods, is not the difference in conductivity, which changes from 4.6 to 7.1 mS/cm regarding those latices and the pH. Table 5 shows no difference between Zeta Potential in the original state at pH 5.0 with different conductivity, and for Zeta Potential at the same pH and the same level of conductivity (7 mS/cm).

The unexpected rating of Zeta Potential is supposed to be caused by the different thickness of the adsorbed ionic layer on the particle surface. The latices of series A in Fig. 19 first show a strong rise in viscosity with increasing ESASS-comonomer content, and then a

depletion. Viscosity was measured in a rotation viscosimeter Fa. Haake model RV100 at a shear rate of $100 \, sec^{-1}$. At lower shear rate there is a more marked maximum while it is reduced to one-half at a shear rate of $400 \, sec^{-1}$. It is assumed that the adsorbed layer of the comonomer of vinyl acetate and ESASS leads to an increase of hydrodynamic volume of the particles and thus to a depletion of Electrophoretic Mobility EM and Zeta Potential. This gel-like envelope of chemically bound or adsorbed polyelectrolytes forms a corona of hairs as discussed by Zimehl and Lagaly [12]. When such anionic macromolecules radiate away from the surface, the electrostatic stabilization is supported by steric stabilization, called electrosteric stabilization [14]. By this model the high value of CCC can be understood, as well as the reciprocal correlation for Zeta Potential, which usually is only discussed for electrostatic stabilization.

For series B no strong viscosity effects can be seen in Fig. 19, but Zeta Potential decreases with the rising content of the emulsifier Hoe S2928 (Table 5). It might be that this emulsifier is active with respect to hydrodynamic volume for Electrophoretic Mobility measurements at low shear rate, but not for viscosity measurements at high shear rate. On the other hand, this anionic emulsifier with poly(ethylene oxide) chains and roughly 23 EO-units, might also be active

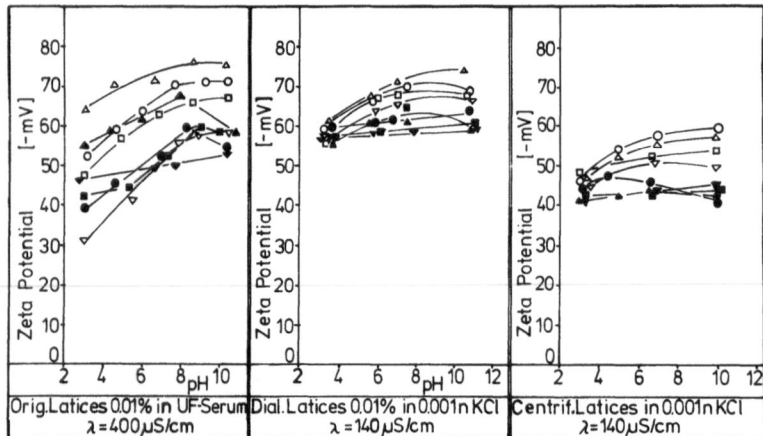

Fig. 20. Dependence of Zeta Potential on pH of poly(vinyl acetate) latices with 0.0% (○); 0.1% (△); 0.2% (□); 0.3% (▽); 0.5% (●); 1% (▲); 1.5% (■); 2% (▼) ESASS-comonomer determined by microelectrophoresis at different levels of latex cleaning (left: original latices in own ultrafiltrate serum; middle: dialyzed latices in 0.001 n KCl solution; right: centrifuged latices in 0.001 n KCl solution with the same conductivity 140 µS/cm)

Table 5. Results of Zeta Potential measurement determined by Mass Transport Electrophoresis (MTE) for 51% solids content or by Micro electrophoresis for 0.01% solids content

Series	Latex No.	Additive [%] ESASS	Hoe S2928	Original latices (51% solids) Visc. at 100^{-1} shear rate [m Pas]	pH	Cond. λ [mS/cm]	Zeta Potential by MTA [−mV] Orig. λ=7 pH 5	Orig. pH 5	pH 7	Diluted latices (0.01% solids) Zeta Potential by Microelectrophoresis [−mV] Original latices in their own UF serum pH 3.5	pH 7	pH 10.5	Dialyzed latices in 0.001 n KCl pH 3.5	pH 7	pH 10.5	CaCl$_2$ pH 7
A	40	0	0	30	5.2	4.6	22.0	21.9	22.5	53	68	71	60	70	68	35
	41	0.1	0	51	5.1	4.6	18.6	18.1	19.2	65	74	76	61	71	73	36
	42	0.2	0	69	5.2	4.6	17.5	16.9	18.3	49	63	67	56	68	68	34
	43	0.3	0	89	5.2	5.0	14.3	15.0	14.4	33	51	58	57	65	66	32
	44	0.5	0	119	5.3	5.2	12.4	13.0	13.2	41	54	55	60	62	63	29
	45	1	0	215	5.4	6.2	10.9	10.8	11.5	56	65	59	55	63	59	27
	46	1.5	0	186	5.4	6.7	7.5	7.5	7.5	42	52	59	57	59	61	27
	47	2	0	148	5.4	7.1	6.6	6.6	7	47	50	53	58	59	59	24
B	40	0	0	30	5.2	4.6	22.7	21.9	−	53	68	71	60	70	68	35
	40/0.5	0	0.5	35	5.0	4.7	18.5	16.5	−	42	51	55	51	58	63	34
	40/1	0	1	40	5.0	4.8	14.8	14.1	−	−	−	−	−	−	−	−
	40/2	0	2	39	5.0	4.9	11.4	10.6	−	39	48	52	45	57	57	34
	40/3	0	3	38	5.0	4.9	9.3	9.0	−	−	−	−	−	−	−	−
	40/4	0	4	38	5.1	5.0	8.7	6.5	−	51	59	61	50	64	61	35

by electrosteric stabilization with a lower steric part than the ESASS-copolymers.

b) Microelectrophoresis

Using a commercial analyzer ZETASIZER IIc (by Malvern Instr. Ltd.), Electrophoretic Mobility and Zeta Potential have been additionally measured upon high dilution. The Electrophoretic Mobility Distribution was almost monomodal. In order to compare latices we measured average values of Zeta Potential for different measuring condition (see Table 5 and Fig. 20):

1. Original latices were measured in their own ultrafiltration serum at a constant level of conductivity (400 µS/cm), (see Fig. 20, left part).
2. Dialyzed latices were measured in 0.001 n KCl solution with HCl or NaOH at the same level of conductivity (140 µS/cm), (see Fig. 20, middle part).
3. Centrifuged latices were measured in 0.001 n KCl solution with HCl or NaOH at the same level of conductivity (140 µS/cm), (see Fig. 20, right part).

With the exception of original and dialyzed latices with 0.1% ESASS-comonomer, the arrangement of

Zeta Potential is approximately the same as for Mass Transport Electrophoresis shown in Fig. 18. The higher values for Zeta Potential in the diluted systems in Fig. 20 were expected from the previous work, as were the lower viscosity of the diluted systems. The same applies to the very low dependence on pH, because the anionic activity of the SO_x^--charges does not really depend on pH.

The lower range for centrifuged latices, and the more narrow range for both dialyzed and centrifuged latices compared with the original latices in their own ultrafiltration sera, correlate with the amounts of surface bound or adsorbed charges determined via SCD in section 2.5. The arrangement with the lower Zeta Potential for latices with higher stability at higher ESASS-content level should have the same basis already discussed for Mass Transport Electrophoresis: the gel-like hairy corona of the ESASS-vinyl acetate copolymer surrounding latex surface. This influence of hydrodynamic volume on Electrophoretic Mobility and Zeta Potential exceeds the influence of the amount of charges. In systems like ours, simple Zeta Potential measurements do not correlate in any case with latex stability determined by practical and other colloid chemistry-based methods, such as measurement of charges via SCD or CCC because of electrosteric stabilization existing in our latex systems.

In the last column of Table 5, Zeta Potential is measured in 0.001 n $CaCl_2$ solution in order to look for a special influence of $CaCl_2$. No significant difference is found for the two series investigated at a lower level of Zeta Potential, compared with results in 0.001 KCl.

2.8 Correlation between Zeta Potential and other stability criteria

In section 2.7, Zeta Potential determined by direct measurements did not correlate with the results of the other methods in sections 2.3−2.6. It does not appear under the conditions of a titration with a cationic polymer like Poly-DADMAC. In Fig. 21 four methods are compared for dialyzed latices with 0,1 or 2% ESASS-comonomer, respectively, during titration with Poly-DADMAC.

Zeta Potential of these three latices initiates at nearly the same level of −60 mV, but the change of Zeta Potential during titration depends strongly on the ESASS-comonomer content. Zeta Potential passes zero at roughly the same consumption of Poly-DAD-MAC, whereas the Streaming Current Detector (SCD)-Potential passes zero as well as turbidity, and particle size d_z rise strongly, thus indicating coagula-

tion at the same point of titration. The isoelectric point of Zeta Potential and the point of zero charge for SCD-titration correlate well with coagulation, indicated by the rise of turbidity and particle size. Beyond the isoelectric point or point of zero charge a certain region of cationic charged latices can be measured by Zeta Potential or SCD-Potential but this unstable range depends on shearing and rate of titration on measurement.

Other cationic additives like ®Hyamine give similar correlations of Zeta Potential with the other stability characterizing methods, if the amounts of destabilizing additives remove the ionic charges toward the isoelectric point or point of zero charge. This is also true for latices with cationic and anionic groups when simply changing the pH; for this we found good correlations between the data of Zeta Potential and the content of charges determined by SCD, and stability vs electrolyte determined by CCC. The measurements will be published in a further paper.

3. Conclusion

For poly(vinyl acetate) latices, the stability during shearing, storage at 70°C, and electrolyte addition, can be improved progressively by copolymerization with ethylene sulfonic acid sodium salt (ESASS), or by addition of a nonylphenol poly-(ethylene oxide) sulfate ammonium salt as emulsifier. The copolymerization with ESASS has been found to be more effective, especially after cleaning of the latices by dialysis or centrifugation.

The different stages of stability during shear and storage at 70°C can be measured by residue on a sieve, or by the solids content after sieving. On the other hand, measurements of particle size and particle size distribution by Aerosol Spectroscopy or Photon Correlation Spectroscopy give a more accurate evaluation.

The determination of stability to electrolyte by turbidity titration with $CaCl_2$ and the evaluation by CCC correlate well with stability during shear and storage at 70°C, as well as with the distribution of charges via Streaming Current Detector (SCD) measurements. Critical Coagulation Concentration is confirmed as a good criterion for latex stability because it correlates well with effective surface charge determined at different levels of cleaning by SCD-polyelectrolyte titration.

Zeta Potential does not correlate with the other criteria of latex stability in our latex system if Zeta

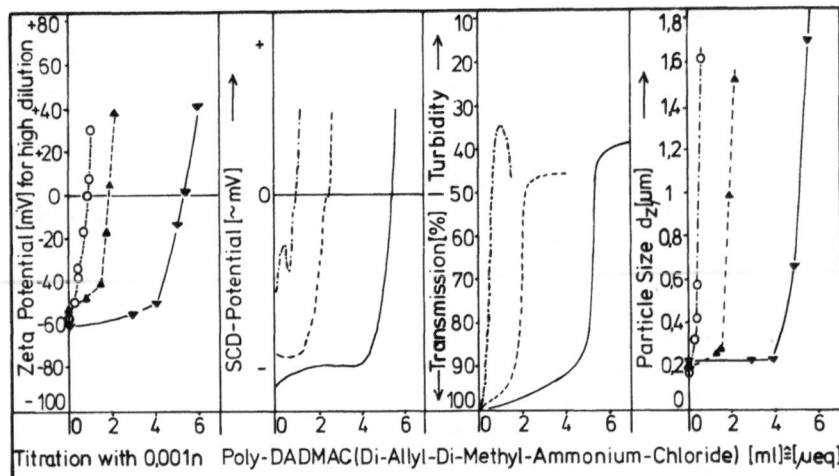

Fig. 21. Correlation of results from four methods: Zeta Potential, Streaming Current Detector (SCD) Potential, turbidity (transmission), and particle size *dz* at different stages of titration with Poly-DADMAC for dialyzed poly(vinyl acetate) latices (series A) with 0 (○); 1% (▲) or 2% (▼) ESASS-comonomer

Potential is simply measured by Mass Transport Electrophoresis or Microelectrophoresis in its own serum or at different pH in 0.001 n KCl solution. A small amount of sulfate or sulfonate groups on the latex particle surface leads to a high Zeta Potential and a small change at different pH. The depletion of Zeta Potential with increasing ESASS or emulsifier content can be explained by a steric part of stabilization, e.g., a gel-like hairy corona of the ESASS-copolymers on the latex surface.

Zeta Potential can be correlated well with the other latex stability criteria if the measuring conditions are changed, e.g., by addition of additives with cationic charges. By proving the capacity of stabilizing charges latices of various stability show Zeta Potential curves which give a good parameter for latex stability, too. On the other hand SCD-Potential shows similar curves and can be measured much more simply than Zeta Potential.

In order to evaluate latex stability from a practical point of view some of the methods investigated have to be used simultaneously since characterization of colloid chemical status seems like a puzzle. For optimizing stability of latices and their application in combination with pigments, characterization of colloid status needs to be as comprehensive and detailed as possible. We hope this paper contributes some clarification about the role of Zeta Potential and we hope it will initiate further work on stability criteria for latices.

References

1. Roth C, Gebhart J, Heigwer G (1976) J Coll Interf Sci 54:265
2. Löhr G, Reinecke R (1980) Angew Makromol Chem 85:181
3. Fischer JP, Löhr G (1986) Org Coat Sci & Technol 8:227, Parfitt GD, Patsis AV (eds). Marcel Dekker Inc, New York
4. Schempp H, Tran HT (1981) Wochenbl Papierfabr 109 (19):726
5. Schempp H, Heß P, Krause Th (1982) Das Papier 36 (10A):V 41
6. Address for commercial SCD: Mütek GmbH, Arzberger Str. 10, D-8036 Herrsching
7. Ottewill RH, Shaw JN (1966) Disc Far Soc 42:154
8. Smoluchowski MV (1917) Z Phys Chem 92:129
9. Cooper WD (1972) Koll Z & Z Polym 250:38
10. Bongards M, Langer G, Werner U (1980) Progr Coll & Polym Sci 67:91
11. Overbeek JThG (1980) Pure & Appl Chem 52:1151
12. Zimehl R, Lagaly G (1986) Progr Coll & Polym Sci 72:28
13. Hunter RJ (1981) Zeta Potential in Colloid Science. Academic Press, London, etc.
14. Napper DH (1983) Polymeric Stabilization of Colloidal Dispersions. Academic Press, London, etc.

Received October 14, 1987;
accepted June 3, 1988

Author's address:

J. P. Fischer
Forschung und Entwicklung GB G/H Polymerphysik
Gebäude G 832
Hoechst AG
D-6230 Frankfurt/Main 80, F. R. G.

Progress in Colloid & Polymer Science　　　　　　　Progr Colloid & Polymer Sci 77:195–200 (1988)

Kinetic analysis of the influence of additives on spontaneous precipitation from electrolytic solutions

H. Füredi-Milhofer, D. Škrtić, M. Marković, and Lj. Komunjer

"Rudjer Bošković" Institute, Zagreb, Croatia, Yugoslavia

Abstract: Kinetic data describing the precipitation of calcium oxalates (calcium oxalate trihydrate (COT) and mixtures of COT and the monohydrate (COM)) in the presence and absence of sodium and potassium chloride, phosphate ions, and/or glutamic acid (Glu) have been employed to obtain information on the rate controlling precipitation processes, e.g., nucleation, crystal growth and aggregation. Precipitation was induced from unseeded high ionic strength solutions at pH 6.0 or 6.5 and the kinetics followed by Coulter counter or calcium selectrode. In selected samples the composition of the precipitates was determined by thermogravimetric analysis and x-ray diffraction. It was found that phosphate ions retard crystal growth but do not influence nucleation or aggregation of the crystals. The neutral electrolytes induced aggregation at concentrations comparable to their critical coagulation concentration. Glu affected the composition of the precipitate by inducing the formation of COM. Its effect on the intensity of precipitation was concentration dependent. At low concentrations ($c(Glu) = 3 \times 10^{-6} - 5 \times 10^{-5}$ mol dm^{-3}) reduction of the rate of crystal growth and simultaneous enhancement of aggregation was observed, while at higher concentrations ($c(Glu) \geq 10^{-4}$ mol dm^{-3}), precipitation of calcium oxalate was enhanced. The effects of the aminoacid were attributed to its interaction at the stage of heterogeneous nucleation. A comprehensive discussion on the effect of additives on precipitation processes is included.

Key words: Calcium oxalate, additives, nucleation, precipitation, crystal growth, aggregation

1. Introduction

The interaction of additives with precipitating solid phases is utilized in many different fields of human endeavour. Thus additives have been used as crystal growth inhibitors in scale inhibition [1, 2], crystallization modifiers in industrial crystallization processes [3, 4], coagulants in waste disposal [5], etc. Also, their possible role in controlling biological [6] and in inhibiting pathological mineralization [7] has prompted extensive research on relevant model systems. It appears, however, that the great influence which even trace amounts of foreign ions, small, and macromolecules exert on the properties of precipitates is still poorly understood. It has been shown [8–10] that the type of the effect depends not only on the particular additive and solid phase but on the experimental conditions (solution composition, temperature, etc.) as well.

In this paper some of our results showing the influence of neutral electrolyte (NaCl), phosphate ions (PO$_4$), and glutamic acid on the nucleation, crystal growth and aggregation of calcium oxalates are represented and discussed. Interpretation is facilitated by recently developed kinetic models [11–13] which allow one to differentiate between successive or simultaneously occurring precipitation processes. Although the model systems used are relevant to urolithiasis research, the results are also of general importance because they clearly demonstrate the manifold role that additives may assume in modifying the rates and mechanisms of processes that control solid phase formation.

2. Experimental

2.1 Materials and methods

Stock solutions were prepared by dissolving analytical grade chemicals (e.g., calcium chloride, phosphoric acid, oxalic acid, glutamic acid, sodium chloride, sodium hydroxide) in triply distilled water and then standardized by conventional chemical analysis. The sodium hydroxide solution was kept in a nitrogen atmosphere. Samples were prepared by mixing known concentrations of solutions of sodium oxalate (pH = 5.0 or 6.5, adjusted with NaOH) and calcium chloride, both made up with sodium chloride to constant ionic strength, as indicated. The additives (PO_4-ions, Glu) were added to the oxalate solution prior to pH adjustment. Solutions used for particle size analysis were purified by filtering through 0.22 μm Millipore filters. All experiments were conducted at (298.0 ± 0.1) K and the suspensions were stirred during the kinetic run. Three different modes of stirring (as indicated) were employed: mechanical (with stirring paddle), magnetic and "mixed" (e.g., magnetic until the first appearance of the precipitate, thereafter, mechanical). Precipitation kinetics was followed by calcium selectrode or Coulter counter; solid phase analysis was performed by TGA and x-ray diffraction.

2.2 Treatment of data

The kinetic data have been interpreted as previously [11–13] in terms of the equations

$$(da/dt)a^{-2/3} = KN_t^{1/3}S^q \qquad (1)$$

and (for dominant crystal growth with $N_t \sim$ const.)

$$(da/dt)a^{-2/3} = K_a(1-a)^p \qquad (2)$$

where N_t is the total number of particles, S is the supersaturation (expressed in Eq. (2) as $1-a$), K and K_a are rate constants, q and p give the order of the reaction. The degree of the reaction a is defined as

$$a = (c_0-c_t)/(c_0-c_s) = V_t/V_{max} . \qquad (3)$$

In Eq. (3) c_0, c_t, and c_s are the calcium concentrations at time 0, t and at equilibrium respectively, while V_t and V_{max} represent the precipitate volume at time t and at equilibrium. The quantities c_0 and c_t were obtained by calcium selectrode; V_t was calculated from the Coulter counter output data [15], c_s and V_{max} were calculated from equilibrium constants as previously [9, 11, 14] defined. Log-log plots of Eq. (1) have been used to distinguish between time periods during which nucleation accompanied by crystal growth (A), crystal growth (B) and growth inhibited by aggregation (C) are rate determining ([11–13] and Figs. 1, 2, 5). If section B can be approximated to linearity, the slope of the corresponding straight line gives the exponent p in Eq. (2) which is used to estimate the intensity of crystal growth inhibition (provided the initial course of the rate vs supersaturation curves is identical, $p_{in} > p_0 > p_{pr}$, where the subscripts designate inhibition, control, and promotion, respectively [13]).

3. Results

In the following experiments precipitation was initiated from unseeded solutions supersaturated in calcium oxalate with initial supersaturations ranging from $S_0 = IP/K_{sp} = 2.7–6.9$ (where $IP = a(Ca) \times a(C_2O_4)$ is the ion activity product and K_{sp} is the thermodynamic solubility product of the precipitating

Table 1. Initial calcium and oxalate concentrations (c_0), ionic strength (I; made up with NaCl), stirring mode and type of additive in the experiments described. Temperature 298 K, pH 5.0 (expt 1) or 6.5 (expts 2–4) solid phase in absence of additive: COT (expts 1–3) or COM+COT (expt 4)

Expt No	c_0(Ca)$\times 10^4$ (mol dm^{-3})	c_0(C$_2$O$_4$)$\times 10^4$ (mol dm^{-3})	I·10 (mol ×dm^{-3})	Additive	Stirring mode
1	260	1.5	2.6	PO_4	mixed
2a	3.0	3.0	0	–	magn
2b	10	10	3.0	NaCl	magn
3	7.0	7.0	3.0	Glu	mixed
4	8.0	8.0	3.0	Glu	mech.

phase). The conditions at which the experiments were conducted are listed in Table 1. The mode of stirring largely determined the nature of the precipitating solid phase in the control systems [17]. In experiments 1–3 the triclinic calcium oxalate trihydrate (COT) precipitated, while in experiment 4 mixtures of the monoclinic calcium oxalate monohydrate (COM) and COT initially appeared. TGA showed that in the latter experiment COM was the dominant growing phase [9].

The influence of a large excess of *phosphate ions* (concentration as in urine, e.g., $(c_0(PO_4) = 2.6 \times 10^{-2}$ mol dm$^{-3})$ on the precipitation of COT is shown in Fig. 1, (experiment 1 in Table 1). In this experiment the starting solution was supersaturated with respect to two solid phases (COT and calcium hydrogenphosphate dihydrate (DCPD)) but since the difference between the induction periods, t_i, preceeding the appearance of the respective solid phase was large enough (e.g., t_i(COT) $\approx$ 10–12 min and t_i(DCPD) $\approx$ 105 min) the kinetics of the precipitation of COT was followed at t_i(COT) $< t < t_i$(DCPD) [16]. Comparison of the V_t vs time curves in Fig. 1a shows that in the presence of PO_4, precipitation of COT commenced later and was less abundant than in the controls. The corresponding rate vs supersaturation curves (Fig. 1b) have similar courses but the slopes of sections B differ significantly (e.g., $p_0 = 3.8$, $p_{(PO_4)} = 4.8$). Applying the criterion $p_{in} > p_0 > p_{pr}$ leads to the conclusion that it is the crystal growth process which is inhibited in the presence of phosphate ions. It is furthermore apparent from Fig. 1b that, both in the presence and absence of PO_4, aggregation starts inhibiting crystal growth at the same critical rate (e.g., $R_{aggr}(1) \approx R_{aggr}(2)$). Further analysis of the kinetic data has shown [16] that there is no significant dif-

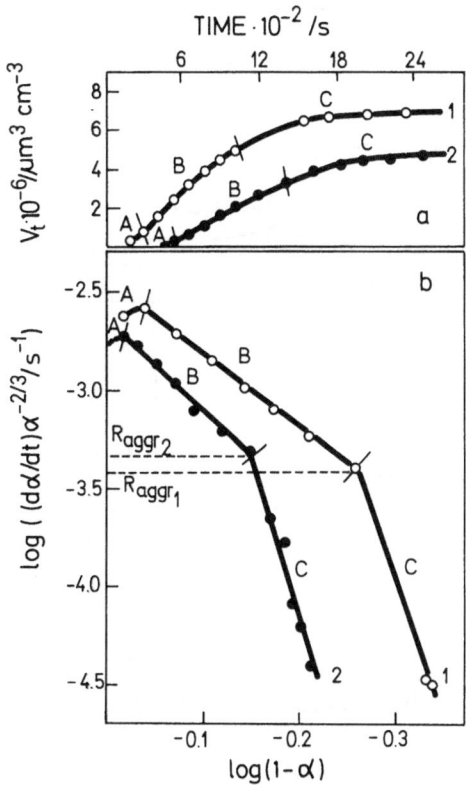

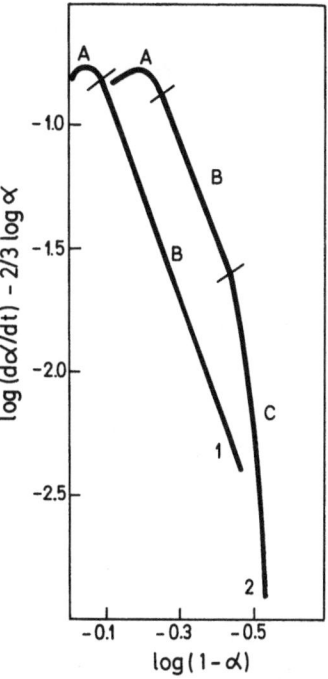

Fig. 1. V_t vs time (a) and rate vs supersaturation (b) curves, (Coulter counter) showing the influence of phosphate ions on the precipitation of COT. Curve 1, without PO_4; curve 2, $c_0(PO_4) = 2.6 \times 10^{-2}$ mol dm^{-3}. Experimental conditions as listed in Table 1 (experiment 1). Sections A (nucleation and growth), B (dominant crystal growth) and C (growth inhibited by aggregation) correspond in time. Slopes p of sections B in (Fig. 1b): $p(1) = 3.8$, $p(2) = 4.8$. R_{aggr} is the rate at which aggregation starts influencing the rate of crystal growth. After [16]

Fig. 2. Rate vs supersaturation curves (Ca selectrode) showing the influence of the concentration of NaCl on the precipitation kinetics of COT. Experimental conditions as in Table 1, experiments 2a (curve 1) and 2b (curve 2). Initial supersaturation $S_0 = 3.92$. Slopes of parts B $p = 3.16$. After [11]

ference in the intensity of growth inhibition (as a result of aggregation) between the PO_4-containing systems and the controls. Also PO_4 had no influence on the type of the solid phase precipitated [17].

The rate vs supersaturation curves shown in Fig. 2 are representative of a series of kinetic precipitation experiments performed at different concentrations of *sodium* or *potassium chloride* [11]. Curves obtained at $c(NaCl) \leq 0.1$ mol dm^{-3} and $c(KCl) \leq 0.07$ mol dm^{-3} were similar to curve 1 in Fig. 2, showing only sections indicating nucleation and concurrent growth (A) and dominant crystal growth (B). Aggregation significantly inhibited crystal growth only at $c(NaCl) \geq 0.2$ mol dm^{-3} and $c(KCl) \geq 0.15$ mol dm^{-3}, resulting in the appearance of sections C in the rate vs supersaturation curves (curve 2 in Fig. 2). No effect on nucleation or on the rate of crystal growth was detected in this series of experiments.

The influence of *glutamic acid* has been demonstrated on two model systems (experiments 3 and 4 in Table 1). In both experiments the total volume, V_t, of the precipitate changed in an irregular way with Glu concentration (Fig. 3), exhibiting a minimum at low concentrations of the aminoacid which was more pronounced in experiment 4 in which the the growing solid phase was COM [9]. At least in one case (experiment 3), enhanced precipitation was observed at higher concentrations of Glu. Also, in both experiments Glu affected the type of the solid phase formed by promoting precipitation of COM, as was ascertained by TGA and x-ray diffraction (Fig. 4). Solid phase analysis performed at different time intervals indicated that this effect may be extrapolated to the earliest stages of precipitation, most probably to the period of heterogeneous nucleation [9, 17]. Characteristic rate vs supersaturation curves showing the influence of Glu on crystal growth and aggregation of COM (experiment 4) are represented in Fig. 5. While section B of curve 2 ($c(Glu) = 1.0 \times 10^{-5}$ mol dm^{-3}) can be approximated to linearity with a relatively high slope ($p = 7.1$ as compared to $p = 3.6-4$ for the controls), curve 3 ($c(Glu) = 2.4 \times 10^{-5}$ mol dm^{-3}, corresponding to the minimum on the V_t vs time curve

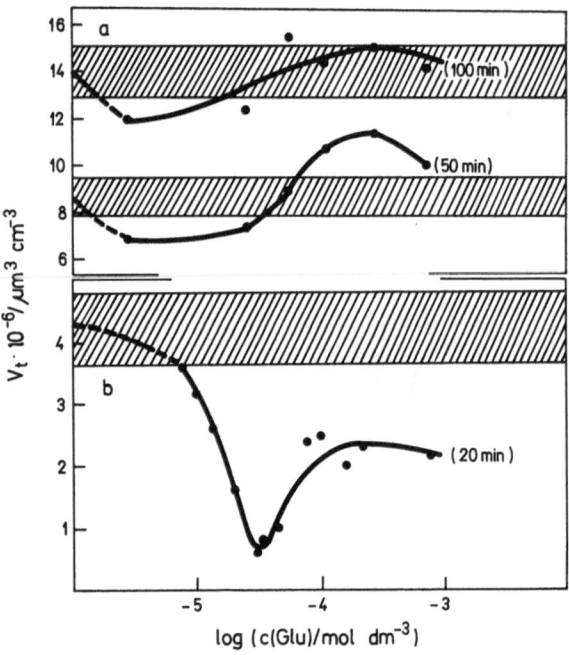

Fig. 3. The influence of the concentration of Glu on the volume of precipitated calcium oxalate (COT (a) and COM+COT (b)) as measured by Coulter counter at the indicated time intervals. Experimental conditions as in Table 1, experiments 3 (a) and 4 (b). Shaded areas correspond to the respective control systems without Glu

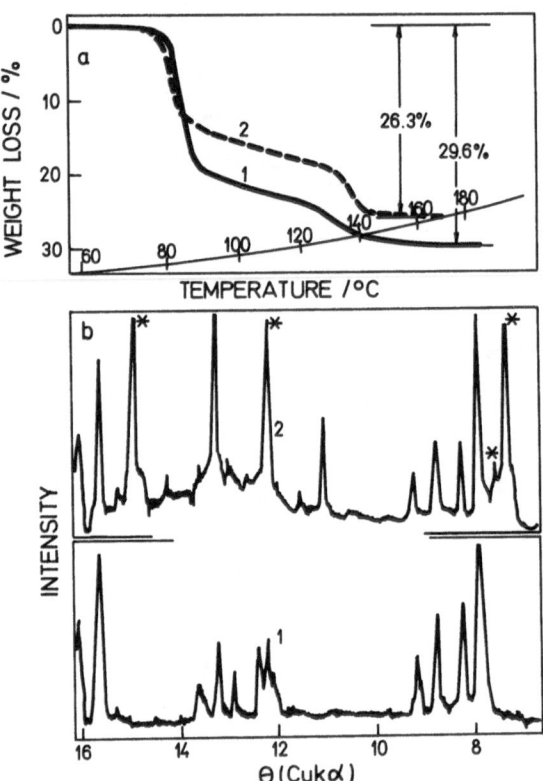

Fig. 4. TGA (a) and x-ray diffraction (b) of calcium oxalate precipitated in the absence (1) and presence (2) of 3.4×10^{-5} mol dm^{-3} Glu. Experimental conditions as in Table 1, experiment 3; precipitates isolated 120 min after sample preparation. Curve 1 corresponds to COT; curve 2 indicates mixtures of COM+COT (stars indicate x-ray diffraction lines characteristic of COM)

shown in Fig. 3b) shows no linear part at all. This result suggests a continuously decreasing number of particles (N_t in Eq. (1)) due to substantial aggregation.

4. Discussion

In the range of supersaturations encountered in the above experiments precipitation is normally initiated by heterogeneous nucleation upon nonspecific impurities or upon seed crystals if such have been added. The number of nuclei formed is determined by the number of effective heteronuclei and amounts to $N \approx 10^5 - 10^7$ per cm^3. If the formation of several phases is possible the less stable phases with higher solubility form first, transforming at a later stage into the thermodynamically most stable precipitate. Many crystallization events encountered in nature or in technology belong to this category.

The following development of the precipitate is brought about by several successive or simultaneously occurring processes (Fig. 6): once formed, nuclei grow into crystals (compact or dendritic, depending on the supersaturation [18]). Simultaneously, or at a later stage of the precipitation aging processes, e.g., ag-

gregation and/or recrystallization (Ostwald ripening, phase transformation) set in and further change the morphology and/or composition of the precipitate.

In real systems the above precipitation processes are seldom clearly separated, usually at least two of them occur simultaneously. Which of the processes will be dominant during a certain time period is determined by their mutual rates; rates may be changed by additives at any stage of precipitate development (Fig. 6). The quality and intensity of this interaction depends on the nature of the additive and on the conditions of precipitation. Several examples given in the "Results" section of this paper will be discussed below.

4.1 Additives as crystal growth inhibitors

The action of additives as inhibitors of crystal growth and dissolution has been demonstrated in a great number of seeded crystal growth experiments and the reduction in the respective rates has been inter-

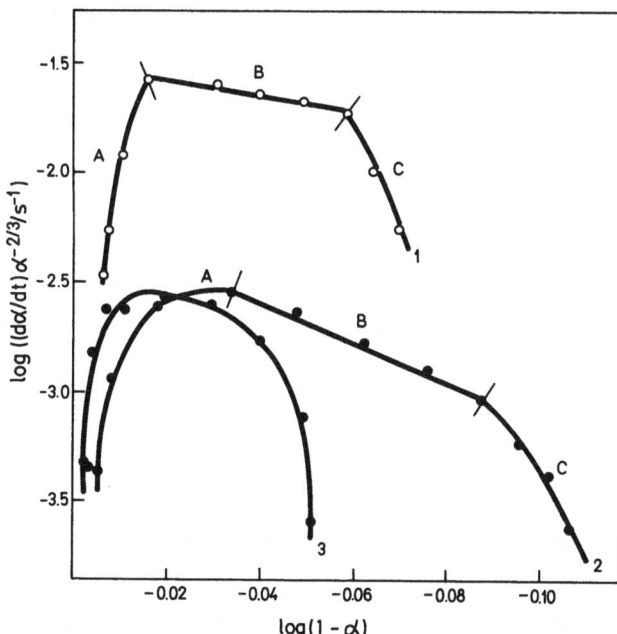

Fig. 5. Rate vs supersaturation curves (Coulter counter) showing the influence of Glu on the precipitation kinetics of calcium oxalate (COM+COT). Experimental conditions as in Table 1, experiment 4. Curve 1 without Glu; curves 2 and 3: c_0(Glu) = 1.0×10^{-5} and 2.4×10^{-5} mol dm^{-3}, respectively. Slopes of sections B: $p(1) = 4.0$, $p(2) = 7.1$. After [9]

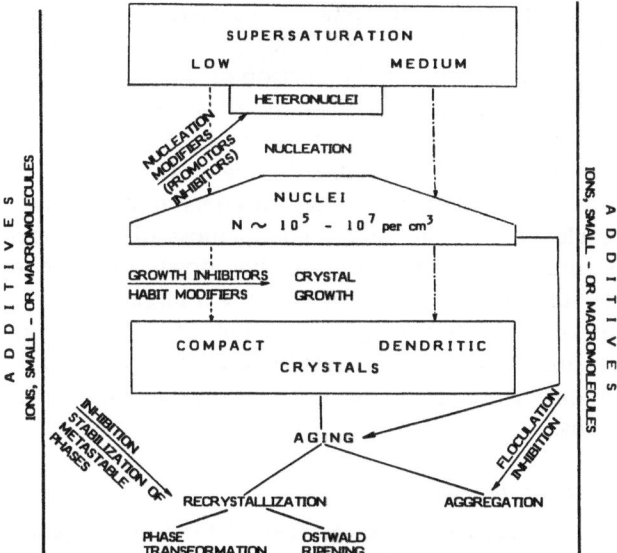

Fig. 6. Schematic presentation showing the possible role of additives in the precipitation of slightly soluble salts initiated by heterogeneous nucleation

preted in terms of the Langmuir adsorption isotherm [19]. In the example given in Fig. 1 this type of interaction is demonstrated on a spontaneously precipitating calcium oxalate. The inhibiting action of PO_4 is not specific to COT but has also been observed on mixtures of COM and calcium oxalate dihydrate (COD), and on this system the adsorption of PO_4 at the crystal/solution interface has also been demonstrated [14].

There is also plenty of evidence that impurities can modify the habit of crystals [20, 21] by adsorbing at specific crystal faces. Thus, an interesting experiment carried out in our laboratory has shown that some di- and tricarboxylic acids adsorb at the (001) face of DCPD thereby promoting the growth of needles in systems in which platelets are normally obtained [21].

Finally, it has been shown [22] that some crystal growth inhibitors (pyrophosphate, polyphosphates) can stabilize metastable hydrates of calcium oxalate by preferential adsorption at the surfaces of growing COM crystals in the course of phase transformation.

4.2 Effect of additives on aggregation

A straightforward example of the action of the simplest additives − neutral electrolytes − as coagu-

lants of a paracolloidal system (e.g., calcium oxalate crystals) is shown in Fig. 2. It appears that the crystals were effectively coagulated at concentrations of NaCl and KCl comparable to the coagulation concentrations of monovalent neutral electrolytes which were defined by the Schulze-Hardy rule ([20] p. 178), and actually determined for silver halide sols [23]. Since, in the absence of NaCl or KCl, the crystals were slightly positive, coagulation was most probably achieved by the chloride ion [11].

4.3 Additives as nucleation modifiers

There is mounting evidence of the action of organic molecules as nucleation modifiers [6, 8−10, 17, 24, 25] affecting the number of the nucleating particles, the morphology, and/or the composition of the precipitating solid phase. It has also been shown [8] that calcium oxalate deposits on the surface of sodium urate crystals only if Glu has been previously adsorbed. Besides the specific nucleation of calcium carbonate crystals by aspartic acid-rich proteins adsorbed at rigid surfaces has been demonstrated [6]. Furthermore, changes in the composition of calcium oxalate precipitates have been induced by other aminoacids [17, 24] and by a surfactant (sodium dodecyl sulphate), [10]).

The effect of Glu on the composition of calcium oxalate precipitates (promotion of the formation of COM, Fig. 4 and [9, 17]) and the enhanced precipitation of the solid phase observed at $c(Glu) > 10^{-4}$ mol dm^{-3} (Fig. 3a) may be interpreted accordingly. The substantial reduction of the crystal growth rate and simultaneous enhancement of aggregation observed at low concentrations of Glu ($c(Glu) = 3 \times 10^{-6} - 5 \times 10^{-5}$ mol dm^{-3}, Figs. 4, 5) may be similarly explained if one considers that enhanced nucleation would have used up most of the supersaturation, significantly slowing down the rate of crystal growth; and – by creating a large number of particles – enhancing aggregation. Such explanation is supported by experimental evidence obtained by Azoury et al. [25] under different experimental conditions.

Acknowledgements

The financial support granted by the Selfmanagement Council for Scientific Research, Croatia, Yugoslavia, and the US-Yugoslav Joint Fund for Scientific and Technological Cooperation in collaboration with the National Institutes of Health, Bethesda, Maryland, USA, (project no. JFP 697), is gratefully acknowledged.

References

1. Cowan JC, Weintritt DJ (1976) Water-Formed Scale Deposits. Gulf Publ Co, Houston, Texas
2. Nancollas GH (1978) Adv Colloid Interface Sci 10:215
3. Botsaris GD (1982) in Jančić SJ, de Jong EJ (eds) Industrial Crystallization 81. North Holland, Amsterdam New York Oxford, pp 109–118
4. Davey RJ (1982) In: Jančić SJ, De Jong EJ (eds) Industrial Crystallization 81. Norht Holland, Amsterdam New York Oxford, pp 123–135
5. Nemerov NL (1971) Liquid Waste of Industry. Theories, Practices and Treatment. Chapter II. Addison-Wesley, Massachusetts, California; London Ontario, pp 97–109
6. Addadi L, Weiner S (1986) Mol Cryst Liq Cryst 134:305–322
7. Schwille PO, Smith LH, Robertson WG, Vahlensieck W (eds) (1985) Urolithiasis and Related Clinical Research, Plenum, New York London, pp 803–908
8. Sarig S (1987) Croat Chem Acta 60:517–530
9. Marković M, Komunjer Lj, Füredi-Milhofer H, Škrtić D, Sarig S (1988) J Crystal Growth 88:118–124
10. Škrtić D, Filipović-Vinceković N (1988) J Crystal Growth 88:313–320
11. Škrtić D, Marković M, Komunjer Lj, Füredi-Milhofer H (1984) J Crystal Growth 66:431–440
12. Marković M, Škrtić D, Füredi-Milhofer H (1984) J Crystal Growth 67:645–653
13. Škrtić D, Marković M, Füredi-Milhofer H (1986) J Crystal Growth 79:791–796
14. Füredi-Milhofer H, Marković M, Uzelac M (1987) J Crystal Growth 80:60–68
15. Marković M, Komunjer Lj (1979) J Crystal Growth 46:701–705
16. Marković M, Füredi-Milhofer H (1988) J Chem Soc Faraday Trans 1, 84:1301–1310
17. Škrtić D, Füredi-Milhofer H, Marković M (1987) J Crystal Growth 80:113–120
18. Füredi-Milhofer H (1980) Croat Chem Acta 53:243–254
19. Nancollas GH, Zawacki SJ (1984) In: Jančić SJ, de Jong EJ (eds) Industrial Crystallization 84, Elsevier, Amsterdam, pp 51–59
20. Walton AG (1967) The Formation and Properties of Precipitates. Interscience Publ, New York London Sidney, pp 166–174
21. Brečević Lj, Sendijarević A, Füredi-Milhofer H (1984) Colloids and Surfaces 11:55–68
22. Nancollas GH (1982) In: Nancollas GH (ed) Biological Mineralization and Demineralization. Springer, Berlin Heidelberg New York, pp 79–99
23. Težak B, Matijević E, Schulz KF, Kratohvil J, Mirnik M, Vouk VB (1954) Faraday Soc Disc 18:63–73
24. Brečević Lj, Kralj D (1986) J Crystal Growth 79:178–184
25. Azoury R, Randolph AD, Drach GW, Perlberg S, Garti N, Sarig S (1983) J Crystal Growth 64:389–392

Received March 7, 1988;
accepted June 5, 1988

Author's address:

Dr. H. Füredi-Milhofer
Laboratory for Precipitation Processes
Department of Technology, Nuclear Energy and Protection
"Rudjer Bošković" Institute
41001 Zagreb, PO Box 1016
Croatia, Yugoslavia

Progress in Colloid & Polymer Science   Progr Colloid & Polymer Sci 77:201–206 (1988)

Crystal growth of lead fluoride using constant composition method
I. The effect of Pb/F activity ratio on the solubility of solid phase

N. Stubičar, M. Ćavar, and D. Škrtić[1])

Laboratory of Physical Chemistry, Faculty of Science, University of Zagreb, and [1])Rugjer Bošković Institute, Zagreb, Yugoslavia

Abstract: The equilibrium solubility boundary has been determined by the light scattering method at 25 °C in a time period up to one month for a very wide range of KF and $Pb(NO_3)_2$ concentrations, i.e., for the activity ratio a_{Pb^2+}/a_{F^-} from 1/0.08, over 1/2, up to $1/10^4$, and the resulting pH from 4.0 to 7.0. Depending on the composition of the supersaturated solution, two distinct crystal phases have been determined to grow, using potentiometric the so called: *constant composition method* (seeded technique). From the neutral solutions with equimolar concentrations, and/or with the excess of KF and given pH = 5.0–7.0, the a-PbF_2 orthorhombic bipyramides have been grown and observed by polarizing microscope. This phase is known and the calculated solubility product $(3.5 \pm 1.2) \times 10^{-8}$, is in agreement with the published value recommended in compiled data of Clever and Johnston, 1980. However, from the acidic supersaturated solution with the excess of $Pb(NO_3)_2$, and given pH = 4.0–5.0, anisotropic needle-like crystals have been formed with an unknown structure. The solubility product of this phase is calculated to be $(1.5 \pm 0.6) \times 10^{-7}$ at 25 °C and $I = 0$.

Key words: Crystal growth, fluoride ion selective electrode, lead fluoride, solubility diagram, solubility product

Introduction

Up to now no detailed study has been attempted of the relation between the composition of supernatant solution, and the morphology and the composition of the solid phase for the electrolytic system: KF–Pb$(NO_3)_2$, in a very wide range of concentration. The investigation of crystal growth's approaching the equilibrium state, or it's steady-state kinetics is, besides a fundamental research point of view, also important for practical reasons, i.e., ecological studies, oxide glass production, metallic fluorides in spectroscopy, etc. The kinetics of crystal growth of lead fluoride has not been extensively investigated, and as far as we know, there are no published data on the growth rate using the *constant composition method* (CCM) [1]. The rate of crystal growth of earth alkaline fluorides has been studied recently by G. H. Nancollas et al.: magnesium fluoride [2], strontium fluoride [3, 4], barium fluoride [5], especially at relatively low supersaturation, and the mechanism of their growth has been proposed [6].

Sometimes the investigations were restricted to the equivalent concentration of precipitating components, as was the case in the study of calcium fluoride growth kinetics [7]. However, in a variety of systems in nature and applied chemistry, the concentrations of components remarkably deviate from stoichiometry, as has been pointed out, [8–12]. A survey on electrolyte crystal growth kinetics [12] and mechanisms [13] was reported recently by A. E. Nielsen.

The aim of our investigation was the experimental determination of the precipitating diagram for lead fluoride at 25 °C and the computation of the activities of all species. This enabled us to define exactly the supersaturated solutions used for seeded crystal growth examination. Specific chemical interactions in solution, as well as at solid/liquid interface, and the amount of all ionic species in supernatant solution, may determine the composition, morphology, size, and particle size distribution (PSD) of grown crystals, giving the appropriate pH of bulk solution.

The kinetic examination of crystal growth of lead fluoride from acidic, neutral or weak alkaline solution by means of potentiostat method (CCM) will be analyzed in the following publication.

Experimental

Chemicals

Analar grade chemicals: KF ("Merck-Alkaloid", Darmstadt-Skopje), $Pb(NO_3)_2$, KNO_3, standard 0.1 mol/dm^3 KOH solution and standard buffers pH = 4.00, 5.00, and 7.00 (all from "Kemika", Zagreb) were used. The solutions were made with twice distilled water, filtered through a 0.2 μm Millipore filter. KF and $Pb(NO_3)_2$ stock solutions were standardized via acidic exchange resin Dowex-50 ("Kemika", Zagreb) and potentiometric titration of the liberated acid, as well as volumetrically by EDTA titration with appropriate indicators. Concentrated KF solutions were stored in plastic bottles to avoid interactions with silicates.

Light scattering measurements

The equal volumes of KF and $Pb(NO_3)_2$ solutions (one with constant, and the others with gradually increased concentration) were mixed and kept in a waterbath thermostat at 25 °C. The appearance of solid phase was checked either tyndallometrically, or by light scattering photometer (Brice Phoenix Model 2000) at appropriate periods of time: 0.5, 1.10 min, 1 h, 1 day, and 1 month after preparation. A cubic cell (Phoenix Cat. No T-101) and light of 546 μm wave length were used. The relative intensity of scattered light of water was $I_{90}/I_0 = (2\pm1)\times10^{-3}$ and systems with values several times larger were considered to contain the solid phase. The precipitation diagram is presented in Fig. 1.

Potentiometric crystal growth measurements

The supersaturated solution was achieved by slow addition of titrants: first 0.1 mol/dm^3 KF, then 0.28 mol/dm^3 $Pb(NO_3)_2$ into 200 cm^3 of water. The fluoride ion activities were measured with a fluoride ion selective electrode ("Metrohm", Switzerland), which was tested before each run. The average slope of Nernstian linear function was calculated to be $(59.2\pm0.4)_{15}$ mV and the non-Nernstian response was below $(1.6\pm0.5)\times10^{-4}$ mol/dm^3 KF. Each experiment of growth was carried out at constant activity of F^- ions in the range from 1×10^{-3} mol/dm^3 (experiment VIII), through 5×10^{-3} mol/dm^3 (exp. IV), to the 8.3×10^{-2} mol/dm^3 (exp. XV), i.e., set potential from 47 mV, through 90 mV, to 165 mV (kept always 2.5 mV higher than the measured EMF; this corresponds to the difference of 2.15×10^{-4} mol/dm^3 KF, which lies in the limit interval of the experimental error).

After checking the constancy of EMF for one hour, 3 cm^3 of seed slurry was added with Justor micropipette, and the automatic titrant addition started with the rate appropriate to the rate of growth. The titrant solutions were prepared in such a way that the activity ratio was $a_{Pb}/a_F = 1/2$; the concentration ratio was 1/3.5. The automatic addition of each solution (KF from "Metrohm Multidosimat 614" and $Pb(NO_3)_2$ from "Radiometer Autoburette ABU-12") was governed by "Radiometer Titrator type TTT", but the signals were recorded separately with two recorders. This adaptation has been made in our laboratory. The potentiostat was checked with 10^{-3} mol/dm^3 KF as a standard. The reference electrode was a saturated calomel electrode ("Iskra", Kranj).

The seeds were prepared by CCM as well and were confirmed to be orthorhombic a-PbF_2 by x-ray diffraction method. The volume of 3 cm^3 of slurry weighted 0.123 g PbF_2 (5×10^{-4} mol/dm^3).

An amount of 10 cm^3 of the sample was withdrawn after each 10, 15 and 25 cm^3 of KF titrant addition usually, at 100, 150, and 250% of growth regarding to the initial amount. Some of the experiments were carried out up to 700% of growth. The samples were analyzed by Coulter Counter technique and the photomicrographs were taken using polarized light and λ plate. The pH of the supernatant solution was measured at the end of the run (after calibration with buffers).

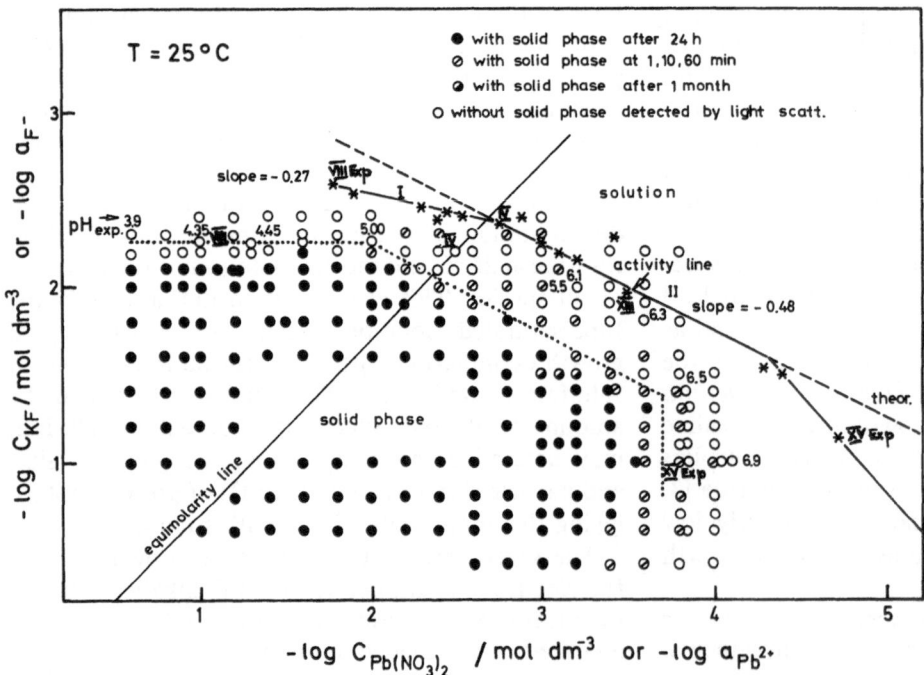

Fig. 1. Precipitation diagram (dotted line) of lead fluoride: log $[KF]_{tot}$ vs log $[Pb(NO_3)_2]_{tot}$, at 25 °C, determined at 1, 10, 60 minutes, 1 day, and 1 month after mixing the components and at a given pH. Activity line (solid line): log a_{F^-} vs log $a_{Pb^{2+}}$, computed from the experimentally measured sets of KF and $Pb(NO_3)_2$ concentrations. Activities of supersaturated solutions for kinetic measurements are denoted by: *. Dashed activity line was calculated for PbF_2 taking into account the solubility product, 3.3×10^{-8}, recommended in [15]

The particle size distributions (PSD) of samples were performed by Coulter Counter MoTA-134 instrument, using 140 μm, and/or 200 μm orifice tubes calibrated with standard monodisperse polyvinyltoluene latexes ($d = 12.45$ μm and 18.6 μm, respectively). The systems were appropriately diluted with original supersaturated solution to avoid the change in size and habit of crystals.

The colour photomicrographs of crystals were taken with a Leitz-Wetzlar universal microscope with automatic camera using polarized light and λ plate to check the anisotropy.

Results and discussion

Precipitation diagram and its analysis

In a series of introductory experiments we tried to collect the data for mapping the areas in the "concentration space" [8] where precipitation occurs at 25°C in the period of time up to 1 month. The results are summarized in Fig. 1 with the axes: logarithm of total cationic vs logarithm of total anionic concentration. The solid/solution boundary for systems in the excess of Pb(NO₃)₂ is rather sharp and parallel to the abscissa. In the excess of KF, however, the transparent precursor – probably lead hydroxide – was formed at early stages of precipitation; which may then be either dissolved or transformed into the white solid lead fluoride. According to Težak [8], the shape of this diagram assumes the complex formation process of both components leading to sparingly soluble lead fluoride formation.

That assumption was confirmed by the computation of the concentration and activities of all different species for given systems at the precipitation boundary. An iterative procedure [14] was employed for a numerical solution of the nonlinear equations, starting from the sets of initial total KF and Pb(NO₃)₂ concentrations and following equilibrium constants of reactions (at zero ionic strength and 25°C) known from the literature [15−23]:

Equilibrium	Equilibrium constants	References
$Pb^{2+} + 2F^- \rightleftharpoons PbF_2^0(aq)$	356.0	[15, 16]
$Pb^{2+} + F^- \rightleftharpoons PbF^+$	18.0	[15, 16]
$PbOH^+ \rightleftharpoons Pb^{2+} + OH^-$	3.02×10^{-8}	[20]
$Pb^{2+} + NO_3^- \rightleftharpoons PbNO_3^+$	3.3	[15, 17]
$Pb^{2+} + 2NO_3^- \rightleftharpoons Pb(NO_3)_2$	3.2	[15, 17]
$KF \rightleftharpoons K^+ + F^-$	4168.0	[20]
$H^+ + F^- \rightleftharpoons HF^0(aq)$	1459.0	[22]
$HF^0(aq) + F^- \rightleftharpoons HF_2^-$	5.01	[22]
$H_2O \rightleftharpoons H^+ + OH^-$	1.01×10^{-14}	
$Pb^{2+} + 3F^- \rightleftharpoons PbF_3^-$	2630.0	[19]
$Pb^{2+} + 4F^- \rightleftharpoons PbF_4^{2-}$	1190.0	[19]

The complexes PbF_3^- and PbF_4^{2-} contributed only 7.4×10^{-3}% and 2.6×10^{-4}%, respectively, even for the greatest KF concentration of 0.1 mol/dm³. The mass balance and electroneutrality have been taken in-

to account and the activity coefficient (y) of species (i) have been computed by successive approximation for the ionic strength from the simple Debye-Hückel equation. The activity product was defined as $AP = a(Pb^{2+})a^2(F^-) = [Pb^{2+}][F^-]^2 y^6$, since

$$y_{Pb^{2+}} = y_{F^-}^4 = y^4 .$$

The activities of constitutive ions for the solutions at the precipitation boundary, i.e., $\log a_{Pb^{2+}}$ vs $\log a_{F^-}$, are presented in the same Fig. 1 and are marked *. The theoretical activity line is shown as a dashed line with a slope of -0.50. That is, it is calculated by assuming that only one solid phase is formed in the whole concentration region, taking into account the activity product 3.3×10^{-8} for PbF₂ [15] and with the classical assumption that $a_{PbF_2}(s) = 1$. As is seen in Fig. 1 in the region of middle concentrations (part II on the activity (solid) line has the slope -0.48; and a corresponding activity product $(3.5 \pm 1.2) \times 10^{-8}$, calculated by the least square method), in pH region $5.1 - 6.5$, is in agreement with the solubility product for PbF₂, recommended by Clever and Johnston [15].

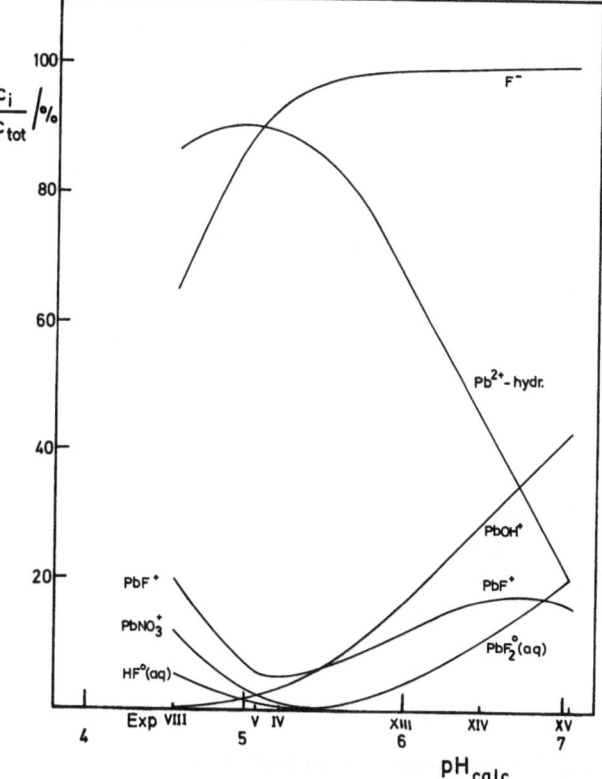

Fig. 2. Calculated composition of supersaturated solutions at the solubility boundary: percentage of the concentration of ionic species normalized to the total concentration of KF and/or Pb(NO₃)₂ as a function of pH calculated

In the region of exceeding $Pb(NO_3)_2$ concentrations, however, the slope of the activity line (part I) is calculated to be -0.27; the ratio of components is then $1/3.7$, and the activity product is $(1.5 \pm 0.6) \times 10^{-7}$ in the pH region $4.0-5.1$, i.e., in the acidic region. It is obvious that there are two distinct concentration regions (and pH regions) corresponding to two different solid phases formed, as we will see later. This then is the reason why we can find the whole branch of various solubility products reported in the literature [15].

Figure 2 presents the composition of supersaturated solutions (for systems at the solubility boundary), i.e., percentage of concentration of a given specie (c_i) in regards to the total concentration of KF and/or $Pb(NO_6)_2$ as a function of pH calculated. The calcu-lated and experimentally measured pH agreed very well (± 0.05 pH). The shift of curves with respect to the abscissa and ordinate may be obtained by using different equilibrium constants from the literature, but the general feature stays the same. In acidic region (experiment VIII), besides Pb^{2+}-hydrocomplexes, about 10% of lead is bound in $PbNO_3^+$ ionic pair, which is probably responsible for the different structure of the solid phase formed in this region. In the pH region above 5.7, i.e., with large excess of F^- ions, the amount of $PbOH^+$ ionic pairs becomes significant (at about pH 7.0 it is approximately 20% of $PbOH^+$ in solution, which causes lead hydroxide formation in this region). The minimum amount of all complexes is at pH $= 5.0-5.4$, i.e., in the region of a small excess of F^- ions (in experiment IV: $a_{Pb}/a_F = 1/2$); the solubili-

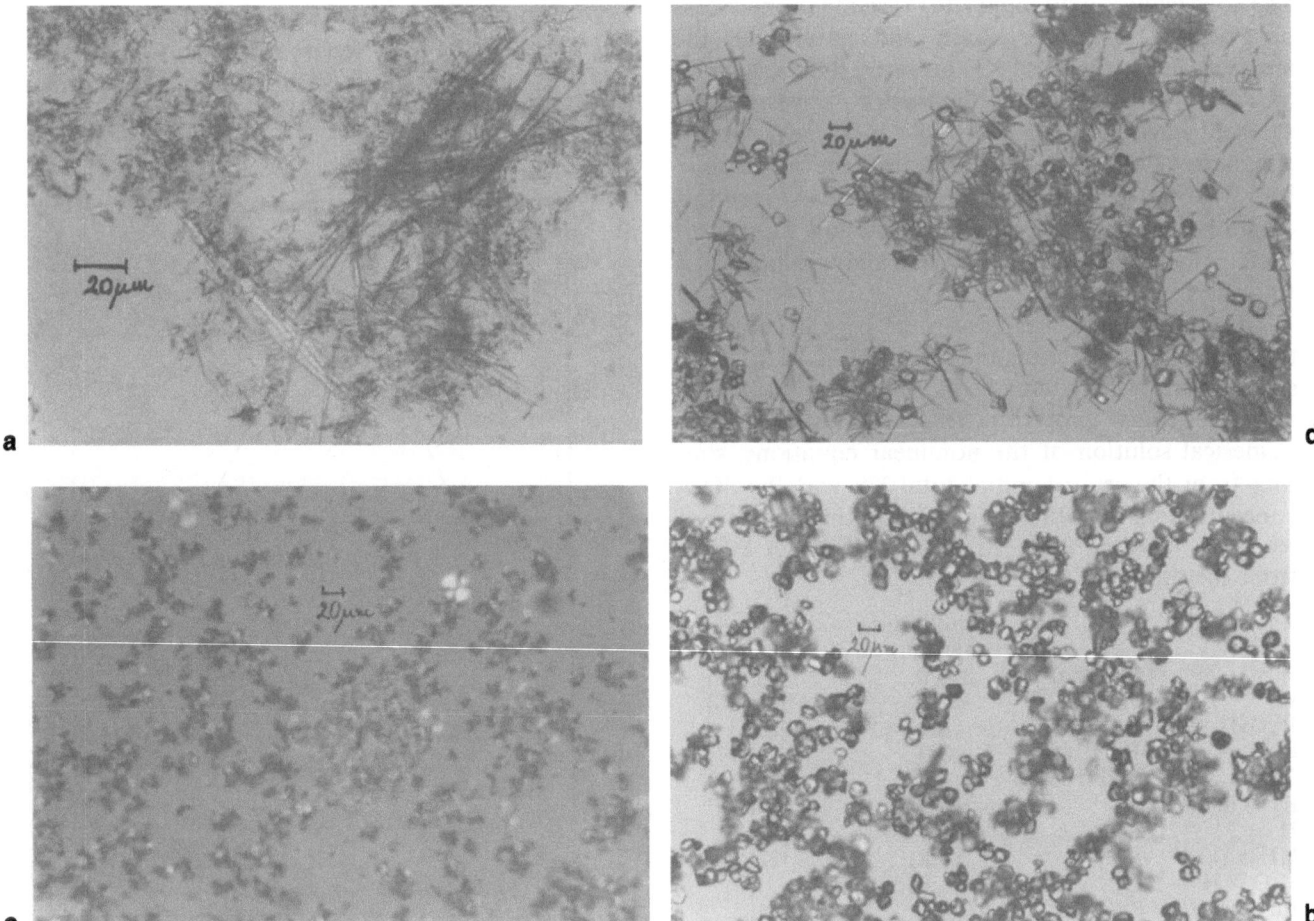

Fig. 3. Photomicrographs of lead fluoride crystal grown in the constant composition supersaturated solution up to 250% of growth regarding the initial amount, taken with polarizing microscope and λ-plate. Concentrations of supersaturated solutions: a) 0.1 mol/dm³ $Pb(NO_3)_2$, 5.5×10^{-3} mol/dm³ KF (experiment VIII); b) 0.01 mol/dm³ $Pb(NO_3)_2$, 5.5×10^{-3} mol/dm³ KF (experiment VI); c) 3.5×10^{-3} mol/dm³ $Pb(NO_3)_2$, 5.5×10^{-3} mol/dm³ KF (experiment IV); d) 2×10^{-4} mol/dm³ $Pb(NO_3)_2$, 0.1 mol/dm³ KF (experiment XV)

ty product for this system is calculated to be 1.8×10^{-8}, that being probably the most reliable activity product for PbF_2.

Size and shape analysis of solid phase

Microscopic observations of solid phase show that two morphologically different solid phases are formed by mixing the components (classical crystal growth), as well as by constant composition crystal growth experiments, depending on the activity ratio of components. Crystals grown from the supersaturated solution where lead (II) nitrate prevails are bundles of needles with some sponge-like precipitate, as is shown in Fig. 3a. These crystals are very settleable because of high ionic strength. As the concentration of $Pb(NO_3)_2$ decreases (with the same KF concentration) the size of needles becomes smaller and at 10^{-2} mol/dm^3 $Pb(NO_3)_2$, (Fig. 3b), a mixture of needles and voluminous bipyramides are found. (The pH in this case was 5.05.) Both kinds, needles and bipyramides, are anisotropic crystals and, as a consequence of partial birefringence for a given crystal thickness, the entire sequence of polarization colours of the visible spectrum were observed (shown also on colour photomicrographs). The small stable bipyramides were formed at pH = 5.2, (experiment IV), and interference figures were observed, e.g. an isogyre cross with yellow and blue fields, such as is characteristic of biaxial crystals (Fig. 3c). Finally, in the solutions with pH up to 7.0, in the large excess of KF, only bipyramidal shaped crystals were seen (Fig. 3d).

The particle size analysis is illustrated in Fig. 4 and the data were analyzed as before [23]. The changes in volume and number size distribution with the percentage of growth are presented for a needle-like crystalline system in the upper part of Fig. 4a, and for a bipyramidal crystalline system in the lower part of Fig. 4b. The increase in the volume diameter during the course of growth is obvious in both cases and is much less pronounced for needle crystals, probably because of their one dimensional growth. Median volume diameters of the system shown in the upper part of Fig. 4 are as follows: 16.8, 19.0, and 20.6 μm (±0.5 μm); for the system in the lower part: 22.5, 29.0, and 36.0 μm. This leads us to the conclusion about the linear increase of median volume diameter with the relative increase of mass precipitated. However, the change of this relationship varies one from each other, which also supports the conclusion about a different growth model in these two cases.

Finally, the structural x-ray diffraction analysis of crystals grown by CCM from equimolar solutions and

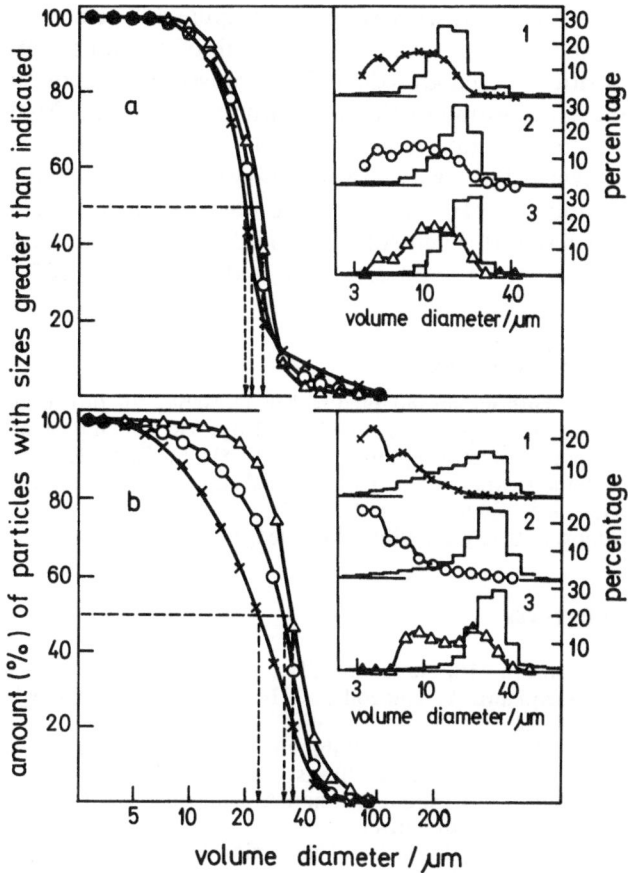

Fig. 4. Number and particle size distribution during the course of seeded growth of lead fluoride grown in the constant composition solution: a) 4×10^{-2} mol/dm^3 $Pb(NO_3)_2$, 5.5×10^{-3} mol/dm^3 KF at 100, 140 and 230% of growth, respectively (1, 2, 3), (experiment XVI); b) 7×10^{-4} mol/dm^3 $Pb(NO_3)_2$, 1.2×10^{-2} mol/dm^{-3} KF at 90, 135, and 200% of growth, respectively (1, 2, 3), (experiment XIII). Arrows denote median volume diameters

from solutions with KF in excess, has shown that pure orthorhombic a-PbF_2 phase was formed. However, the needle-like crystals have a different structure and that will be the subject of a separate article.

Conclusions

In order to characterize the heterogeneous equilibrium and the kinetics of solid phase formation in a lead (II)nitrate-potassium fluoride system at 25 °C, various equilibria must be considered, including the hydrolysis of the lead nitrate solution, as well as the complex formation. Despite the fact that we started always with the same carefully prepared crystal suspension of orthorhombic a-PbF_2, and with a narrow size distribution, we got quite different crystal shapes, sizes, and PSD, as a result of constant composition seeded growth, and this proved dependent on the composition of the supersaturated solution. From a neutral solution at pH = 5.0–7.0, the previously known a-PbF$_2$ was grown; however, from the acidic solution at pH = 4.0–5.0, the unknown solid phase was revealed. Their solubility products were determined to be: $(3.5 \pm 1.2) \times 10^{-8}$ and $(1.5 \pm 0.6) \times 10^{-7}$, respectively.

Acknowledgements

The authors are indebted to Mr. D. Babić, for providing the computer program; to Mr. Z. Dojnović, for improving the potentiometric equipment, and to Ms. M. Beuk, for tyndallometric measurements. Financial support from the Self-Managed Council for Scientific Research of Croatia, Yugoslavia, is gratefully acknowledged.

References

1. Tomson MB, Nancollas GH (1978) Science 200:1059
2. Yoshikawa Y, Nancollas GH (1983) J Cryst Growth 64:222
3. Bochner RA, Abdul-Rahman A, Nancollas GH (1984) J Chem Soc Faraday Trans I 80:217
4. Hamza SM, Abdul-Rahman A, Nancollas GH (1985) J Cryst Growth 73:245
5. Barone JP, Svrjcek D, Nancollas GH (1983) J Cryst Growth 62:27
6. Yoshikawa Y, Nancollas GH, Barone J (1984) J Cryst Growth 60:357
7. Møller H, Madsen L (1985) J Cryst Growth 71:673
8. Težak B (1966) Disc Faraday Soc 42:175
9. Füredi H (1967) In: Walton AG (ed) The Formation and Properties of Precipitates. Interscience, New York, p 152
10. Füredi-Milhofer H, Walton AG (1981) In: Parfitt GD (ed) Dispersions of Powders in Liquids. Applied Sci Publishers, London, N. Jersey, IIIth ed, p 203
11. Nielsen AE, Christoffersen J (1982) In: Nancollas GH (ed) Biological Mineralization and Demineralization, Dahlem Konferenzen. Springer, Berlin, Heidelberg, New York, p 37
12. Nielsen A, Toft JM (1984) J Cryst Growth 67:278
13. Nielsen A (1984) J Cryst Growth 67:289
14. Cruise DR (1964) J Phys Chem 68:3797
15. Clever HJ, Johnston FJ (1980) J Phys Chem Ref Data 9:751
16. Mesarić Š, Hume DN (1963) Inorg Chem 2:788
17. Mironov VE (1961) Zh Neorg Khim 6:659
18. Ivett RW, DeVries TJ (1941) J Am Chem Soc 63:2821, ibid (1947) 69:1644
19. Talipov ShT, Kutumova OF, loc. cit. (1959) Chem Abstr 53:21277g
20. Sillen LG, Martell AE (1964) Stability Constants. Chem Soc, Spec Publ No 17:256, ibid No 25:95
21. Gmelin's Handbuch der anorganischen Chemie (1969) Blei Teil C, 8. Auflage. Verlag Chemie, 47:272
22. Högfeldt E (1982) Stability Constants of Metal-Ion Complexes, Inorganic Ligands, Part A, IUPAC. Pergamon Press, Chem Data Series, 1st ed, No 21:187, 36, 120, 196, 100
23. Škrtić D, Marković M, Komunjer LJ, Füredi-Milhofer H (1984) J Cryst Growth 66:431

Received February 25, 1988;
accepted June 8, 1988

Authors' address:

Dr. N. Stubičar
Laboratory of Physical Chemistry
Faculty of Science
University of Zagreb, POBox 163
41001 Zagreb, Yugoslavia

Progress in Colloid & Polymer Science   Progr Colloid & Polymer Sci 77:207–210 (1988)

Crystal growth in gels – a survey

G. Sperka

Institut für Physikalische und Theoretische Chemie, Graz University of Technology, Graz, Austria

Abstract: This paper gives a short survey of the principal features of crystal growth in gels. Since there are many variations in the experimental setups only the three basic types, to which all others can be traced back, are described in some detail. The substances which can be grown by the gel technique have been categorized into three groups fulfilling different boundary conditions. No attempt has been made to review all the substances grown in gels, or the existing literature in this field. The crystallization criteria and the influence of the gel itself are outlined. Due to the restricted possibilities of structural investigations, emphasis is laid on the possibilities of simulating gel structures on computers, using simple random growth models.

Key words: Crystal growth, gels, basic types, silica gels, gel structure

Introduction

In almost all fields of solid state science there is a strong necessity for the use of high quality single crystals. In many cases it is not possible to grow crystals of suitable size and quality using conventional methods, like crystal growth from melts or solutions. The method of crystal growth in gels – or gel method – is often a very powerful alternative.

Although this method is rather old (it was originally introduced by Liesegang [1, 2]), it has not been in use for many years, until book by Henisch [3] caused its renaissance.

In this contribution a short survey of the principal features of crystal growth in gels is given. No attempt will be made to review all the substances which can be grown by the gel technique or the published literature; interested readers should refer to one of the cited reviews [3, 4].

Possibilities of the gel method

The gel method is suitable to grow crystals of substances which fulfill one of the following conditions:
a) they have water soluble reaction components but are insoluble in water (two-layer, three-layer, and U-tube method)
b) they have water soluble reaction components and are themselves water soluble. The product must be insoluble in a solvent not reacting with the gel or other compounds present in the gel, so that it can be crystallized (solubility reduction method)
c) the substance forms a water soluble complex which can undergo a decomplexion reaction whose product is insoluble in water or a suitable solvent not reacting with the gel (chemical decomplexion method)

One of the main disadvantages of the gel method is that it is impossible to grow single crystals of good quality larger than about 1 cm in diameter for most substances, and in some cases, the biggest crystals do not exceed some 3 mm, such as with $BaSO_4$ [5].

Furthermore, it is not possible to predict size and quality of the growing crystals; the growth conditions have to be optimized for every new growth system.

On the other hand gel growth is an inexpensive and powerful method to grow crystals of substances which cannot be grown by usual techniques for various reasons, e.g., thermal instability or different high-temperature phases.

Since the crystals grow at ambient temperatures (usually the temperature does not exceed ca. 315 K) there are practically no thermal strains in the lattice, and the number of defects in the crystals is very small [6]. All nuclei are spatially separated and can be observed in practically all stages of their growth. In addition, the gel prevents convection currents and turbulence. By remaining chemically inert the gel protects the crystals in all stages of their growth. For several purposes it is very useful that many substances can easily be doped, as will be shown in the experimental section.

Experimental

The primary effect of the gel is to slow down the diffusion rates of the reacting solutions. This is a first restriction for the materials to be used as gels. Since most of the organic gels — like gelatin gel or agar agar — have typical pore sizes one order of magnitude larger than silica gel [7], the vast majority of them are not suitable for crystal growth because the nucleation densities are too high. Only in some cases where the diffusing solutions undergo reaction with the silica gel, e.g., as in some fluorides [8], are organic gels used.

Essentially all the various experimental equipments described in the literature can be traced back to three basic types:

a) Two-layer-technique: Primarily used for the growth of ambient insoluble substances, like calcium tartrate tetrahydrate [9]. It consists simply of a reaction vessel, in most cases just a test tube partially filled with gel. One component of the desired product is yet incorporated in the gel. In the case of calcium tartrate tetrahydrate the tartaric acid acts as a proton donor for the formation of the gel and as one reaction component. After the gel has set a solution of the other component is placed carefully on the surface of the gel (Fig. 1 a)

b) Three-layer-technique: Between the gel with one reacting component and the solution of the second component a "neutral" gel (termed so because it does not contain any reactant) is placed. The diffusion is slowed down once more, and the critical concentration product, which is responsible for the formation of crystals to a first approximation, is reached very slowly. Therefore very insoluble substances are grown with the three layer technique (Fig. 1 b)

c) U-tube-technique: "neutral" gel is poured into a U-tube. After the gel has set solutions of the reaction components are poured on top of the gel on both sides of the U-tube. The advantage of this method is the easy control of the diffusion length (Fig. 1 c).

There are many variants of these three basic techniques concerning size, form, solution reservoirs, etc. The most common experimental setups are reviewed in the work of Arora [4]. Doped crystals, which are useful for many purposes, e.g. spectroscopic studies, are easily grown by the gel technique. The desired amount of dopant is added to one of the solutions of the reacting compounds. The disadvantage is that the concentration of the in-doped ions is not uniform over all crystals but depends on the position of the crystals in the gel [9].

Crystallization in gels

As mentioned, the primary effect of the gel is to slow down the diffusion of the reacting solutions. Usually the Fick laws are sufficient to describe the diffusion in gels. In the one dimensional case an expression of the form

$$\frac{\delta c}{\delta t} = D \cdot \frac{\delta^2 c}{\delta x^2} \qquad (1)$$

can be derived [10].

There is no analytical solution of this equation since all the growth systems are finite. For complete analytical solutions the systems dealt with have to be at least semi-infinite. However, numerical solutions under time-dependent boundary conditions are simple [11, 12].

For practical purposes an equation of the form

$$x^2 = K \cdot t \quad \text{with} \quad K = \text{const.} , \qquad (2)$$

were x denotes the diffusion length, has turned out to be sufficient. The measurement of diffusion coefficients is done with radioactive tracers or organic indicators [13].

For practical reasons we shall restrict the following discussion to U-tube systems. If the criterion that crystallization starts when the concentration product exceeds a critical value K'_s is correct, first precipitation should always take place in the middle of the diffusion length, provided the diffusion coefficients are the same. This is in gross contrast to the observations of Garcia-Ruiz and Miguez [14], who observed a dependence of the position of the first precipitate on the reservoir concentrations A_R and B_R. Furthermore, the position of the first precipitate depends on the ratio of the diffusion coefficients. It is therefore necessary to define a probability of nucleus formation in the form

$$P_N = \exp\left[\frac{M}{(\log s)^2}\right] \qquad (3)$$

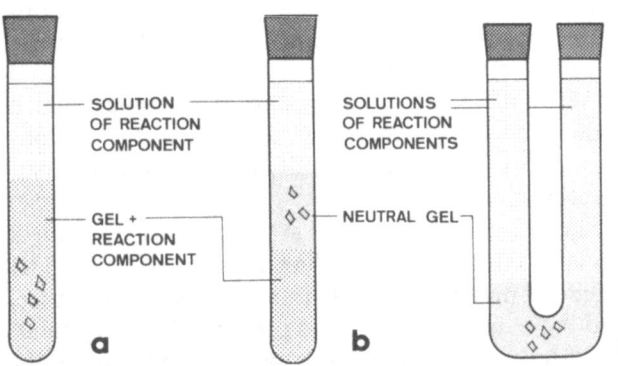

SOLUTION OF REACTION COMPONENT

SOLUTIONS OF REACTION COMPONENTS

GEL + REACTION COMPONENT

NEUTRAL GEL

a   b   c

Fig. 1. The three basic techniques of crystal growth in the gels: a) two-layer technique, b) three-layer technique, c) U-tube technique

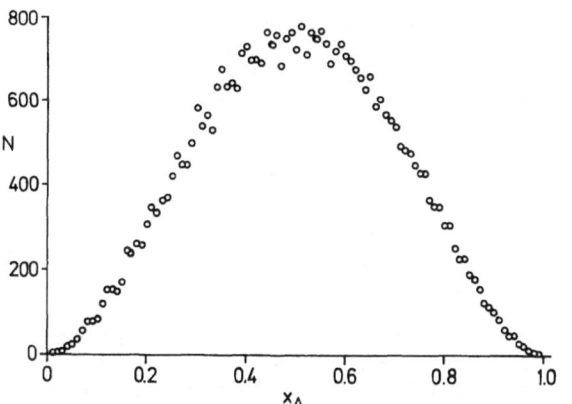

Fig. 2. Result of a simulation in a two-dimensional lattice of 12 000 cells. N denotes the number of $ABBA$ items found for this $[A]/[B]$ ratio

where s is the supersaturation defined as

$$s = AB/(AB)_s . \tag{4}$$

$(AB)_s$ denotes the concentration product at saturation. The term $(\log s)$ may be considered as the driving force of the saturation process.

From the simplest and most schematic geometrical model it becomes clear that the $[A]/[B]$ ratio influences the place of the first precipitation. Figure 2 shows the result of a simulation in a rectangular matrix of 12 000 cells, populated randomly with A and B items. $ABBA$ configurations were regarded as nuclei. Without taking this simple two-dimensional model too seriously it is at once plausible that the nucleation probability is fairly high over a rather wide range of $[A]/[B]$ ratios.

Influence of the gel

The above discussion has not included any influences of the gel material itself. It is known that some properties of the gel such as pH-value, density, etc., do influence the quality of the grown crystals. Since the gel state lies between liquid and solid states, the methods for determining the structural properties are somewhat restricted. The most common method is Scanning Electron Microscopy (SEM). Strong indications point to the validity of the Random Network Theory [15]. However, there is some short-range ordering but no indications for long-range ordering in gels [16, 17]. Experimental observations show that the quality of the crystals is higher according to increased homogenity of the pore size distribution of the gel.

This pore-size distribution depends, as does the specific surface of the gels [18], mainly on the pH-value, and *not* on the density of the gel [9, 19]. There is no really convincing theoretical interpretation for this behavior.

Simulations of gel structures

As pointed out in the preceding section it is desirable to obtain a deeper understanding of the gel structure. Simple random growth models allow the simulation of gel structures on computers [20]. The most successful model is the "clustering for clusters" [21, 22], which starts with N_0 particles on a (hyper-cubic) lattice of the (euclidian) dimension d and the lattice constant L. The particles are allowed to diffuse. If they collide, dimers and polymers are formed (which, of course, have a slower rate of diffusion), until a large, rigid cluster is formed.

In the "chemically limited" clustering of clusters model [23] the bonding probability is reduced and the formed bonds do have a distinct lifetime T (the resulting cluster is therefore non-rigid). This model provides a good depiction of the principal features of the gel formation, the formation of clusters and the gelation threshold. The main disadvantage of the model is that it requires much computer memory due to the explicit assumption of the diffusion of all particles and clusters. The Static Gel Model [9] does not take the diffusion into account and represents momentary pictures of the gel formation. As in the case of the clustering of clusters it starts with N_0 particles on a for convenience two-dimensional lattice. These points serve as starting points for bonds. The number of starting points and the bond lengths are free parameters and can be correlated, e.g., with pH-value and density. The model runs through all the surface points and looks for "allowed" points, (allowed points have only a certain number of neighbors in the surrounding cell). The number of neighbors and the size of the surrounding cell are chosen at the beginning as additional parameters; from the allowed points new bonds are drawn and a chosen number of new points (monomers) is set on the lattice. This is repeated until there are no more allowed points. The results are similar to those obtained with the clustering of clusters model, but can easily be calculated on a personal computer. There is a very good qualitative correlation between calculated structures and those obtained from SEM [9]. This can be used to choose the optimum conditions to achieve a homogenous pore size distribution in the gels.

Conclusions

The method of crystal growth in gels is a very powerful tool to obtain high quality single crystals from substances not easily grown by conventional techniques. In this contribution the principal features have been outlined. The three basic growth procedures have been described and some of the advantages and problems have been given. The boundary conditions for substances which can be grown in gels have been systematized into three categories. Since the methods of investigating gel structures are limited, an attempt has been made to simulate gel structures using a personal computer.

Acknowledgement

The author thanks Prof. H. P. Fritzer for valuable discussions.

References

1. Liesegang RE (1906) Z anorg Chem 48:364
2. Liesegang RE (1907) Z phys Chem 59:444
3. Henisch HK (1973) Crystal Growth in Gels. Pennsylvania State University Press
4. Arora SK (1981) Prog Cryst Growth Charact 4:345
5. Patel AR, Bhat ML (1972) J Cryst Growth 12:288
6. Patel AR, Arora SK (1976) J Mater Sci 11:843
7. Blank Z, Reimschüssel AC (1974) J Mater Sci 9:1815
8. Leckebusch R (1974) J Cryst Growth 23:74
9. Sperka G (1987) Thesis, Graz University of Technology
10. Dennis JC (1968) J Chem Ed 45:432
11. Henisch HK, Garcia-Ruiz JM (1986) J Cryst Growth 75:195
12. Henisch HK, Garcia-Ruiz JM (1986) J Cryst Growth 75:203
13. Wunderlich W (1982) Cryst Res Techn 17:987
14. Garcia-Ruiz JM, Miguez F (1982) Estudios Geol 38:3
15. Bell RJ, Dean P (1972) Phil Mag 25:1381
16. Bursill LA, Thomas JM (1981) J Phys Chem 85:3007
17. Bursill LA, Mallinson LG, Elliott SR, Thomas JM (1981) J Phys Chem 85:3004
18. Zachariasen WH (1932) J Am Chem Soc 54:3841
19. Blank Z, Drake LC (1947) J Coll Sci 2:399
20. Family F, Landau DP (eds) (1984) Kinetic of Aggregation and Gelation. North Holland Physics Publishing, Amsterdam
21. Meakin P (1983) Phys Rev Lett 51:1119
22. Kolb M, Botet R, Jullien R (1983) Phys Rev Lett 51:1123
23. Jullien R, Kolb M (1984) J Phys A 17:L639

Received March 3, 1988;
accepted June 8, 1988

Author's address:

Dr. Gunter Sperka
Insitut f. Physikalische and Theoretische Chemie
Technische Universität Graz
Rechbauerstraße 12
8010 Graz, Österreich

Progress in Colloid & Polymer Science   Progr Colloid & Polymer Sci 77:211–216 (1988)

Polymer Solutions and Polymers

Electro-optical study of a polyelectrolytic graft copolymer

M. Tricot

Laboratoire de Chimie Macromoléculaire et Chimie Physique, Université de Liège au Sart-Tilman, Liege, Belgium

Abstract: Electro-optical experiments have been performed on a poly (styrene-co-N-ethyl-4-vinylpyridinium bromide) graft copolymer in water-methanol mixtures and in the absence of added salt. A positive electric birefringence was observed for the graft copolymer while the birefringence of the corresponding polyvinylpyridinium homopolymer was negative. This change of the sign of the optical anisotropy was discussed on the basis of the orientation of the pyridinium ring with respect to the main backbone and hence to the applied electric field.

We determined the electric polarizability parameter and the relaxation time of the birefringence. It was shown that the graft copolymer does not display the so-called polyelectrolytic effect, i.e. the pronounced extension of the chain with dilution. The conformation of the graft copolymer remains rather compact and much less extended than the corresponding homopolymer. The influence of the solvent composition was also investigated and the effects discussed in terms of preferential solvation of the styrene units by methanol.

Key words: Electro-optics, polyelectrolytes, polyvinylpyridinium, graft copolymers, Kerr effect

Introduction

In many studies of polystyrene-poly-N-ethyl-4-vinylpyridinium bromide copolymers, Selb and Gallot [1–3] showed that block copolymers exhibited association phenomena in water/methanol/lithium bromide mixtures. The degree of association is a function of the molecular characteristics of the block copolymer and of the composition of the binary solvent mixture. On the contrary, these authors observed that the graft copolymers [4, 5] remain in a molecularly dispersed state, whatever the composition of the solvent medium. The absence of intermolecular associations in the latter case can be explained on the basis of the protection of the polystyrene sequence by the numerous hydrophilic grafts [5]. Let us recall that both methanol and water are precipitants for the polystyrene sequence, but are good solvents for the vinylpyridinium grafts.

We previously studied the solution behavior of poly-2-and-4-vinylpyridinium halides by means of electro-optical and hydrodynamic methods, with a peculiar emphasis on the very dilute solution range and low ionic strengths smaller than 5×10^{-3} M [6–9]. We investigated the influence of molecular weight, length of the alkyl lateral chain, nature of the counterion and ionic strength. Attention has been called to the change of optical anisotropy from negative values in water and methanol to positive values in dimethylformamide and nitromethane [10]. This change has been explained by variations of the extension of polyvinylpyridinium chains in relation to the ionization degree of pyridinium sites and their subsequent influence on the electrostatic repulsion between charged sites [7]. From a hydrodynamic point of view it was also proved that polyelectrolytes in very dilute solution and in the absence of added salt behaved as wormlike chains whose degree of rigidity can be quantitatively determined by means of the persistence length [8].

The purpose of the present work is to gain further insight into the solution behavior of the polystyrene-polyvinylpyridinium graft copolymers with the aid of the electric birefringence method. We will determine the optical anisotropy, electric polarizability, and relaxation time of the graft copolymer and compare these parameters with the same obtained for a homopolymer. The influence of polymer concentration and solvent composition (water-methanol mixtures) will be investigated in detail and we will work without added salt, i.e, under experimental conditions that differ markedly from those of Selb's and Gallot's previous studies.

Experimental (materials and method)

1. Synthesis and characterization of the samples

The preparation of the graft copolymer has been exhaustively described [4, 11]; the principle of the procedure can be summarized as follows: the polymerization of 4-vinylpyridine is performed in tetrahydrofuran-dimethyl-formamide media at −70°C, using diphenylmethylsodium as initiator. The graft copolymer is formed by carbanionic deactivation of the living polyvinylpyridine onto partially chloromethylated polystyrene [11]. Fractionation is then carried out by using a mixture methanol-tetrahydrofuran as solvent and heptane as precipitant. Quaternization of the pyridinic units by ethylbromide in dimethylformamide at 50°C yields graft copolymers whose degrees of quaternization ranges from 90 to 99% [4, 5].

A graft copolymer (G) was given to us by Selb and Gallot and designed as sample NG22 in their papers [4, 5]. Its weight-average molecular weight is equal to 8.35×10^5, whereas the molecular weights of polystyrene sequence and of the polyvinylpyridinium grafts are equal to 5.9×10^4 and 2.25×10^4. The average number of polyionic grafts per copolymer molecule is of about 35. A poly-N-ethyl-4-vinyl-pyridinium bromide homopolymer (H), having a $\bar{M}_w$ value of 3.8×10^5, was also studied. The degrees of quaternization are of about 100% for the homopolymer and 93% for the graft polymer.

2. Methods

Electric birefringence and birefringence relaxation experiments have been performed at 20°C and at a wavelength of 550 nm with a previously described instrument which involves a fast transient recorder and a minicomputer for the storage and the treatment of the signals [12, 13]. The pulse generator (Cober Electronics, Stamford, Connecticut 06902) is able to deliver single rectangular pulses at 2.5 kV, which corresponds with our Kerr cell's (electrodes separation of 1.5 mm) applied electric field of 16 kV/cm. For the reversing pulse method, the association of two such generators yields two consecutive pulses at 2.5 kV of reversed polarity with a transition time lower than 1 µs.

In order to avoid previously observed unreproducible association phenomena, the graft copolymer has been first dissolved in pure methanol before water was added.

Results and discussion

1. Determination of the electric-optical parameters and of the relaxation time

The optical anisotropy effect, i.e., the electric birefringence in the present case resulting from the orientation of the particles in a pulsed electric field may be expressed as

$$\Delta n = \Delta n_s \Phi \qquad (1)$$

where Δn is the steady state birefringence for a given electric field E, Δn_s is the birefringence at saturation (at infinite field) and Φ is an orientation function reflecting the degree of orientation of the particle [14].

The orientation function Φ depends upon both permanent and induced dipole terms β and γ. Of the polyions studied, we will be concerned only with a purely induced dipole for the orientation mechanism. We will appeal to the single orientation function $\Phi(\gamma)$ expressed in [14, 15], with

$$\gamma = \Delta a E^2 / 2kT \qquad (2)$$

where Δa is the excess of electric polarizability ($\Delta a = a_a^E - a_b^E$) along the great axis a and the transverse axis b of the particle.

At low electric fields, the Kerr law is obeyed, i.e., the birefringence varies linearly with the square of the electric field strength. If K_B represents the Kerr constant

$$\Delta n_{E \to 0} = K_B E^2 = \left(\frac{\Delta n_s \Delta a}{15 kT} \right) E^2 . \qquad (3)$$

At high electric fields a linear variation of Δn vs $1/E^2$ should be observed for the orientation of an induced dipole.

For the analysis of the field strength dependence of the birefringence, we applied a procedure which requires two calculation methods [15]:

a) First, a graphic extrapolation is performed by plotting the Δn values vs E^{-2}, in order to obtain Δn_s. Then, the Kerr constant values issued from the measurements at very low field strengths allow the determination of Δa, according to relation [3]

b) Secondly, using a multiparametric curve fitting procedure a fitting of Δn vs E curves is computed with the aid of the orientation function $\Phi(\gamma)$ for an induced dipole. This implies that Δn_s and a set of two (sometimes three) electric polarizability terms Δa_i affected by their respective contributions ω_i should be kept as adjustable parameters. The mean polarizability is then given by $\overline{\Delta a} = \Sigma \omega_i \Delta a_i$.

It was shown that the Δn_s values derived from the graphic and computed fitting procedures are generally in good agreement (within 1−2%), whereas the Δa values differ by 2−4% [15]. Only the mean Δn_s and Δa values given by the two procedures will be retained for further discussion.

After the end of the electric pulse, the oriented particles submitted to the brownian motion return to a random orientation state. The analysis of the exponential decay curve of the birefringence

$$\Delta n = \Delta n_0 e^{-t/\tau} \qquad (4)$$

yields the relaxation time τ which is directly related to the dimension and the conformation of the particle.

2. Orientation mechanism and sign of the optical anisotropy

One of the best ways to see the orientation mechanism of a particle under the action of an external electric field is the reversing pulse method [16, 17] which consists of applying two consecutive pulses of opposite polarity and of short duration to the particle suspension or solution. Whatever the nature of the solvent and the polymer concentration, we did not detect any change in the birefringence value when the electric field is reversed. This was observed not only with the graft copolymer, but also with a block copolymer having a short polystyrene sequence. The absence of transient in the electro-optical signal at the reversal of the electric field proves the absence of a permanent dipole moment contribution [16]. This also excludes the presence of any slow-induced dipole moment, so the orientation mechanism retained is the fast-induced dipole moment due to the polarization of the counter-ionic atmosphere. This is in complete agreement with earlier results on polyions such as

sodium polystyrenesulfonate [15, 18] and polyvinyl-pyridinium salts [6].

But the main feature we retained from our preliminary observations is a change in the electric birefringence (and of the optical anisotropy), which is positive for the graft copolymer and negative for both the homopolymer and block copolymer. Such a change of the optical term was previously observed in the case of polyvinylpyridinium salts in organic solvents [7] and polyvinylimidazolinium salts in the presence of platinum compounds [19]. This change is clearly depicted in the field strength dependences of the electric birefringence in water/methanol mixtures at a concentration of 0.1 g/l (Fig. 1).

Two factors could be responsible for this positive anisotropy:

a) the so-called form birefringence linked to the difference between the refractive indices of the polymer ($n_p \simeq 1.6$) and the solvent ($n_s \simeq 1.33$). This form anisotropy, which is always positive, has been used to explain the positive birefringence detected with the poly-N-butyl-4-vinylpyridinium bromide in nitromethane solution [7]. Owing to a possible overestimation of this form's contribution [20], we must be cautious about considering this positive term in a discussion of the sign of birefringence.

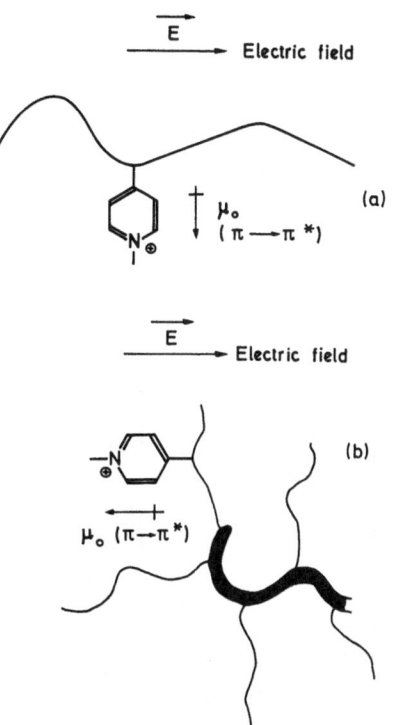

Fig. 1. Field strength dependences of the electric birefringence of the copolymer G (open signs) and the homopolymer H (dark signs) at a concentration equal to 0.1 g/l, and in water/methanol mixtures ($V/V\%$): 90/10 ($\circ$, $\bullet$); 65/35 ($\square$, $\blacksquare$); 35/65 ($\triangledown$, $\blacktriangledown$); 10/90 ($\triangle$, $\blacktriangle$); pure methanol ($\diamond$, $\blacklozenge$)

Fig. 2. Schematic representation of the orientation of pyridinium lateral group with respect to the chain backbone for the homopolymer (a) and the graft copolymer (b)

b) for polyions with aromatic lateral groups such as polyvinylpyridinium salts [21] and sodium polystyrenesulfonates [22], a negative dichroism has been observed in the UV region (290–220 nm), in good agreement with an orientation of the main $\Pi \to \Pi^x$ transitions in a direction more or less perpendicular to the long axis of the polyion (Fig. 2a). Owing to the peculiar shape of the graft copolymer, it seems reasonable suspect that many of the pyridinium rings could adopt an orientation more or less parallel to the orientation axis of the whole particle and, more or less parallel to the external electric field (Fig. 2). Such an assumption is confirmed by the measurement of the reduced dichroism $\Delta A/A$ at a low concentration ($C = 0.05$ g/l). ΔA represents the difference of the absorbances in the polarization directions parallel and perpendicular to the field. The absorption spectrum of the graft copolymer displays two peaks at 228 and 259 nm. The reduced dichroism vanishes at the former wavelength, but it is positive and very small at 260 nm ($\Delta A/A = +0.02$). This positive value of UV dichroism is in good agreement with the positive birefringence in the visible region.

3. Influence of the polymer concentration

The field strength dependences of the birefringence and the relaxation time (at $E \simeq 6$ kV/cm) have been measured as functions of concentration in a 90/10% vol water/methanol mixture. The Δn_s, Δa and τ values of the polymers G and H, calculated as mentioned above, are compared in Fig. 3. It appears that the classic "polyelectrolytic effect", i.e., the sharp increase of electric polarizability and the relaxation time (similar to that of reduced viscosity) with the dilution, is observed for the homopolymer. Let us recall that this effect arises from the electrostatic repulsion between neighbor-charged sites, which in-

creases when the concentration and/or the ionic strength decrease. This polyelectrolytic effect is not observed with the graft copolymer: Δa increases only slightly with dilution in the range of $5-8 \times 10^{-32}$ $F \cdot m^2$, whereas τ remains almost constant, ranging between 15 and 19 µs (Fig. 3). At $c = 0.05$ g/l, $\tau = 15$ µs (G) and $\tau = 115$ µs (H); despite a molecular weight twice smaller, the homopolymer has a relaxation time of about eight times larger than the copolymer.

The chain length, assuming a rigid-rod model, can be roughly estimated with the aid of the relation

$$L^3 = \frac{18 kT\tau (\ln 2p - 0.8)}{\Pi \eta_0} \tag{5}$$

where p is the axial ratio and η_0 the viscosity of the solvent. The axial ratio can be estimated on the basis of the viscosity and partial specific volume, according to Simha [23]. In the 90/10% water/methanol medium, the reduced viscosity η_s/c measured at $C \sim 0.05$ g/l (in a classical dilution viscometer), and the partial specific volume $\bar{v}$ (measured by picnometry), are equal to 3200 cm³ g^{-1} and 0.70 cm³ g^{-1} for the homopolymer, and to 250 cm³ g^{-1} and 0.67 cm³ g^{-1} for the graft copolymer. Using the function of Simha, axial ratio values of 315 and 78 have been found for the samples H and G. Attention must be focused on a possible overestimation of these axial ratios, owing to the well-known electroviscous effect arising in viscosity experiments performed on polyelectrolytes in the absence of added salt [24]. Using Eq. (5), chain lengths equal to 2450 Å and 1140 Å have been calculated for the samples H and G. The latter value proves the very compact structure of the graft copolymer; the mean distance separating each polyvinylpyridinium graft is of the order of 32 Å.

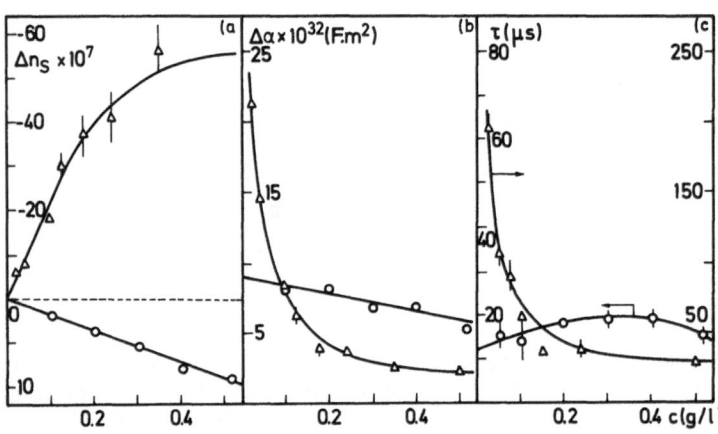

Fig. 3. Concentration dependences of the birefringence at saturation (a), the electric polarizability (b), and the relaxation time (c) for homopolymer $H(\triangle)$ and graft copolymer $G(\bigcirc)$ in a 90/10 vol% water/methanol mixture

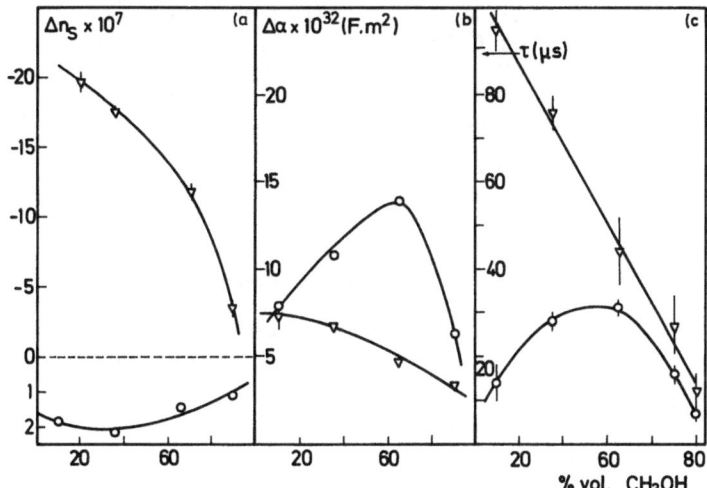

Fig. 4. Variations of the birefringence at saturation (*a*), the electric polarizability (*b*), and the relaxation time (*c*) in function of the solvent composition (water/methanol mixtures) for homopolymer *H* (▽) and copolymer *C* (○) at $C = 0.1$ g/l

Similar conclusions can be drawn from the comparison of the electric polarizability data: at $C = 0.06$ g/l, the Δa values are equal to 23×10^{-32} F·m²(*H*) and 8×10^{-32} F·m²(*G*). The former value is in fairly good agreement with previous results for polyvinylpyridinium salts [6]; the latter corresponds to that found for a polystyrene sulfonate sample with a molecular weight of 6.7×10^4 [15]. It is then concluded that the conformation of the graft copolymer does not change under the effect of dilution. In this highly hydrophilic medium which is in fact a precipitant of the polystyrene sequence, the latter adopts a rather compact conformation which has been identified to a "pseudomicellar" structure [5].

4. Influence of the solvent composition

For the homopolymer, the progressive change of the nature of the medium from water to methanol causes a continuous decrease of the Δn_s, Δa, and τ parameters (Fig. 4). This is expected from the decrease of the dielectric constant (water: $\varepsilon = 78$, CH₃OH: $\varepsilon = 32$) and of the electrostatic repulsion between charged sites. We showed previously that in protic solvents (water and aliphatic alcohols) there is a linear correlation of the Kerr constant of poly-4-vinylpyridinium salts with the dielectric constant [7]. In pure methanol, both polymers *G* and *H* have almost the same relaxation time, of about 10 ± 2 μs, and hence similar degrees of extension, despite their differences in structure and molecular weight.

Peculiar variations of Δa and τ have been observed in the case of the graft copolymer which exhibits maximum values when the solvent composition reaches about 60% in methanol (Fig. 4). The former assumption was to consider an association phenomenon,

yielding polymolecular micelles, such as those identified with the block copolymers [1 – 3]. However, Selb has shown that no intermolecular association takes place with the graft copolymer [5]. The polystyrene backbone is indeed completely protected by a rather large number of hydrophilic grafts (35 for copolymer *G*). The existence of the maximum can be explained by considering preferential solvation effects. In water-rich media, the copolymer can be represented by a conformation "in star", and the highly hydrophobic polystyrene sequence is very contracted [5]. As the methanol content increases, the preferential solvation of the styrene units by methanol causes an expansion of the central polystyrene backbone, so that the whole particle tends to a conform "in comb", in which the polyelectrolytic sequences have much more mobility (Figs. 3, 4, of [5]). This explains the observed increase of Δa and τ (Fig. 4). When the percentage of methanol exceeds 70%, the decrease of extension of polyvinyl-pyridinium grafts becomes preponderant; both Δa and τ decrease and reach, in a 90/10 medium, values similar to those of the homopolymer.

Acknowledgement

Joseph Selb and Yves Gallot (Centre de Recherches sur les Macromolécules, Strasbourg, France) are gratefully acknowledged for the samples, helpful comments, and discussion.

References

1. Selb J, Gallot Y (1980) Makrom Chem 181:2605 – 2624
2. Selb J, Gallot Y (1980) Makrom Chem 182:1491 – 1511
3. Selb J, Gallot Y (1981) Makrom Chem 182:1513 – 1524
4. Selb J, Gallot Y (1980) Makrom Chem 181:809 – 822
5. Selb J, Gallot Y (1981) Makrom Chem 182:1775 – 1786

6. Tricot M, Houssier C, Desreux V (1976) Eur Polymer J 12:575–588
7. Tricot M, Houssier C, Desreux V (1978) Eur Polymer J 14:307–315
8. Tricot M, Houssier C, Desreux V, van der Touw F (1978) Biophys Chem 8:221–234
9. Tricot M (1980) In: Goethals EJ (ed) Polymeric amines and ammonium salts. Pergamon Press, Oxford, New York, pp 229–238
10. Tricot M, Debeauvais F, Houssier C, Desreux V (1975) Eur Polymer J 11:589–595
11. Selb J, Gallot Y (1979) Polymer 20:1273–1280
12. Houssier C, O'Konski CT (1981) In: Krause S (ed) NATO Adv Study Inst Ser, Ser B, 64:309–340
13. Houssier C (1974) Laboratory Practice 562–563
14. O'Konski CT, Yoshioka K, Orttung WH (1959) J Phys Chem 63:1558–1565
15. Tricot M, Houssier C (1982) Macromolecules 15:854–865
16. Tinoco I, Yamaoka K (1959) J Phys Chem 63:423–427
17. Matsumoto S, Watanabe H, Yoshioka K (1970) J Phys Chem 74:2182–2188
18. Yamaoka K, Ueda K (1980) J Phys Chem 84:1422
19. Houssier C, Tricot M (1979) In: Jennings BR (ed) Electro-optics and Dielectrics of macromolecules and colloids. Plenum Press, New York, London, pp 247–257
20. Harrington RE (1970) J Am Chem Soc 92:6957–6965
21. Tricot M, Houssier C, Desreux V (1973) Biophys Chem 3:291–296
22. Matsuda K, Yamaoka K (1982) Bull Chem Soc Jpn 55:69–76
23. Simha R (1940) J Phys Chem 44:25–34
24. Vink H (1970) Makromolekulare Chem 131:133–145

Received February 15, 1988;
accepted June 8, 1988

Author's address:

M. Tricot
Laboratoire de Chimie Macromoléculaire
et Chimie Physique
Université de Liège au Sart-Tilman
4000 Liège, Belgium

Progress in Colloid & Polymer Science   Progr Colloid & Polymer Sci 77:217–220 (1988)

Rheological properties of solutions of a colloid-disperse homoglucan from *Schizophyllum commune*

E. Steiner[1]), H. Divjak, W. Steiner, R. M. Lafferty, and H. Esterbauer[1])

Institute of Biotechnology, Microbiology and Waste Treatment, Graz, University of Technology, Graz, Austria,
[1])Institute of Biochemistry, University of Graz, Graz, Austria

Abstract: The rheological properties of a homoglucan produced by *Schizophyllum commune* were investigated in water and dimethylsulfoxide (DMSO). Aqueous solutions of the polymer showed non-Newtonian character at concentrations of about 0.1% (w/v), and thixotropic behavior was observed. The apparent viscosity of aqueous solutions was not influenced by high salt concentrations and also showed high stability from pH = 2–12. The solutions showed Newtonian character in DMSO at polymer concentrations of about 0.1% (w/v). Moreover, the dependency of the apparent viscosity on the polymer concentration was investigated in water and DMSO. The intrinsic viscosities of the polymer in these two solvents were determined by graphic extrapolation of the reduced viscosity to polymer concentration = 0. From these intrinsic viscosities the average molecular weight of the polysaccharide in water and DMSO was calculated.

Key words: Homoglucan, Schizophyllum commune, pseudoplasticity, thixotropy, intrinsic viscosity, random coil, triple helix

Introduction

A polysaccharide was isolated from a culture broth of a fungus identified as *Schizophyllum commune*. This extracellular homoglucan is supposed to be schizophyllan, composed of repeating units consisting of three β-1,3-linked D-glucose residues and added to one of those, a single-D-glucose molecule attached through a β-1,6-linkage [1–3]. Schizophyllan has been studied in Japan [2, 4, 5] because of its unique rheological properties resulting from the triplehelical structure in water and from the random coil structure of the single chains in DMSO.

The main aim of our studies on the complex rheological behavior of the isolated polysaccharide was to achieve a better characterization of the isolated polymer.

Materials and methods

The biopolymer used in our studies was produced by a fungus which was isolated in Bangladesh. The strain was identified by Centraalbureau voor Schimmelcultures as *Schizophyllum commune*. Fr.: Fr. Fermentation was carried out in a 100-liter fermenter (Braun, 100D, Braun Melsungen AG) at 30°C, pH = 5±0.5, 150 rpm. The composition of the medium per liter was as follows: 50 g glucose, 6 g KH_2PO_4, 10 g caseinpeptone and 3 ml mineral solution. The polymer was harvested after a fermentation time of 350 h.

The polysaccharide was isolated by precipitation with ethanol from the culture filtrate and purified by means of ion exchange resins and repeated precipitation with ethanol, followed by freeze-drying from aqueous solution. After the total hydrolysis of the purified polymer only glucose could be detected by HPLC and GC analysis.

The polymer solutions for the rheological characterization were prepared by solving the freeze-dried polymer in water or DMSO at 40°C; unsolved particles were removed by filtration. The concentration of the polymer in these solutions was determined gravimetrically.

Rheological properties of the solutions were measured by a rotation viscosimeter (Fa. Contraves, Rheomat 115, Searl type, DIN 125), and a low shear viscosimeter (Fa. Contraves, LS2, Rheomat 30, Couette type).

Results and discussion

Solutions of the isolated homoglucan show a complex rheological behavior. The viscosity of aqueous solutions decreases with increasing shear rates which indicates the non-Newtonian character of pseudoplasticity. This is demonstrated for different polymer concentrations from 0.02% to 0.1% (w/v) and shear rates

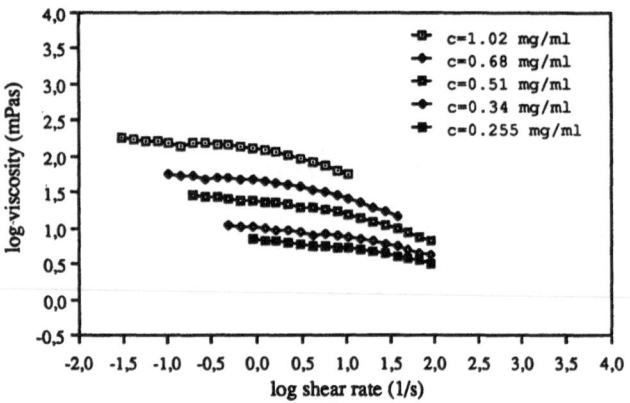

Fig. 1. Dependency of the apparent viscosity of aqueous homoglucan solutions on the shear rate (20 °C, LS 2)

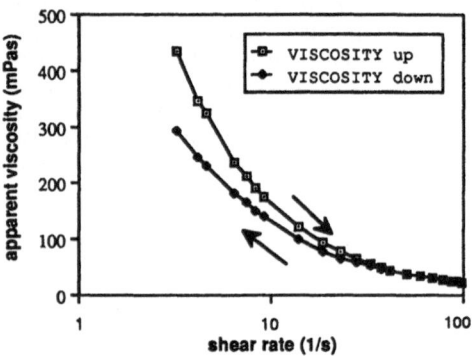

Fig. 2. Thixotropic behavior of an aqueous polymer solution ($c = 3.21$ g/l, 25 °C, DIN 125)

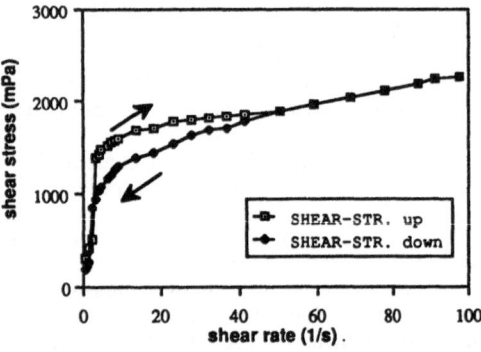

Fig. 3. Correlation between shear rate and shear stress for a thixotropic polymer solution ($c = 3.21$ g/l, 25 °C, DIN 125)

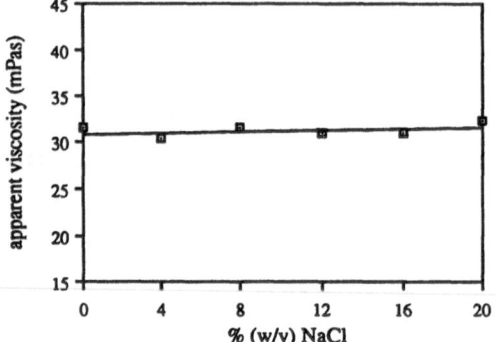

Fig. 4. Influence of high salt concentrations on the apparent viscosity of an aqueous polymer solution (1% w/v, 25 °C, DIN 125)

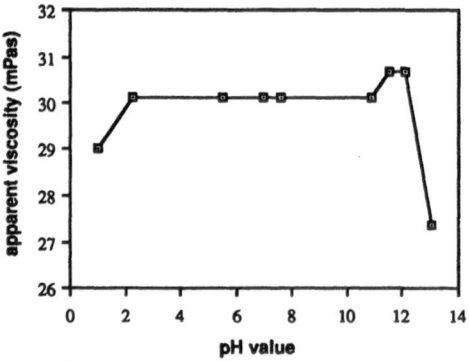

Fig. 5. Influence of the pH value on the apparent viscosity of an aqueous polymer solution (1% w/v, 25 °C, DIN 125)

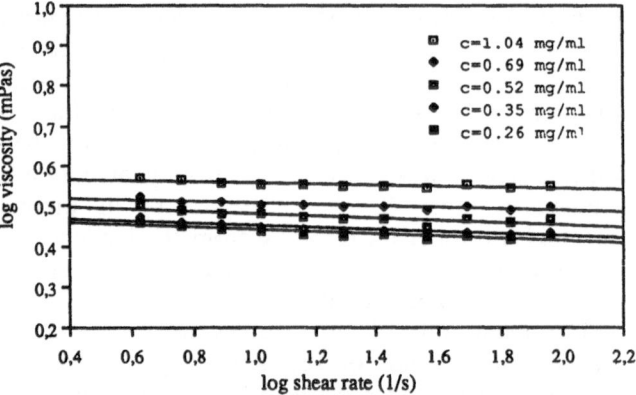

Fig. 6. Dependency of the apparent viscosity on the shear rate of polymer solutions in DMSO (20 °C, LS 2)

varied from $3.12 \cdot 10^{-2}$ s^{-1} to 91.0 s^{-1} in Fig. 1 (measured with LS 2 at 20 °C). This pseudoplasticity might be due to the anisotropy of the rigid triple helix structure of the polymer.

Additional measurements revealed the thixotropic behavior of aqueous solutions, which means that

sheared polymer solutions reach their zero-shear-viscosity after a time shift. Graphic representations of thixotropy for the polymer solution ($c = 3.21$ g/l) are shown in Figs. 2, 3 (measured with Rheomat 115 at 25 °C). This thixotropic effect can be explained by entanglements of the polymer molecules; these en-

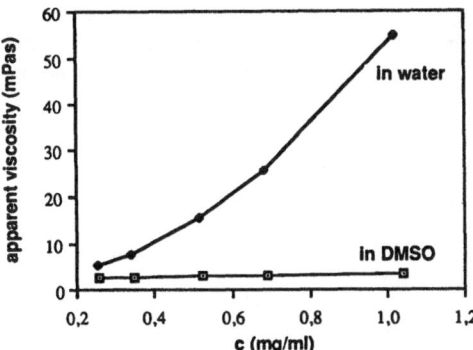

Fig. 7. Dependency of the apparent viscosity on the polymer concentration in water and DMSO (20 °C, $D = 10.6 \, \text{s}^{-1}$, LS 2)

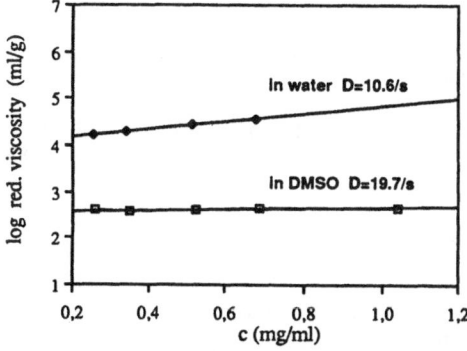

Fig. 8. Graphic determination of the intrinsic viscosities of the polymer in water and DMSO

tanglements are destroyed under shear-stress and restored after a regeneration-time [1, 6].

The apparent viscosity of aqueous polysaccharide solutions (1%, w/v) is not influenced by high salt concentrations up to 20% (w/v) NaCl, when measured at 63,5 rpm (Rheomat 115) and 25 °C (Fig. 4). Furthermore, the polymer showed a very high stability over a large range of pH values (pH = 2–12). If the pH-value of the glucan solution is larger than pH 12, the apparent viscosity decreases irreversibly because of alkaline depolymerisation (Fig. 5). These properties are of great interest for possible technical and industrial use and are in good agreement with other research groups [7, 8].

Contrary to aqueous solutions, the polysaccharide showed Newtonian character in DMSO which was verified up to a polymer concentration of 0.1% (w/v). That means that the viscosity does not depend on the shear-rate, which was varied from $4.24 \, \text{s}^{-1}$ to $91.0 \, \text{s}^{-1}$ (Fig. 6 shows measurements with LS2 at 20 °C). This Newtonian behavior can be explained by the destruction of the triple-helix in DMSO and the formation of single random coils which show far less anisotropy

and shear orientation than the triple-helix in aqueous solution [9, 10].

Figure 7 represents the comparison of the dependency of the apparent viscosity on the increasing polymer concentration in water and DMSO ($D = 10.6 \, \text{s}^{-1}$, 20 °C, LS2). The diagram shows the non-Newtonian behavior of aqueous polymer solutions in contrast to the Newtonian character of the polymer in DMSO.

The intrinsic viscosities at 20 °C of the polymer in water and DMSO were determined according to the Martins' formula by graphic extrapolation of the reduced viscosity to polymer concentration = 0 ($D = 10.6 \, \text{s}^{-1}$ for aqueous solutions, $D = 19.7 \, \text{s}^{-1}$ for measurements in DMSO). The extrapolation to zero shear rate was not necessary for the polymer in DMSO, as these solutions showed Newtonian behavior; for aqueous solutions this extrapolation was not carried out.

The results from these viscosity measurements are presented in Table 1. The graphic extrapolations for the determination of the intrinsic viscosity in water and DMSO can be seen in Fig. 8.

Evaluated intrinsic viscosities:

water: $10\,115 \, \text{ml g}^{-1}$
DMSO: $370 \, \text{ml g}^{-1}$

The rheological properties of the polymer studied were comparable to those obtained from investigations on schizophyllan samples by other researchers [9, 11, 12]. Therefore, we used their models for the calculation of the average molecular weight of the polysaccharide.

With the help of the Staudinger-Mark-Houwink-equation ($[\eta] = K \cdot M^a$) the calculated molecular weight for the polysaccharide in DMSO was $1.2 \cdot 10^6 \, \text{g/mol}$ when using the data from Norisuye ($K = 2.314 \cdot 10^{-2}$; $a = 0.69$) [9].

Table 1. Apparent viscosities measured for different polymer concentrations at 20 °C (LS2) in water and DMSO, and the reduced viscosities calculated from these data

Water ($\eta = 1.002 \, \text{mPas}$)			DMSO ($\eta = 2.424 \, \text{mPas}$)		
$c \cdot 10^3$ (g/ml)	η (mPas)	η_{red} (ml/g)	$c \cdot 10^3$ (g/ml)	η (mPas)	η_{red} (ml/g)
0.255	5.178	16343	0.26	2.663	396
0.34	7.539	19188	0.35	2.74	385
0.51	15.50	28370	0.52	2.939	418
0.68	25.74	36306	0.69	3.138	434
			1.04	3.551	453

The average molecular weight in water was $4.9 \cdot 10^6$ g/mol determined by Yanaki's diagram [13].

These average molecular weights for the isolated polymer correspond with those for schizophyllan [9, 12]. Nevertheless, the ratio of the molecular weight of the polymer in DMSO to the molecular weight in water (1 : 4) is higher than expected assuming the triple helix in water turns into three single random coils in DMSO. This might be due to the uncertainty factor that arose when we had to use data for schizophyllan to determine the molecular weights.

To be sure that the polysaccharide we studied is identical with schizophyllan, further investigations on the structure of the polymer are necessary and will be carried out subsequently.

Acknowledgement

Measurements with the Low shear viscosimeter LS2 were made possible kind permission of Dr. V. Ribitsch (Institute of Physical Chemistry, University of Graz).

This work was partially supported by the Austrian "Ministerium für Wissenschaft und Forschung".

References

1. Miyajima T, Yoshizumi S, Kikumoto S, Takahashi H (1970) Seito Gijutsu Kenkyukaishi 22:35–44
2. Kikumoto S, Miyajima T, Yoshizumi S, Fujimoto S, Kimura K (1970) Nippon Nogei Kagaku Kaishi 44(8):337–342
3. Kikumoto S, Miyajima T, Kimura K, Okubo S, Komatsu N (1971) Nippon Nogei Kagaku Kaishi 45(4):162–168
4. Enomoto H, Einaga T, Teramoto A (1984) Macromolecules 17:1573–1577
5. Yanaki T, Ito W, Tabata K, Kojima T, Norisuye T, Takano N, Fujita H (1983) Biophysical Chemistry 17:337–342
6. Kojima T, Tabata K, Ikumoto T, Yanaki T, Itoh W (1980) J Polym Sci 18:547–558
7. Cottrell I (1980) American Chemical Society, Washington DC:251–270
8. Komatsu N (1969) GANN 60:137–144
9. Norisuye T, Yanaki T, Fujita H (1980) J Polym Sci: Polym Phys Ed 18:547–558
10. Sato T, Norisuye T, Fujita H (1981) Carbohydrate Research 95:195–204
11. Carriere C, Amis E, Schrag J, Ferry J (1985) Macromolecules 18:2019–2023
12. Enomoto H, Einaga Y, Teramoto A (1985) Macromolecules 18:2695–2702
13. Yanaki T, Norisuye T, Fujita H (1980) Macromolecules 13:1462–1466

Received February 15, 1988;
accepted June 3, 1988

Authors' address:

W. Steiner
Institute of Biotechnology, Microbiology
and Waste Treatment
Graz, University of Technology
Schlögelgasse 9
8010 Graz, Austria

Progress in Colloid & Polymer Science

Progr Colloid & Polymer Sci 77:221–226 (1988)

Kinetics of swollen surface layer formation in the diffusion process of polymer dissolution

L. Lapčík, L. Valko, M. Mikula, V. Jančovičová, and J. Panák

Slovak Technical University, Bratislava, Czechoslovakia

Abstract: The diffusion process of polymer dissolution in liquid solvents and the kinetics of the surface swollen layer formation have been studied. The following fundamental kinetic parameters of the dissolution and swelling processes have been presented: the rate of dissolution u_{OC} and its relevant thermodynamic and kinetic parameters ($\Delta H_d^{\neq}$ and $\Delta H_{D,d}^{\neq}$); the induction period of dissolution t_Q and its apparent activation energy values in the swelling process ($\Delta H_s^{\neq}$); and, in the diffusion process connected with swelling ($\Delta H_{D,s}^{\neq}$). Both the kinetic and thermodynamic activity of individual solvents for given polymers (carboxymethylcellulose (CMC), hydroxyethylcellulose (HEC) and polyvinyl chloride (PVC)) have been expressed through the internal quantities $RA_{\delta,s}$ and $RA_{\delta,d}$, i.e., expansion work of the polymeric coil accomplished by action of the internal pressure P_i.

Key words: Polymer solutions, diffusion process, surface layer formation, thermodynamic and kinetic parameters, swelling processes, carboxymethyl cellulose, hydroxyethyl cellulose, polyvinylchloride

Introduction

The diffusion of liquid solvents into polymer foils can be evaluated on the basis of parameters describing the formation of a surface diffusion layer. In some cases investigated it has been recognized that a polymer matrix swells due to the solvent effect, the developed surface diffusion layer being denoted as a surface swollen layer [1]. In this connection an important aspect that might partially influence the correct interpretation of the experimental results obtained should be taken into consideration: optical methods use the velocities of optical interphase movement in the study of diffusion of solvent into the polymer substance. The optical interphase is usually viewed as the front of the diffusion boundary of molecules (index R) where the maximum gradient value of their concentration is assumed. This assumption, however, is not always strictly complied with. It can be admitted – also in terms of polymer dissolution – that in the front of the diffusion boundary the transport takes place through infiltration; the flow (index R) can then be characterized as the penetration of individual molecules, as in the case of permeation of the organic compounds vapors [1]. It follows that the maximum concentration gradient cannot be located in the front of the boundary (index R) but only in that certain point (which should be very close) behind this region. For this reason the optical interphase movement is not expected to be exactly identical to the movement of the front of the diffusion boundary of molecules.

The influence of these factors can be determined during the analysis of the empirical parameter A_δ which appears in the equation for the temperature dependence of thickness of the surface layer in this form [1]

$$\delta = \delta_0 \exp\left(-A_\delta/T\right) \tag{1}$$

where δ = the total thickness of the surface layer, δ_0 = pre-exponential factor, A_δ = empirical parameter, T = absolute temperature.

The initial theoretical analysis of the variable A_δ, in which the possibility of influencing the above-mentioned phenomenon is not included, was carried out by Valko and Lapčík [2, 3].

The theoretical analysis of physical importance of the empirical constant A_δ

In performing the analysis the equation is based on that for the rate of the optical interphase movement [4]

$$u_{RC} = D_{solv}/\delta \cdot \phi_\xi^{solv} \qquad (2)$$

where u_{RC} = the velocity of the optical interphase movement, ϕ_ξ^{solv} = solvent concentration (a molar fraction) in the optical interphase, D_{solv} = diffusion coefficient, δ = thickness of the swollen layer.

The temperature dependence u_{RC} can be expressed on the basis of the activated complex theory as follows [5]

$$u_{RC} = u_{RC}^0 \exp(-\Delta G_s^{\neq}/RT) \qquad (3)$$

where $\Delta G_s^{\neq}$ = the free activation enthalpy in the diffusion process of swelling, and u_{RC}^0 is defined by the equation

$$u_{RC}^0 = \varkappa e \lambda^2 \frac{k_B T}{h \delta_0} \qquad (4)$$

where $\varkappa$ = transmission coefficient, e = Euler number, λ = the distance of two consecutive positions on the flow coordinate, k_B = Boltzmann constant, h = Planck constant.

The effective activation energy for diffusion is given

$$E_D = RT^2(\partial \ln D/\partial T) \qquad (5)$$

and similarly for the activation energy of the swelling process

$$E_s = RT^2(\partial \ln u_{RC}/\partial T) . \qquad (6)$$

The mutual coherence between those equations can be achieved by modification of Eq. (2). After logarithmic and derivative calculations it is

$$\partial \ln u_{RC}/\partial T = \partial \ln D/\partial T - \partial \ln \delta/\partial T - \partial \ln \phi_\xi^{solv}/\partial T \qquad (7)$$

In considering the general validity of the equation applied under the constant pressure [6]

$$(\partial \ln \phi_\xi^{solv}/\partial T)_p = \frac{\Delta H_{dil}}{RT^2} \qquad (8)$$

where ΔH_{dil} is the dilution heat under the constant

pressure, and with respect to Eqs. (1), (5), and (6), and after inserting it into Eq. (7), it is possible to make a deduction

$$E_{s,P} = E_{D,P} - RA_{\delta,s} - \Delta H_{dil,s} \qquad (9)$$

where $E_{s,P}$ and $E_{D,P}$ are apparent activation energies under the constant pressure and, $RA_{\delta,s}$, $\Delta H_{dil,s}$ are variables corresponding to the swelling process.

Analogically, under the constant volume the following equation validity may be considered

$$(\partial \ln \phi_\xi^{solv}/\partial T)_V = \frac{\Delta U_{dil}}{RT^2} \qquad (10)$$

where ΔU_{dil} is the dilution heat under the constant volume expressed as a change of internal energy.

Then the following form may be expressed as

$$E_{s,V} = E_{D,V} - \Delta U_{dil,s} \qquad (11)$$

where $E_{s,V}$ and $E_{D,V}$ are apparent activation energies under the constant volume and, $\Delta U_{dil,s}$ is a variable corresponding to the swelling process, because in the case of isotropic swelling of the polymer chains network it holds true that

$$(\partial \ln \delta/\partial T)_V = 0 . \qquad (12)$$

The product of $R \cdot A_{\delta,s}$ can then be expressed as

$$R \cdot A_{\delta,s} = (E_{D,P} - E_{D,V}) - (E_{s,P} - E_{s,V}) \\ - (\Delta H_{dil,s} - \Delta U_{dil,s}) . \qquad (13)$$

The term in the last bracket of this equation can be written as

$$\Delta H_{dil,s} = \Delta U_{dil,s} + P'\Delta V_{dil,s} \qquad (14)$$

where P' is pressure and the expression $\Delta V_{dil,s}$ represents the difference between the value of molar segments volume of the polymer chains in the optical interphase V_{RC} and the molar solid-phase volume V_M (at the appropriate temperature)

$$\Delta V_{dil,s} = V_{RC} - V_M . \qquad (15)$$

For a very condensed system the validity of this equation is assumed

$$\Delta H_p = \Delta U_P + P'\Delta V . \qquad (16)$$

After applying Eq. (16) to Eq. (14) and using the Taylor expansion we have

$$\Delta H_{\mathrm{dil}} = \Delta U_{\mathrm{dil},s}$$
$$+(\partial U/\partial V)\Delta V_{\mathrm{dil},s}+\frac{1}{2!}\,(\partial^2 U/\partial V^2)\Delta V_{\mathrm{dil},s}^2$$
$$+\frac{1}{3!}\,(\partial^3 U/\partial V^3)\Delta V_{\mathrm{dil},s}^3+\ldots+P'\Delta V_{\mathrm{dil},s}\ . \tag{17}$$

By substitution for Eq. (13) and by application of the equation derived in an earlier paper [2] for the quantity $R\cdot A_{\delta,s}$ without considering the concentration change in the optical interphase it may be received

$$R\cdot A_{\delta,s} = (P_i+P')(\Delta V_D^{\neq} - \Delta V_s^{\neq})$$
$$+V_M/\beta\ \sum_{j=0}^{\infty}\ (-l)^j\ \frac{j}{j+2!}$$
$$\cdot[(\Delta V_D^{\neq}/V_M)^{j+2} - (\Delta V_s^{\neq}/V_M)^{j+2}] \tag{18}$$

where P_i = internal pressure, β = isothermal compressibility, $V_D^{\neq}$ and $V_s^{\neq}$ = activation volumes of the appropriate diffusion and swelling processes.

After making some modifications it is

$$R\cdot A_{\delta,s} = (P_i+P')(\Delta V_D^{\neq} - \Delta V_{\mathrm{dil},s} - \Delta V_s^{\neq})+(V_M/\beta)$$
$$\cdot\sum_{j=0}^{\infty}\ (-l)^j\ \frac{(j)!}{(j+2)!}\cdot[(\Delta V_D^{\neq}/V_M)^{j+2}$$
$$-(\Delta V_s^{\neq}/V_M)^{j+2} - (\Delta V_{\mathrm{dil},s}^{\neq}/V_M)^{j+2}], \tag{19}$$

which is an expression showing the meaning of the empirical constant A_δ or of the product of $R\cdot A_\delta$ during the swelling or dissolving of polymers.

From a physical point of view the expressions $R\cdot A_{\delta,s}$ and $R\cdot A_{\delta,d}$ can be understood as the voluminous work accomplished by polymer coils being affected by internal pressure and associated with their transition from a solid polymer phase to a swollen gel of the surface swollen layer (index s), or from the surface swollen layer to the solution in the case of polymer dissolution (index d).

Experimental

The process of the surface swollen layer formation as well as the rate of dissolution were studied with the film-shaped polymer samples prepared by casting from the concentration solutions in suitable solvents: PVC/THF, CMC/H$_2$O, HEC/H$_2$O, and by evaporating the solvent. The examina-

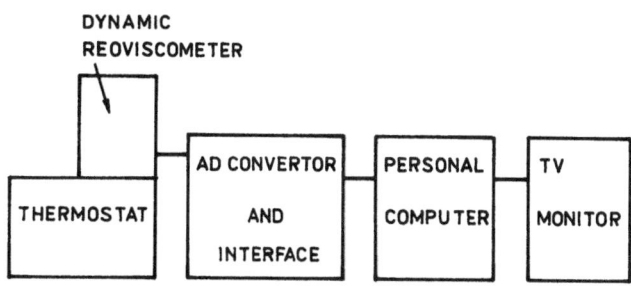

Fig. 1. Schematic of equipment designed for the kinetic study of dissolution of solid polymers

tion was carried out in experimental devices of three types, two of which have been detailed in our previous work [7, 8]. Briefly, the direct measurement of thickness of the surface swollen layer was made in the wedge interferometer; the kinetics of dissolution was studied in the apparatus for optical registration of the concentration change [5], as well as in that for investigation of the viscosity change in the co-existing liquid phase by using a rotary viscometer, being directly attached to the personal computer through the appropriate interfaces (Fig. 1).

Results and discussion

The typical examples of kinetic curves of the dissolution diffusion process of polymers in the glassy state are for some of the samples identical and independent of the applied method of investigation (the optical or viscosimetric one), which is obvious from (Figs. 2, 3). The curves are characterized by the induction period t_Q, during which the surface swollen layer

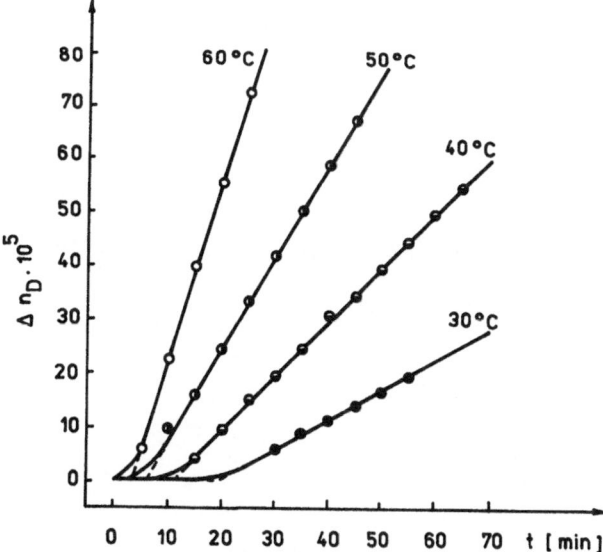

Fig. 2. Kinetic curves of dissolution of polyvinyl chloride in the cyclohexanone ($\bar{M}n = 60000$)

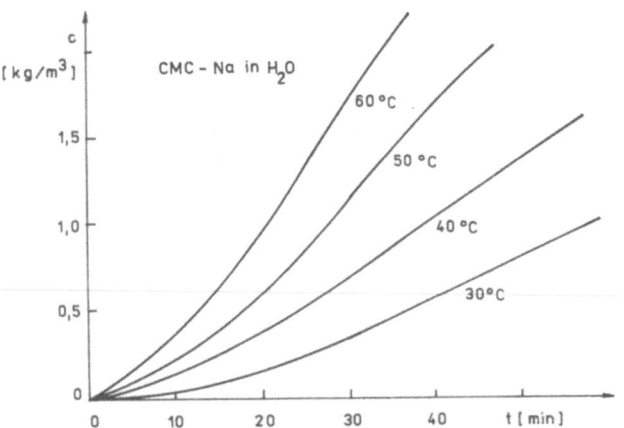

Fig. 3. Kinetic curves of dissolution of carboxymethylcellulose (CMC) in water ($\bar{M}n = 110000$)

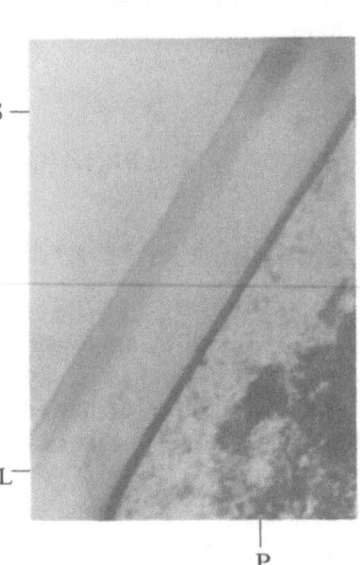

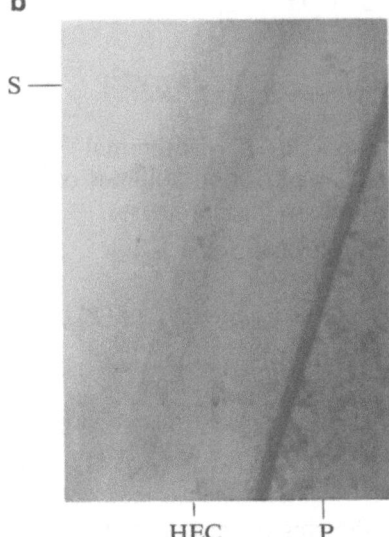

Fig. 5. Surface swollen layer of (a) CMC and (b) HEC (1 cm ~ 0.1 mm)

of defined structure and thickness (Figs. 4−6) is formed on the surface of the solid polymer sample.

The shape of the concentration solvent field in this layer can only regarded as linear in the first approximation, namely in the case where the values of differential diffusion coefficients for both substances, polymer and solvent, differ essentially (in several orders) one from another. When the difference between those variables is smaller and the proper dissolution is associated with the significant thermodynamic effects (the high value of dissolution heat), then the exponential character of the solvent's concentration distribution in the surface swollen layer becomes manifest. The situation is further complicated due to the originated "crazing effect" which leads to the splitting of optical interphase − "the surface swollen layer/solid polymer". This fact reflects upon the temporal variation observed in the rate of

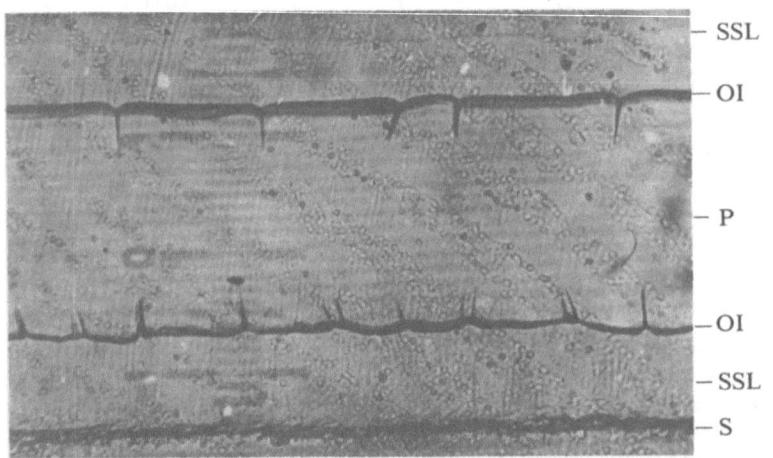

Fig. 4. Surface swollen layer of polyvinylchloride (PVC). The figure demonstrates the microcrack formation in the optical interphase − "the solid polymer/surface swollen layer", (the surface swollen layer, optical interphase, polymer, and solvent)

Table 1. Kinetic parameters of swelling and dissolution for polymers investigated

Polymer	$\bar{M}_n$	$H_s^{\neq}$ [kJ/mol]	$H_d^{\neq}$ [kJ/mol]	$R \cdot A_{\delta,d}$ [kJ/mol]	t_Q^0 [s]	u_{RC}^0 [cm/s]	Solvent
PVC-1	85 000	58.2	51.5	−7	$2.3 \cdot 10^{-7}$	854	Cyclohexanone
-2	60 000	49.9	49.0	0	$3.0 \cdot 10^{-6}$	525	Cyclohexanone
-3	46 000	47.3	45.2	−2	$5.8 \cdot 10^{-6}$	191	Cyclohexanone
-4	38 000	43.6	36.4	−7	$1.2 \cdot 10^{-5}$	10	Cyclohexanone
-5	31 000	36.4	31.4	−5	$1.2 \cdot 10^{-4}$	2	Cyclohexanone
CMC-Na	110 000	17.5	28.3	11	$9.8 - 10^{-1}$	0.25	Water
HEC	100 000	28.0	32.0	4	$3.0 \cdot 10^{-2}$	3.2	Water

solvent incorporation into the solid sample (e.g., the typical distances of two consecutive positions expressed as a change of internal energy on the sorption curves) or, at the same time, upon the "contour crazing" related to the development of microcracks observed microscopically.

The value of activation parameters describing the diffusion process of swelling and dissolution of polymers under investigation, are shown in Table 1. The values are considerably influenced by the polarity of the basic structural polymer unit, a monomer.

As it is evident from Table 1, the values of activation parameters $\Delta H_s^{\neq}$ and $\Delta H_d^{\neq}$ for PVC are largely dependent upon the molar weight and they increase with its increasing value, which may be ascribed to the mutual chain entanglement caused mainly by the higher amount of branched structures in the higher-molar-weight fractions. It is remarkable that the values of activation parameters $\Delta H_s^{\neq}$ and $\Delta H_d^{\neq}$ for CMC are, in comparison to the identical parameters for

PVC, substantially lower due to the smaller chain segment flexibility of CMC compared to PVC. This can also be confirmed by the density of CMC which is approximately 40% lower than that of PVC. The role of hydrogen bonds is less important in this case than in that of PVC, inasmuch as the sterically large cyclic elements of a water-soluble cellulose derivative minimize the probability of forming the higher amount of hydrogen bonds between the chains of the fundamental skeleton. Consequently, the voluminous expansion work for CMC is supposed to be relatively higher than for PVC. However, it is striking that while in the case of PVC the activation energy values of swelling are higher than those of dissolution, it is the reverse in the case of CMC. Obviously, this phenomenon relates to a significant role of the constitution of solvatation envelope and to the easier aggregability of dipole water structures in general, and to the hydroxyl groups of the polymer chain in particular. One may express an assumption, (and the diffusion anomalies observed by other researchers support the assumption), that the solvatation of the CMC and HEC chains results in the decreasing flexibility of the linear cellulose-derivative skeleton and in the increasing of the potential barrier which must be overcome in the diffusion transport. The vacation mechanism supposes the free volume formation for segmental displacement from the lower interstitial position n to the consecutive positions $n+1$ (in agreement with Eyring's theory) which take place in the direction of the polymer flow coordinate. The apparent molar weight enhancement of the solvated cellulose-derivative chain in the swollen layer has, by contrast with the initial conditions of the unswollen gel, an effect equal to that expressed by a different molar weight associated with the change of molecular weight of PVC. This can be acknowledged by a visual observation suggesting that the thickness of the surface swollen layer of the cellulose derivative is several orders higher than the swollen surface layer of PVC.

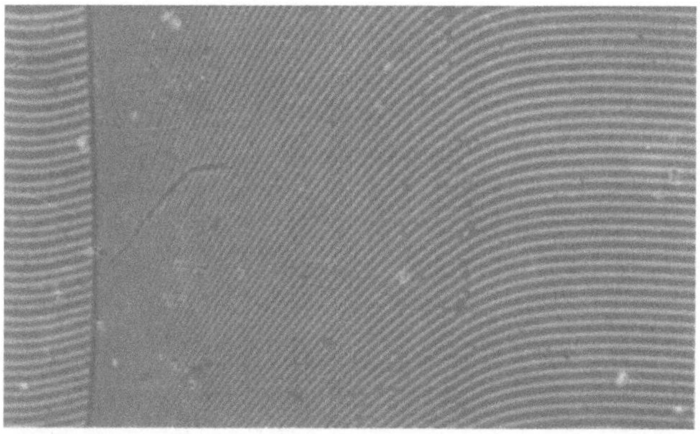

Fig. 6. Concentration field of PVC in the surface swollen layer expressed as the change in the index of refraction (cyclohexanone, 25 °C)

Progress in Colloid & Polymer Science, Vol. 77 (1988)

References

1. Überreiter K, Asmussen F (1957) J Polym Sci 23:75
1a. Überreiter K, Asmussen F (1962) J Polym Sci 27:187
2. Valko L, Lapčík L (1967) Zbornik prác CHTF SVŠT 47
3. Lapčík L, Kellö V (1974) IUPAC Macromol Symp Preprints (Madrid) 2:645
4. Valko L (1961) Chem zvesti 15:1
5. Lapčík L, Valko L (1971) Polym Sci A-2 9:633
6. Roff WJ, Scott JR (Ed) (1971) "Fibres, Film, Plastics and Rubbers". Butterworths, London, p 282
7. Lapčík L, Očadlík J, Kellö V (1972) Chem Papers 26:18
8. Lapčík L, Panák J, Kellö V, Polavka J (1976) J Polymer Sci Polym Phys Ed 14:981

Received February 26, 1988;
accepted June 9, 1988

Authors' address:

L. Lapčík
Slovak Technical University Radlinskeho 9
CS-81237 Bratislava, Czechoslovakia

Progress in Colloid & Polymer Science

Progr Colloid & Polymer Sci 77:227–233 (1988)

Investigations of the permselectivity
of styrene-butadiene block copolymer membranes *) **)

A. Ferdinand[1]) and J. Springer

Institut für Technische Chemie der Technischen Universität Berlin, Fachgebiet Makromolekulare Chemie

Abstract: The gas permeability of styrene-butadiene block copolymers has been examined for mixtures of CO_2 and CH_4. The permeated gas has been analyzed using a mass spectrometer as detector. The diffusion coefficient of CH_4 depends on the composition of the gas mixture and on the composition of the block copolymer. In comparison to the pure CH_4 the permeation of CH_4 in mixtures with CO_2 increases with increasing CO_2-content and increasing butadiene content in the block copolymer. Besides the plastification of the polymer by CO_2, the diffusion rate of CH_4 increases according to the high solubility of CO_2 in polystyrene.

Key words: Styrene-butadiene block copolymers, percolation, gas transport (permeability, diffusion, solubility), permselectivity

1. Introduction

Much technological interest exists about the separation processes of gas mixtures using membranes. Normally such processes require less energy than conventional methods, and consequently are less expensive. The basis for these processes is the different retention of polymers for different gases. Scientists have measured the permeabilities of membranes using pure gases and have defined the proportion of the permeability coefficients as to the ideal permselectivity. Often these ideal values cannot be compared with real permselectivities, because interactions between polymers and gases change with different gas mixtures and have a strong influence on the gas transport. The permeation rate of a gas can be strongly influenced by the presence of a second gas of different physical properties. Therefore, the exact determination of permselectivity of polymers for gas mixtures is only possible with equipment whose analyzers can detect different gases. This publication describes the construction and function of an apparatus for the determination of the real permselectivity using a quadrupole mass spectrometer as analyzer. The first measurements are done

with membranes consisting of styrene-butadiene (SB-) block copolymers, because the transport of pure gases is well known in such multiphase polymer systems [1]. The description and the denotation of the polymers have been specified. The number given in the denotation of the polymer describes the content of butatiene in the block copolymer in weight percent.

2. Apparatus and procedure of measurement

In the case of free-pore membranes the gas transport is described by the solution diffusion mechanism. Therefore the three characteristic values are the permeability coefficient P, the diffusion coefficient D and the solubility coefficient S. These values can be determined by the two fundamental different kinds of measurement, the sorption and the transmission [2, 3]. In his detailed publication, Felder [4] has discussed the advantages and disadvantages of these two methods for permeation experiments with pure gases. Our own studies [1 a] have shown that the differential transmission measurement is especially suitable for the investigation of permselectivity. For this kind of measurement the membrane to be tested is put into a special cell. The pressure difference existing between the two sides of the membrane causes the transport of the test gas and the gases or molecules are forced to move to the detector after their diffusion through the

---

*) Dedicated to Professor Dr. G. W. Becker on the occasion of his 60th birthday

**) Presented in part at the 33rd Annual Meeting of the Kolloid-Gesellschaft, Graz, Austria, September 14–16, 1987

[1]) Present address: Hoechst AG, Werk Kalle, Wiesbaden

membrane and their desorption by the help of a gas stream, or the suction of a pump. The resulting sigmoidal curve represents the actual permeation flux J as a function of time t. The use of an analyzer able to detect different gases nearly simultaneously allows us to continuously determine the permeation curves of different gases in a mixture. In contrast to a gas chromatograph as an analyzer, a mass spectrometer has the advantage that no carrier gas can disturb the gas transport, and in the case of gas mixtures no loss of time occurs during the measurement. A quadrupole mass spectrometer (QMA) (Balzers, Liechtenstein), was installed. The apparatus was constructed for high vacuum (Fig. 1) because using the QMA is only possible for pressures $\leq 10^{-4}$ mbar. The normal pressure during measurement is between 10^{-6} and $5 \cdot 10^{-5}$ mbar.

Gas is supplied in order to build up the desired pressure at the upper side of the membrane in the thermoregulated cell (pressure side), which must be free of gas at the beginning (Fig. 1b). At the other cell side (vacuum side) a turbo molecular pump creates the high vacuum (Fig. 1a, P). Each of the permeated molecules enters the suction of the pump and is detected by the QMA (Fig. 1a, 2). Because of the increasing permeation flux the QMA-signal I increases until a steady state is reached. This state is indicated by the stable signal I_s, which is a measure for the permeability coefficient. The diffusion coefficient can be calculated using different mathematics that describe the nonsteady state.

Measuring and evaluating is as follows. To calibrate the steady state signal I_s a special cell (Fig. 1a, 4) is evacuated into the system via a micro-leak. The leak was installed near the permeation cell. Each time during the evacuation t' the signal of QMA $I_{t'}$ indicates a pressure decrease $-(dp/dt)_{k,t'}$, in the calibration cell (index k). The amount of this decrease represents the number of molecules entering the high vacuum system at this time (Fig. 2). Therefore, a certain molecule flux $(dn/dt)_{t'}$ belongs to each signal $I_{t'}$ (n being the number of moles). In order to evaluate the permeability coefficient the molecule flux has to be expressed as a volume stream under standard conditions (STP). Using the ideal gas law the permeability coefficient can be calculated

$$P = \lim_{t' \to \infty} P_{t'} = -(l/A\,\Delta p)(V_k T_0 / T_k p_0)(dp/dt)_{k,\,t'} \quad (1)$$

l is the thickness of the membrane, A its area, Δp the existing pressure difference at the membrane, V_k the volume of the calibration cell, T_k its temperature and $T_0 = 273$ K resp. $p_0 = 1\,013$ mbar.

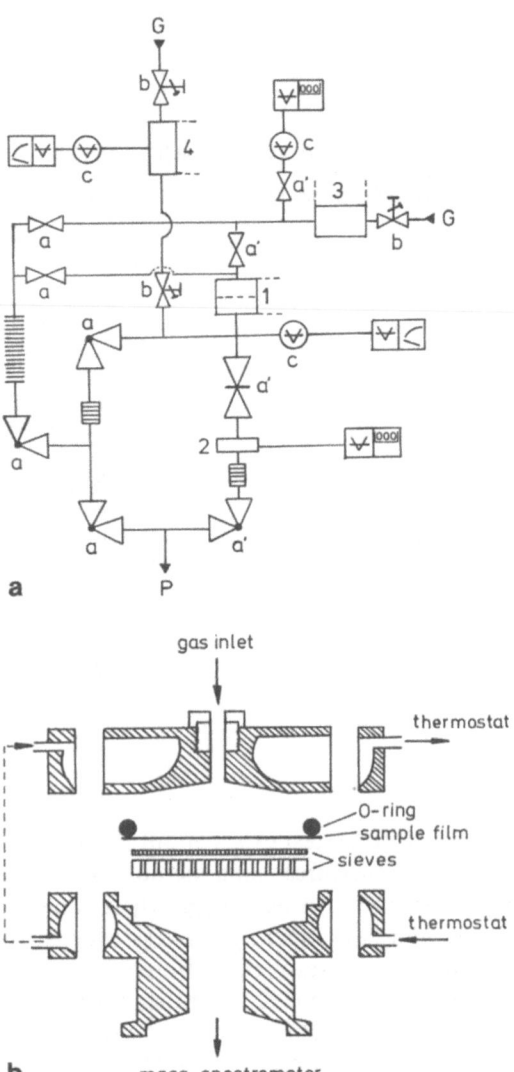

a

b

Fig. 1. High vacuum apparatus (A) and permeation cell (B). 1 − measuring cell, 2 − mass spectrometer, 3 − gas supply (thermoregulated), 4 − calibration cell (thermoregulated), a − cut-off valves, b − fine regulation valves, c − pressure gauge, G − gas inlet, P − vacuum pump

The diffusion coefficient results from the permeation curve in the nonsteady state. The differential equation of Fick's second law can be solved in two different ways with the help of assumptions made by the solution diffusion mechanism, either trigonometrically by separation of the variables time and place, or by removing one variable using a Laplace transformation. The solutions will be simple in case of possible extrapolations to long or short times.

The diffusion coefficient can be calculated by the use of these solutions and the plots as shown in Fig. 3.

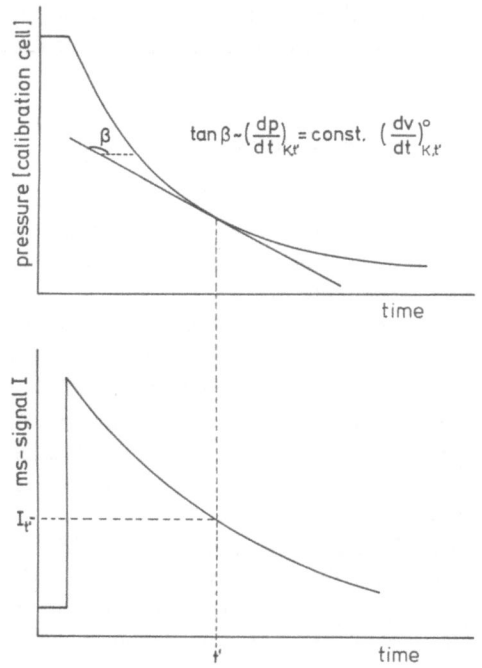

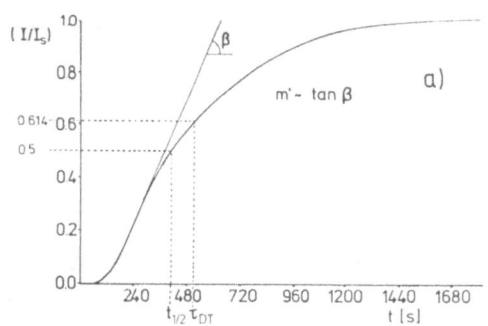

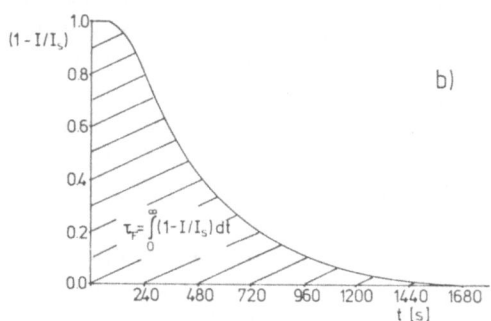

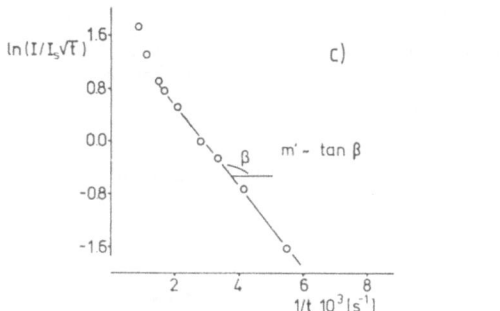

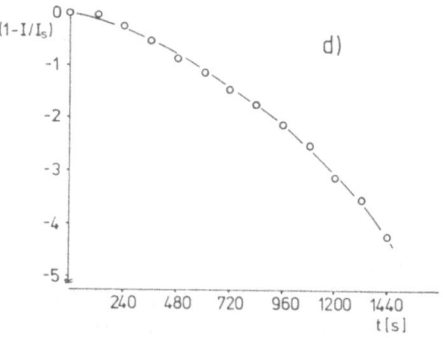

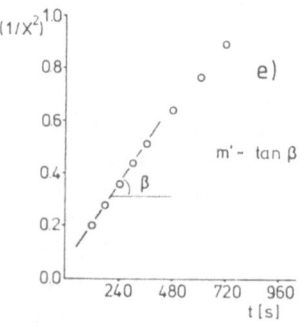

Fig. 2. Schematic plot of a calibration curve for the mass spectrometer (ms)

$(dv/dt)^0_{k,t'}$ = volume stream with standard conditions

Ziegel et al., [5] (Fig. 3a)

$$D_Z = l^2/7.119 t_{1/2} \qquad (2)$$

Ziegel et al., [5] (Fig. 3d)

$$D_{ZJ} = -l^2 m'/\pi^2 \qquad (3)$$

Felder et al., [6] (Fig. 3b)

$$D_F = l^2/6\tau_F \qquad (4)$$

Dung Tu, [7] (Fig. 3a)

$$D_{DT} = l^2/6\tau_{DT} \qquad (5)$$

Rogers et al., [8] (Fig. 3c)

$$D_R = -l^2/4m' \qquad (6)$$

Pasternack et al., [9] (Fig. 3e)

$$D_{PI} = l^2 m'/4 \qquad (7)$$

Pasternack et al., [9] (Fig. 3a)

$$D_{PI\text{-}I} = 0.176 l^2 [d(J/J_s)/dt] \qquad (8)$$

Equations (7) and (8) results from the combination of the second derivation of the error function and of

Fig. 3. Different plots for the calculation of the diffusion coefficients (Eqs. 2–8); experimental data for SBS 67, *Ar*, $T = 298$ K

an equation for the time dependence of the reduced flux J/J_s. The expression $l^2/4Dt$ has been substituted by x^2. In Fig. 3 only experimental data are shown; they will be discussed in section 4.

3. Gases and experimental data

The experimental procedure for the measurement of pure gases has been described elsewhere [1 b]. Based on these investigations the gases and conditions have been chosen for the measurement of permselectivity.

The gas mixtures have been produced as described in the literature [10]. At the pressure side in the permeation cell the total pressure has been set in such a way that the partial pressure of the investigated gas always is at 400 mbar. In this case, a direct comparison is possible between these results and those of pure gases. In order to guarantee a constant composition of the gas mixture at the pressure side the gas mixture flows continuously over the membrane. During the measurement the rate of the gas flow has been set 100 times higher than the maximally expected permeation rate. In order to test the influence of this procedure permeation measurements were made with pure gases that stay or flow at the pressure side. Both ways have produced the same results so that no anomalies occur because of the flow.

4. Test of the evaluation methods for determination of the diffusion coefficient

In order to evaluate a permeation curve according to the Eqs. (2)–(8) the plots as described in section 2 have been constructed as shown for instance in Fig. 3. As this example states, all methods stand the test except that of Ziegel (Eq. 3, Fig. 3d). In this case the linear correlation between $\ln(1-I/I_s)$ and t is not fulfilled.

The diffusion coefficients for Ar for SBS 67-films are liste in Table 1 for different temperatures. Using the method of Rogers et al., (Eq. 6) one gets always the highest values in comparison with the other evaluation methods. Further tests with other gases also show that calculations according to Rogers always yield too high diffusion coefficients. Therefore, this method does not seem suitable for the calculation of the diffusion coefficient. For this purpose the arithmetical mean value of the diffusion coefficients $\bar{D}$ given by the methods of Ziegel (Eq. 2) and Dung Tu (Eq. 5) proves to be good, although these results might be less exact in comparison with other methods as a consequence of the evaluation of points. The disadvantage is reduced

by forming the mean value. Therefore, the calculation of all diffusion coefficients was carried out after this method.

5. Measurements of the permselectivity

Figure 4 shows the real diffusion coefficients of the gases CO_2 and CH_4 measured on gas mixtures and depending on their composition. Four different block copolymers were investigated.

The copolymers having a high styrene content (SBO and SB 12.5) show the same transport behavior as the pure gases (CO_2 and CH_4) and as their mixtures. The calculation according to Ziegel (Eq. 2, $\bar{D}_Z$) and Dung Tu (Eq. 5, $\bar{D}_{DT}$) gives nearly the same diffusion coefficient, therefore for this polymer the mean values are only presented in Fig. 4.

With increasing butadiene content (SB 25, SBS 67) the diffusion rate of CO_2 remains the same in the case of pure gas and of gas mixtures. In contrast, the diffusion coefficient of CH_4 strongly depends on the composition of the gas mixture. A decreasing CH_4-content resp. an increasing CO_2-content causes the increase of the diffusion coefficient of CH_4. In an extreme case (SBS 67), the values of CH_4 become higher than those of CO_2. Parallel to this increase there is an increasing difference between the results of the calculation methods used according to Ziegel and Dung Tu (Eq. 2, resp., Eq. 5). The diffusion coefficients $\bar{D}_Z$ and $\bar{D}_{DT}$ differ so much that the formation of a mean value is not suitable. The $\bar{D}_Z$-values are always lower than the $\bar{D}_{DT}$-values. The difference increases with increasing CO_2 content in the gas mixture.

The important influence on this effect is the existence of the continuous polybutadiene (PB-)phase in the membrane which changes the copolymer composition. The increase of CH_4-diffusion rate starts at the

Table 1. Diffusion coefficient of Ar for SBS 67-membrane

T [K]	D_τ	10^6		[cm² s⁻¹]	D_R	D_{PI}	D_{PII}	D_{DT}
		$D_{\bar{D}}$	D_Z	D_F				
303	2.4	1.91	1.99	1.90	2.38	1.82	1.94	1.89
308	–	2.26	2.35	2.29	2.77	1.99	2.38	2.28
313	3.2	2.37	2.52	2.23	3.60	2.34	2.35	2.39
318	–	2.67	2.82	2.71	4.02	2.40	2.76	2.61
323	4.4	3.26	3.34	3.56	4.52	2.89	3.34	3.18

Diffusion coefficients of Ar for a SBS 67-membrane
D_τ integral method
$D_{\bar{D}}$ differential method, mean values from the different evaluation methods with D_R

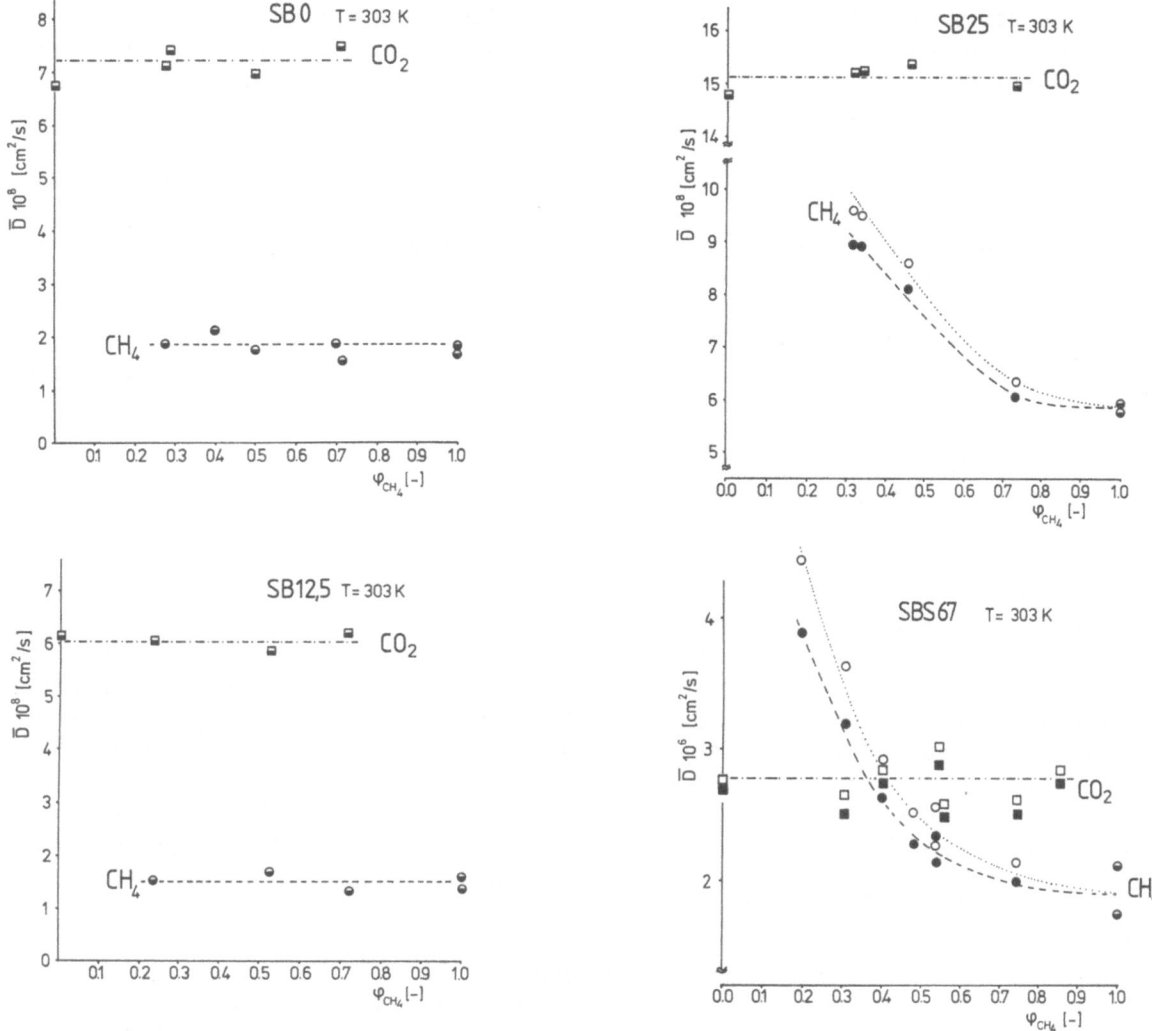

Fig. 4. Real diffusion coefficients $\bar{D}$ of CO_2 and CH_4 as a function of the composition of the gas mixture and the composition of the copolymer, $T = 303$ K. $\square \, \circ$, $\bar{D}_{DT}$ of CO_2 resp. CH_4; $\blacksquare \, \bullet$, $\bar{D}_Z$ of CO_2 resp. CH_4; $\square \, \ominus$, mean values from $\bar{D}_{DT}$ and $\bar{D}_Z$ of CO_2 resp. CH_4

point when the volume fraction of PB becomes higher than the percolation point. In the case of SB block copolymers this point has been determined as $\varphi_{PB}^c = 0.23$ by using gas permeability measurements [1 b]. Because of the obvious and important influence of a continuous PB-phase two theories explain this anomalous behavior of CH_4:

a) A plastification of the PB transport phase is caused by the presence of CO_2 in the gas mixture. This effect has been described in the literature for other polymers [11, 12]. Therefore, there will be an increase of the effective free volume by increasing the CO_2-concentration in the membrane and, as a consequence, there will be an increase of the diffusion rate of CH_4. The difference between the $\bar{D}_Z$

and $\bar{D}_{DT}$-values noticed above supports this theory.

Concerning the beginning of measurement, Dung Tu evaluates the resulting curve later than Ziegler does (Fig. 3a). During this time difference the permeation continues to approach the steady state. The result is a higher concentration of dissolved CO_2 molecules and therefore a higher degree of plastification at the evaluation point of Dung Tu in comparison to that of Ziegel. In consequence the $\bar{D}_{DT}$-values are higher than the $\bar{D}_Z$-values.

The diffusion coefficient of CO_2 is nearly independent of the composition of the gas mixture and so no plastification is caused by CH_4. This result is similar to that of Stern et al., [12], who investigated the permselectivity of polyethylene membranes using mixtures of CO_2/C_2H_4 and CO_2/C_3H_8. They found that the plastification increases with rising C_3H_8-content and so the diffusivity of CO_2 becomes higher. However, there is no similar influence on the diffusivity of C_3H_8 by increasing the CO_2-content in the gas mixture.

b) It is also possible to explain the described effect with the better solubility of CO_2 in polystyrene (PS) in comparison to that of CH_4 [1b]. According to the previous investigations [1] it is well known that the gas transport in SB block copolymers mainly occurs in the PB-phase, while the PS-phase is nearly impermeable. The idea is that the gas molecules are adsorbed in or at the PS-regions during the permeation process in the SB-membrane. Then the whole diffusion time is determined by the diffusion rate in the PB-phase and the mean sorption time in or at the PS-regions. This mechanism is valid for all gases, but the influence will be very strong for that gas which dissolves well in one of the components of the block copolymer. For the gas mixture of CO_2 and CH_4 with a high content of CO_2, the high number and good solubility of CO_2-molecules cause a high filling of the PS-sorption places; as a result the sorption of CH_4 molecules will be reduced. Compared to SB 25 and SBS 67, this effect is stronger in the instance of lower PS-content, resp. of a lower number of sorption places. The CH_4-molecules diffuse without longer lasting retention at or in the PS-phase through the membrane. Consequently the diffusion time of CH_4 decreases. A lower diffusion time means a higher diffusion coefficient.

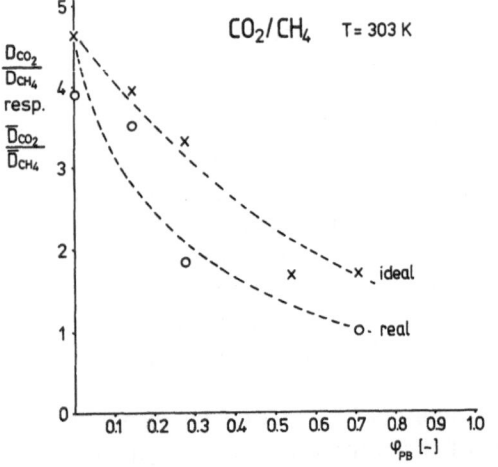

Fig. 5. Ideal and real selectivities for CO_2/CH_4 mixtures as a function of the composition of the copolymer, $T = 303$ K

Whether "plastification" or "retention" exert the main influence on the diffusion of CO_2/CH_4 mixtures in SB-block copolymers cannot be determined on the basis of the results discussed.

Both the real and ideal selectivity ($\bar{D}_{CO_2}/\bar{D}_{CH_2}$ resp. D_{CO_2}/D_{CH_4}) are functions of the composition of the block copolymer (Fig. 5). A direct comparison indicates that the described anomaly causes a decrease of the real selectivity with a higher PB-content and these selectivities are based on the relation of diffusion coefficients measured with 50/50 CO_2/CH_4-mixtures. Only this composition of a gas mixture allows the described direct comparison to the ideal selectivities.

6. Summary

The determination of the permselectivity of polymer membranes against gas mixtures is possible using the differential transmission technique. For this purpose an apparatus was constructed with a mass spectrometer as analyzer. The evaluation methods for diffusion coefficients proposed by Ziegel et al., Dung Tu, Felder et al. and Pasternack et al. give good corresponding results; the methods by Rogers et al., and Ziegel/Jost, do not.

Investigations of permselectivity of SB-block copolymers for CO_2/CH_4 gas mixtures show that the ideal and real selectivities do not agree in the case of PB-volume fractions which are higher than the PB-percolation point.

Regarding this polymer composition and in comparison with pure gas, the diffusion coefficients of CH_4 increase with decreasing CH_4 content in the CO_2/CH_4-mixture and with increasing PB-content in the SB-membrane.

The two possible theories discussed:

– CO_2 molecules dissolved in the membrane increase the free-volume of the polymer, especially of the PB-phase (plastification) and cause the increase of the CH_4-diffusivity.

– The high solubility of CO_2 in PS reduces the solubility of CH_4-molecules in the PS-phase (retention), and therefore the diffusion rate of CH_4 increases in the membrane.

Acknowledgement

This work was sponsored by the Deutsche Forschungsgemeinschaft, to which we express our gratitude.

References

1a. Ferdinand A (1986) Dissertation TU Berlin, D 83
1b. Ferdinand A, Springer J (Publication in preparation)
1c. Ferdinand A, Springer J (1986) In: Sedlacek B (ed) Polymer Composites/Morphology of Polymers. Walter de Gruyter, Berlin New York
2. Felder RM, Huvard GS (1980) In: Morton L (ed) Methods of Experimental Physics – Polymers Part C; Physical Properties. Academic Press, London New York, p 315
3. Lomax M (1980) Polym Test 1:105
4. Felder RM (1978) J Membr Sci 3:15
5. Ziegel KD, Frensdorff HK, Blair DE (1969) J Polym Sci A-2,7:809
6. Felder RM, Spence RD, Ferrell JK (1975) J Appl Polym Sci 19:3193
7. Dung Tu (1978) Dissertation, TH Stuttgart
8. Rogers WA, Buritz RS, Alpert D (1954) J Appl Phys 25:868
9. Pasternack RA, Schimscheiner JF, Heller J (1970) J Polym Sci A-2,8:467
10. Wunschel H (1980) Linde-Berichte Nr. 47
11. Fang SM, Stern SA, Frisch HL (1976) Chem Eng Sci 30:773
12. Stern SA, Mauze GR, Frisch HL (1983) J Polym Sci Phys 21:1275

Received February 15, 1988;
accepted June 5, 1988

Authors' address:

J. Springer
Institut für Technische Chemie
der Technischen Universität Berlin
Fachgebiet Makromolekulare Chemie
Straße des 17. Juni 135
1000 Berlin 12, FRG

Further electronmicroscope observations on polyethylene
II. An artifact makes amorphous surface layers visible
on crystal lamellae in melts*)

G. Kanig

Kunststofflaboratorium, BASF Aktiengesellschaft, Ludwigshafen am Rhein, F.R.G.

Abstract: The granule structure that occurs when a polyethylene melt is stained with chlorosulfonic acid by a rapid technique, e.g. at 150 °C, cannot be ascribed to nodules. In Part I of this study, it was unmasked as an artifact caused by microdemixing [1]. However, demixing can be prevented by first crosslinking the melt by electron beams. In this case, the melt appears to be homogeneously black after staining. If melting of crystalline polyethylene is arrested exactly at 134 °C by the rapid method, not only can the granule structure of the original melt be recognized in the ultrathin section, but also very thin black layers can be seen on the surfaces of the lamellae. Their character is apparently that of a crosslinked melt and they, therefore, conform more to the switchboard model than to the regularly adjacent re-entry model.

Key words: Electronmicroscopy, polyethylene, fine structure, artifact and switchboard model

A. Introduction

In Part I of this study [1], the granule structure was exposed as an artifact, a discovery made when polyethylene melts were stained directly with chlorosulfonic acid by the rapid method; the granule structure could then be seen in ultrathin sections under the electron microscope. Consequently, it cannot be taken as evidence for the presence of nodules in polymer melts, as other authors have done in interpreting similar granule structures obtained by preparing various polymers by other methods.

In the further course of the study, the curious discovery was made that even an artifact might prove quite useful. In this case, it allows the amorphous surface layers on the crystal lamellae to be distinguished from the contiguous melt, as will be demonstrated below.

B. The amorphous surface layers on the crystal lamellae and the contiguous melt

(a) Experimental

About 15 years ago, it was described how the chlorosulfonic acid method could render visible under the electron microscope both the thin crystal lamellae and the thin interlayers that remain amorphous in partially crystalline polyolefins [10–15]. In the photograph of an ultrathin section shown in Fig. 1a, the cross-sections of the crystal lamellae can be recognized as thin white strips, and the amorphous interlayers as thin black strips. The cross-sections are particularly defined if the microtome direction is perpendicular to the lamellae during preparation of the ultrathin specimens, but are unclear if the cut is made in other directions, as can be seen in the upper and lower parts of Fig. 1a. Obviously, the only strips that can be taken for quantitative evaluation are those that are clearly defined.

In the further course of the study, it was discovered that the staining method could be successfully applied not only at low temperatures, i.e. room temperature upwards, but also at temperatures above 100 °C [14, 15]. This is made possible by rapid treatment with chlorosulfonic acid. Thus, treating correspondingly small specimens of about 0.003 g for one second or somewhat longer allows intermediate stages to be fixed in rapidly changing structures, i.e. those encountered in annealing stretched polyethylene fibres at 100, 120 or 130 °C [15]. A depth of penetration by the chlorosulfonic acid of only a few micrometres suffices for the preparation of sufficient ultrathin sections

---

*) Paper presented at the "Kolloid-Tagung" held in Graz, 14–16 September, 1987.

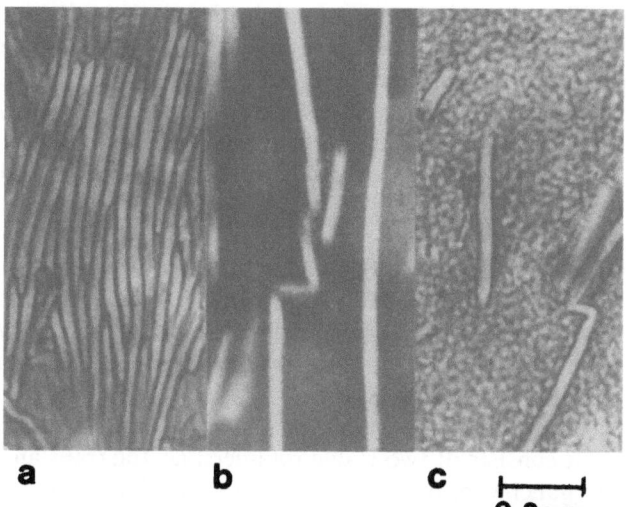

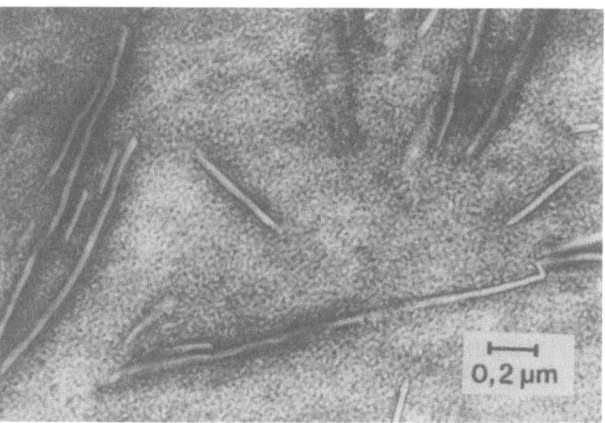

Fig. 2. Larger zone from the specimen in Fig. 1c

Fig. 1. Ultrathin sections of linear polyethylene (LPE; $M_W \approx 100\,000$). (a) Initial specimen; annealed under nitrogen for four weeks at 125 °C; then stained for five hours at 60 °C with liquid chlorosulfonic acid and subsequently treated at room temperature with uranyl acetate solution [12]. (b) Sample of (a) heated to 134 °C and allowed to stand for 30 minutes at this temperature; subsequently irradiated with electrons (200 Mrad) and stained as described in (a) above. (c) Sample of (a) after 30 minutes at 134 °C treated direct for about one second with liquid chlorosulfonic acid and subsequently stained as described in (a) above

after the minute samples have been embedded, e.g. in Araldite.

When the rapid staining method was applied to study crystallization and melting in polyethylene [12], a fine granule structure appeared in the ultrathin sections of the polymer melts. The initial suspicion that it was an artifact was confirmed in Part I of this paper [1]. It could be regarded as microdemixing in the copolymer melt produced by the chlorosulfonic acid treatment (Fig. 1c). The artifact's appearance made it advisable at that time to relinquish the study of crystallization and melting by this method for the time being. Other objections to the method were that adequate electron irradiation (200 Mrad) in a Van de Graaff generator not only fixed individual stages, e.g. melting, but also suppressed the microdemixing responsible for the granule structure. As a result, the melt would appear homogeneously black in the photographs of the ultrathin sections subsequently obtained by conventional staining techniques (Fig. 1b) [10–15].

In view of the long exposure to radiation, viz. about two hours, a suitable starting material must be found for the melt series. Figure 1a shows a linear polyethyl-

ene (LPE; $M_W \approx 100\,000$) that was heated for 16 hours at 160 °C under nitrogen in a sealed tube, annealed for four weeks at 125 °C, and subsequently cooled slowly to room temperature [12]. If a sample were to be taken afterwards, heated slowly to 134 °C, kept at that temperature for 30 minutes, irradiated with electrons as described above, and subsequently stained, the resulting structure would be that shown in Fig. 1b. Some thick lamellae can still be recognized in this photograph; their melting points ought to lie between 134 °C and 135 °C, because no lamellae at all were detected at 135 °C [12].

If a minute quantity of the sample taken after 30 minutes at 134 °C was to be stained directly for about one second with liquid chlorosulfonic acid and not irradiated with electrons beforehand, the melting process ought to be arrested at roughly the same stage as that described above. This assumption finds its confirmation in a comparison of Figs. 1b and 1c. Furthermore, Fig. 1c shows that the remaining isolated lamellae are accompanied by two thin black surface layers of about 6 nm thickness that can evidently be related to the black interlayers in Fig. 1a. The surface layers and the interlayers respond to staining in the same way as the free melt does after prior crosslinkage by electron beams (Fig. 1b). There is no microdemixing, such as is evident in the zones representing the melt in Fig. 1c or Fig. 2, and the layers appear homogeneously black and crosslinked from the outset. It is interesting to note from a comparison of Figs. 1b, 1c and 2, how useful the artifact has proved to be: it allows a thin amorphous surface layer, which behaves as a crosslinked melt, to be distinguished from a contiguous mobile melt whose presence is betrayed by the microdemixing.

(b) The amorphous thin layers and the switchboard model

P. J. Flory suggested that polymer melts, e.g. LPE, consisted of interpenetrating, coiled chain molecules [2]. In 1972, this hypothesis was convincingly proved by low-angle neutron scattery with the aid of deuterized material [3]. It can also be demonstrated that the space occupied by the coiled chain remains practically unchanged during partial crystallization in the melt [4–7, 12].

Thus, unreserved support is given to the switchboard model [8, 9] for the thin amorphous interlayers and surface layers (Figs. 1 a, 1 c) and not to the regularly folded model. According to the switchboard model, the thin black layers (Figs. 3 a, 3 b) consist of coil residues composed of loops and chain ends of various lengths. The interlayers also contain chain segments of various lengths that connect the adjacent lamellae, i.e. tie molecules.

It can be seen from Fig. 3 that long chain ends or large loops are obtained if neighbouring lamellae are melted away from the tie molecules that connect them. In Fig. 3 a, the interlayers are attached to the crystal surfaces on both sides (as is shown by dots in Fig. 3), but the surface layer in Fig. 3 b is anchored on only the one side. In the latter case, there appears to be a smooth transition into the mobile melt. The mobility of the long chain ends and large loops immersed in the melt is not much less than that of the free chains; in other words, they will behave similarly and participate in microdemixing.

As a result of the double anchorage of the smaller loops to the surfaces of the lamellae, as shown in Fig. 3 b, the layer of about 6 nm thickness formed on these surfaces behaves as a crosslinked melt and prevents microdemixing during staining (Fig. 1 c). The surface layer also embraces the lower anchored parts of the long chain ends and loops (Fig. 3 b).

The entanglements caused by interpenetration of the coiled chains may exert an additional crosslinking effect. They must not only have been retained in the thin amorphous layers during crystallization (Figs. 1 a, 1 b, 2, 3) but must have, in fact, multiplied in these layers, because their original fraction in the crystal lamellae was shifted into the surface layers during crystallization [20].

The interlayers in Figs. 1 a and 3 a are composed of the surface layers of neighbouring crystal lamellae. They are penetrated by the chain segments that join the lamellae and are responsible for the strength of the partially crystalline material.

It may be assumed that the coiled LPE chains in the melt consists of two stable rotamers, i.e. the trans and the gauche conformations, which are continuously interconverting and remain in equilibrium. In this case, a given temperature-dependent trans-gauche ratio can represent a mean degree of coiling in the melt [13, 29–33]. Since the two positions differ in partial volume as well as in energy, their ratio is also reflected in the density.

Many studies have revealed that the density of the amorphous interlayers is practically identical to that of the free melt, if the temperature is the same [21, 22]. However, this entails that the mean degree of coiling of the coil residues from which the thin layers are composed is the same as that of the free melt. This is a further argument in favour of the switchboard model.

Two recent publications [16, 23] have pointed out that the switchboard model implies that the difference in density will prevent unrestricted linkage between the amorphous interlayers and the neighbouring crystal lamellae, if the chain stems are perpendicular to the surfaces. Therefore, a link could not be established unless the switchboard arrangement were to be combined with frequent regularly adjacent re-entry. However, if the stems are inclined at a sufficiently large angle from the vertical, the amorphous interlayers can quite readily be linked, despite differences in density. This has been proved by x-ray studies [8, 15, 24–26]. Thus the switchboard model remains entirely relevant. Recently, the inclination of the stems has been made directly visible under the electron microscope as well [12, 13]. Unfortunately, even adherents of the switchboard model ignore this inclination and continue inserting perpendicular stems in the lamellae of their models, e.g. [7, 25].

A frequently voiced opinion is that single crystals grown in solutions have regularly adjacent re-entry on the surfaces [23, 27]. This also is no longer tenable. Figure 4 shows an ultrathin section of a stained LPE preparation composed of lamellar single crystals [28].

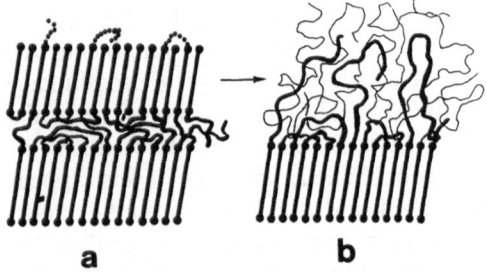

a   **b**

Fig. 3. (a) Loops and tie molecules anchored in the amorphous interlayer in conformance with the switchboard model. (b) If the neighbouring lamellae are melted, the tie molecules in (a) are converted into long chain ends or large loops

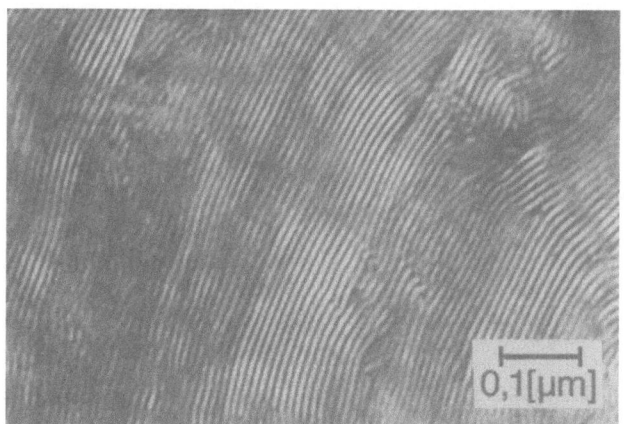

Fig. 4. Ultrathin section of linear polyethylene (LPE; $M_W \approx 150000$) single crystals obtained from solution, compressed at 200−300 bar and room temperature [28]

The crystals were grown in a 0.1% solution of LPE ($M_W \approx 150000$) in xylene at 79 °C. After filtration, washing and drying, they were compressed at 200−300 bar at room temperature to form a transparent film. An ultrathin section of this film is shown in Fig. 4. It can be seen that the structure, in principle, is the same as that in Fig. 1a, i.e. it consists of thin crystal lamellae and very thin interlayers of about 4 nm thickness. An obvious conclusion would be to attribute the homogeneous blackening of the layers to the coil residue anchored to the surfaces of the lamellae. When they are stained with chlorosulfonic acid, they behave in the same way as a crosslinked melt, in which microdemixing cannot occur. The interlayers are composed of the two surface layers on the neighbouring lamellae and, in this case, consist merely of loops and chain ends, because the tie molecules joining the lamellae, as in Figs. 1a and 3a, are not present.

C. Conclusion

In two previous publications [12, 13], details were given on the blocking effect exerted by surface layers on crystal growth. It was demonstrated that the actual reason for this was that the gauche-trans transformation in the tauter, twice-anchored loops and tie molecules are blocked by excessive activation energy. This leads to termination of the all-trans growth in the thickness direction of the lamellae, because the trans-positions are no longer replenished while lateral growth proceeds. Consequently, abnormal crystalline forms, e.g. hollow pyramids and lamellae are obtained. For this reason, the term "autogenous activated

sorption" has been adopted in analogy to the formation of abnormal crystalline forms, e.g. in metals, as a result of activated sorption of oxygen, hydrogen, water, etc. In this case, the gauche position represents the foreign substance [13].

References

1. Kanig G (1987) Coll and Polym Sci 265:855
2. Flory PJ (1953) Principles of Polym Chem; Cornell University, New York; and (1969) Statistical Mechanics of Chain Molecules, Wiley, New York
3. Kirste RG, Kruse WA, Schelten J (1972) Makromol Chem 162:299
4. Wignall GD, Ballard DGH, Schelten J (1976) J Macromol Sci-Phys B 12:75
5. Schelten J, Ballard DGH, Wignall GD, Longmann G, Schmalz W (1976) Polymer 17:751
6. Schelten J, Wignall GD, Ballard DGH, Longmann G (1977) Polymer 18:1111
7. Yoon DY, Flory, PJ (1977) Polymer 18:509
8. Flory PJ (1962) J Am Chem Soc 84:2857
9. Fischer EW, Schmidt GF (1962) Angew Chem 74:551
10. Kanig G (1973) Kolloid-Z Z Polymere 251:782
11. Kanig G (1974) Kunststoffe 64:470
12. Kanig G (1982) Coll and Polym Sci 260:356
13. Kanig G (1983) Coll and Polym Sci 261:993
14. Kanig G (1977) Coll and Polym Sci 255:1005
15. Kanig G (1980) J Crystal Growth 48:303
16. Di Marzio EA, Guttman CM (1980) Polymer 21:733
17. Bonart R (1964) Kolloid-Z Z Polymere 199:136
18. Blackadder DA, Roberts TL (1969) Makromol Chem 126:116
19. Odell JA, Grubb DT, Keller A (1978) Polymer 19:617
20. Flory PJ, Yoon DY (1978) Nature 272:226
21. Zachmann HG (1964) Forschr Hochpolymer Forschg 3:581
22. Fischer EW, Goddar H, Schmidt GF (1967) Polymer Letters 5:619
23. Peterlin A (1980) Macromolecules 13:777
24. Bonart R (1964) Kolloid-Z Z Polymere 199:136
25. Blackadder DA, Roberts TL (1969) Makromol Chem 126:116
26. Odell JA, Grubb DT, Keller A (1978) Polymer 19:617
27. Hay JN (1981) Polymer 22:718
28. Illers KH, Kanig G (1982) Coll and Polym Sci 260:564
29. Flory PJ (1956) Proc Roy Soc (London) 234 A:60
30. Starkweather HJ Jr, Boyd RM (1960) J Phys Chem 64:410
31. Smith RP (1966) J Polym Sci A-2:869
32. Miller AA (1978) Macromolecules 11:859
33. Zetta L, Gatti G (1972) Macromolecules 5:535

Received February 26, 1988;
accepted June 2, 1988

Author's address:

Prof. Dr. G. Kanig
Kunststofflaboratorium
BASF Aktiengesellschaft
6700 Ludwigshafen am Rhein
F. R. G.

Progress in Colloid & Polymer Science   Progr Colloid & Polymer Sci 77:238–241 (1988)

Element specific electron microscopy of polymeric materials

M. Kunz, U.-R. Heinrich, M. Möller, and H.-J. Cantow

Institut für Makromolekulare Chemie, Universität Freiburg, Hermann-Staudinger-Haus, Freiburg, F. R. G.

Abstract: Electron Spectroscopic Imaging (ESI) was performed on a polyethylene specimen and on the poly(styrene-b-2-vinylpyridine) blockcopolymer to obtain information about the microstructural organization. Polyethylene samples were treated with $ClSO_3H$ or RuO_4 prior to cutting ultrathin specimens. The local distribution of different staining agents was investigated by electronic micrography of specific, inelastically scattered electrons. Ultrathin sections of poly(styrene-b-2-vinylpyridine) were quaternized with CH_3I-vapour to yield specific contrasts.

Key words: Electron Spectroscopic Imaging (ESI), element distribution, morphology, polyethylene, poly(styrene)-b-(2-vinylpyridine)

Introduction

In the last 10 years microstructural aspects of polymeric materials have met with increasing interest. Knowledge of the microscopic morphology and the corresponding molecular composition is essential for a better understanding of the particular properties of polymeric materials. The localization of distinct elements or specific functional groups is especially important. The combined analysis of microscopic structure and chemical composition is possible by various electron microscopic techniques, such as EDX or AES [1]. Progress in the evaluation of local element composition has been achieved by the introduction of a new electron microscope, the ZEISS-EM 902. In conventional transmission electron microscopy the image is formed by the sum of elastically, inelastically and nonscattered electrons. The ZEISS-EM 902 allows highly resolved images to be obtained either with elastically or with inelastically scattered electrons. Specific imaging with inelastically or elastically scattered electrons is performed by a prism-mirror-prism-type spectrometer integrated into the optical column of a high resolution transmission electron microscope. This type of imaging was first reported by Ottensmeyer [2]. The electron microscope has been described in detail elsewhere [3, 4]. In bright-field imaging, inelastically scattered electrons are excluded, yielding images of enhanced contrast and sharpness. The energy loss of inelastically scattered electrons varies for different elements. Thus, by excluding the elastically scattered electrons, the inelastically scattered electrons can give information about the composition of the sample.

Polyethylene

Figure 1 a shows a nonstained specimen of a linear ultrahigh molecular weight polyethylene (Hostalen GUR 412, Hoechst AG) with $M_W > 2 \times 10^6$, $M_W/M_n = 5$, and $\varrho = 0.936 \, g/cm^3$. The degree of crystallinity is about 61% and thus rather low. For the application of the ESI-technique very thin sections of the samples (less than 50 nm) are necessary. Because of the plasticity of the material at room temperature the polyethylene was cut by a cryomicrotome (Reichert-Jung) at $-170\,°C$. The sections exhibited a golden interference colour, and were therefore not suitable for ESI-application. Reacting the samples with $ClSO_3H$ according to Kanig [5], or with RuO_4 before cutting, resulted not only in staining but also in hardening. Samples which were treated by $ClSO_3H$ prior to microtoming could be cut with success at $-120\,°C$.

Figure 1 b represents a specimen of the polyethylene which was oxidized with $ClSO_3H$ before cutting. After microtoming, the sections were treated with $UO_2(OAc)_2$ for two seconds and analyzed in the electron microscope. The picture shows the semi-

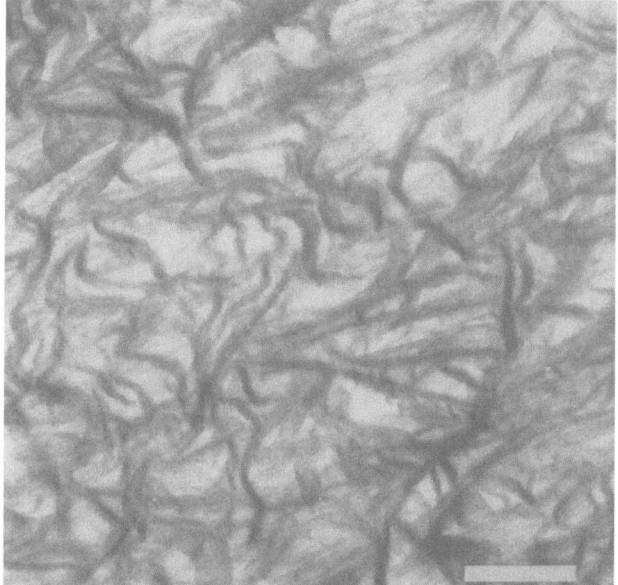

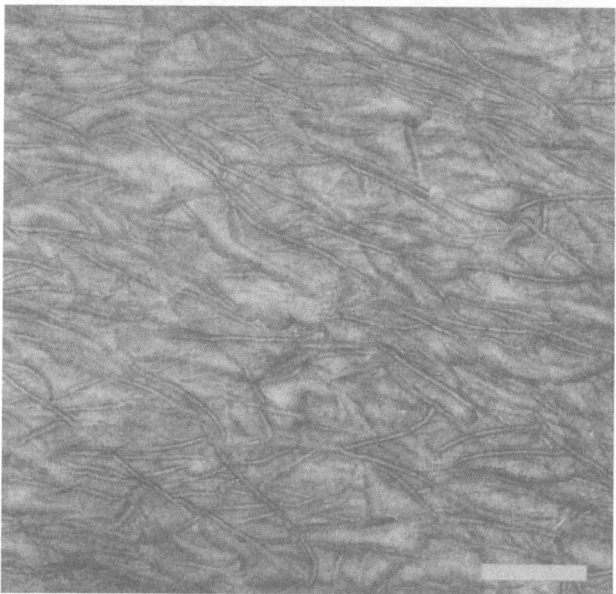

Fig. 1. Elastic bright-field images of linear high molecular weight polyethylene (M_w 2×10^6, $\varrho = 0.936$ g/cm³); a) untreated sample; scale bar = 300 nm; b) sample treated with ClSO₃H and UO₂(OAC)₂; scale bar = 500 nm

crystalline morphology, where the crystals are embedded in a matrix of amorphous polymer. Due to the intense scattering of the uranium atoms, the contrast has been drastically improved compared to the unstained cryocut sample. The staining agent is concentrated in the area between the crystalline and the amorphous phase, as has already been reported by Kanig [6].

Figure 2 shows the polyethylene sample treated with ruthenium trichloride and sodium hypochlorite, which produces RuO₄ in situ [7]. The contrast and resolution of the image formed by the elastically scattered electrons (Fig. 2a) are of extraordinarily good quality. Again, the strongly oxidizing RuO₄ preferentially attacks the area between the isotropic amorphous and the crystalline phase. A possible explanation for this interaction is given elsewhere [8].

The electron micrograph of the inelastically scattered electrons at $\Delta E = 220$ eV shows the distribution of the chlorine (Fig. 2b), which is incorporated in the staining particles. The areas between the amorphous and crystalline parts of the sample, which were dark in the elastic bright-field image (Fig. 2a), now appear bright, showing an increased concentration of chlorine. The electron micrograph of the inelastically scattered electrons at the carbon-edge ($\Delta E = 300$ eV) in Fig. 2c shows an inverse contrast relative to the chlorine specific image seen in Fig. 2b. The crystalline polyethylene lamellae are visible as bright lines because the carbon density in the crystal is higher than in the amorphous regions. Imaging at the oxygen ionization edge ($\Delta E = 532$ eV) shown in Fig. 2d shows results similar to the chlorine specific image in Fig. 2b. This result implies that the staining particles consist of ruthenium chlorides and oxides.

Poly(styrene-b-2-vinylpyridine) blockcopolymers

The copolymer was synthesized by sequential anionic polymerization of styrene and 2-vinylpyridine in toluene with butyllithium as initiator. The number average molecular weight was determined to be $M_w = 114.000$ and the molecular weight distribution to be $M_w/M_n = 1.04$. Bulk materials were prepared by slow solvent evaporation at 50 °C from 5 wt.% solutions from N,N-dimethylformamide (DMF). Ultramicrotomy of the dry polymer films was performed at room temperature. Quaternisation of the vinylpyridine nitrogen was achieved by exposure of the slices to methyliodine vapour. Details are described elsewhere [9, 10].

The application of ESI in ultrastructural analysis of the blockcopolymer with a composition of 80% styrene and 20% 2-vinylpyridine is demonstrated in Fig. 3. Comparison of the conventional global bright-field image (Fig. 3a) with the electron micrograph obtained by imaging only elastically scattered electrons shows enhanced contrast (Fig. 3b). In the images the phase separation of the two different blocks is clearly visible. The dark areas represent rods of poly (2-vinylpyridine), which are sliced at different angles and are arranged in a regular packing as well as in a

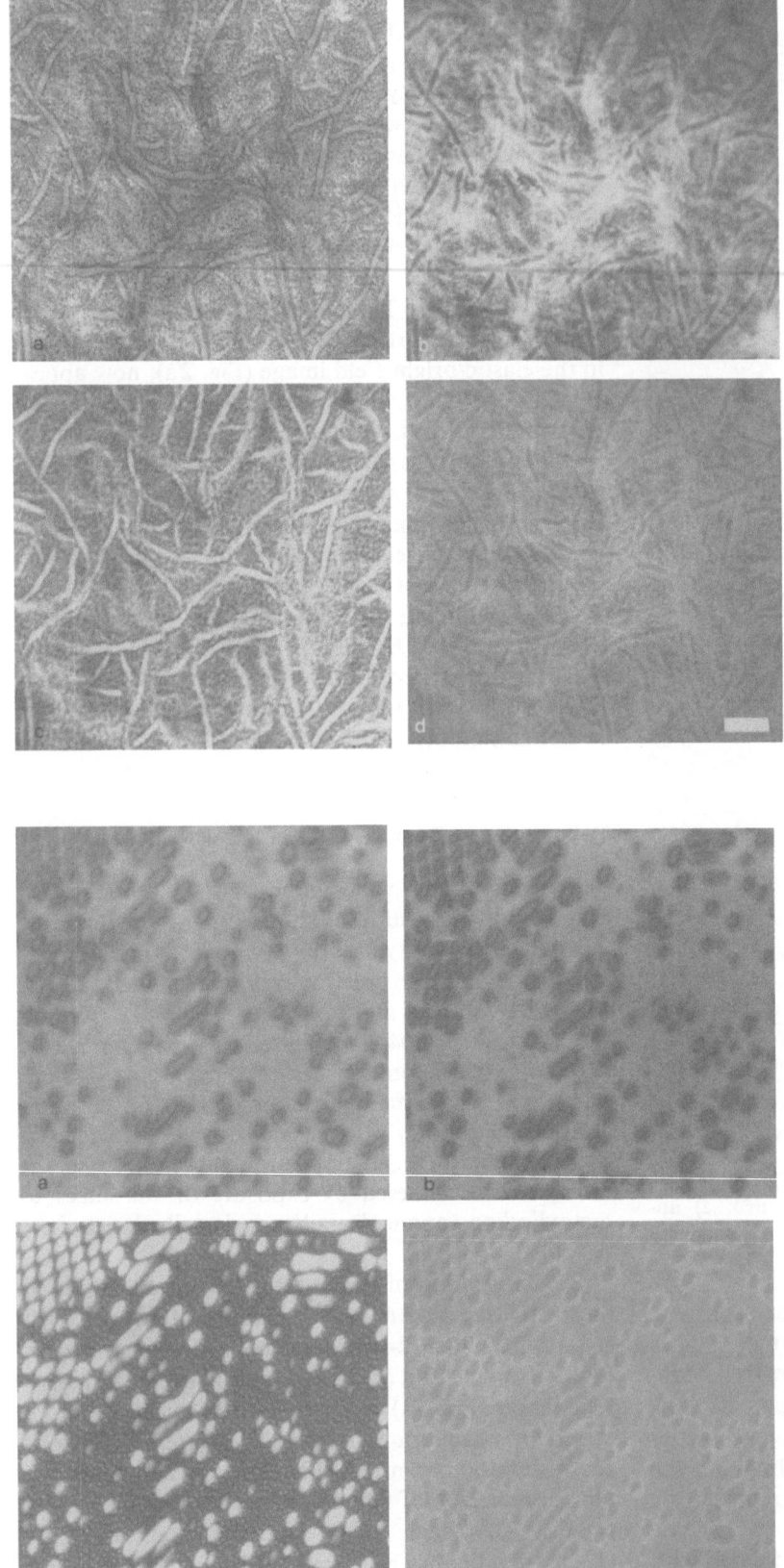

Fig. 2. Electron micrographs of linear high molecular weight polyethylene (M_w 2×10^6; $\varrho = 0.936$ g/cm^3); sample stained with RuCl$_3$/NaOCl; a) elastic bright-field image; b) chloride specific image ($\Delta E = 200$ eV); c) carbon specific image ($\Delta E = 304$ eV); d) oxygen specific image ($\Delta E = 552$ eV); scale bar = 125 nm

Fig. 3. Ultramicrotomed specimen of a polystyrene-block-poly(2-vinyl-N-methylpyridinium iodide); sample evaporated from DMF at 50°C; $M_n = 114\,000$; a) global bright-field image; b) elastic bright-field image; c) iodine specific image ($\Delta E = 206$ eV); d) carbon specific image ($\Delta E = 304$ eV); scale bar = 300 nm

disordered array. The homogeneity of the polymer composition and of the narrow molecular weight distribution appears to contradict the heterogeneity of the observed structure. These observations are understandable if one looks at the iodine distribution in Fig. 3 c. Using the inelastically scattered electrons of the iondine-edge ($\Delta E = 209$ eV), small bright dots appear in the less ordered regions. This is verified by the carbon specific image ($\Delta E = 304$ eV) which shows an inverse contrast due to the higher carbon content in the styrene domains. Obviously, the domain formation is controlled kinetically, resulting in a bimodal distribution of the domain sizes.

Conclusion

Sharp, highly resolved images with excellent contrast can be obtained by the combination of distinct chemical staining techniques and the visualization of the elemental distribution by the newly developed electron microscope. New information offered by electron specific imaging aids in understanding much of the controversial data reported on the structure of semicrystalline polymers, block- and graft copolymers, reversible associations in polymeric systems and polymer blends.

Acknowledgements

We thank Mrs. Christine Aehnelt for the excellent photographic work. Financial support was provided by the Deutsche Forschungsgemeinschaft (Sonderforschungsbereich 60) and by the Minister für Wissenschaft und Kultur von Baden-Württemberg (Schwerpunkt 29).

References

1. Maher DM (1979) In: Hren JJ, Goldstein JI, Joy DC (eds) Introduction to Analytical Electron Microscopy. Plenum Press, New York, p 259
2. Ottensmeyer FP, Andrew JW (1980) Ultrastruct Res 72:336
3. Egle W, Kurz D, Rilk A (1984) Magazine for Electron Microscopists (Zeiss) 3:4
4. Egle W, Rilk A, Bihr J, Menzel M (1984) Electron Microsc Soc Am Proc 42:566
5. Kanig G (1973) Kolloid-Z u Z Polymere 251:782
6. Kanig G (1975) Prog Colloid Polym Sci 57:176
7. Montezinos D, Wells GB, Burns JL (1985) J Polym Sci, Polymer Lett Ed 23:421
8. Kunz M, Möller M, Heinrich U-R, Cantow H-J (1987) Makromol Chem Symposia Series, San Sebastian, in press
9. Kunz M, Möller M, Cantow H-J (1987) Makromol Chem, Rapid Commun 8:401
10. Möller M, Lenz RW (in preparation) J Polym Sci, Polym Chem Ed

Received November 6, 1987;
accepted June 2, 1988

Author's address:

M. Kunz
Institut für Makromolekulare Chemie
Universität Freiburg
Hermann-Staudinger-Haus
Stefan-Meier-Str. 31
D-7800 Freiburg

Author Index

Subject Index

DEGIORGIO, V. (Guest-editor)

Trends in Colloid and Interface Science II

1988. X, 320 pp.
Progress in Colloid & Polymer Science, Vol. 76
Editors: H.-G. KILIAN; G. LAGALY
Hard cover DM 168,–. US$ 98.00.
ISBN 3-7985-0777-5 (Steinkopff Verlag); ISBN 0-387-91355-1 (Springer-Verlag New York)

Trends in Colloid and Interface Science II contains the proceedings of the 1st European Colloid and Interface Society Meeting, held in Como, Italy, in 1987. The book's three main sections, "Colloids," "Amphiphile Solutions," and "Interfaces," provide an up-to-date account of new developments in the structure and stability of colloidal dispersions, colloidal semiconductors, fractal aggregates, micelles, microemulsions, lyotropic phases, biological amphiphiles, monolayers, adsorption on colloids such as clays, and mineral extraction.

CHUDÀCEK, I. (Guest-editor)

Relationships of Polymeric Structure and Properties

1988. 212 pp.
Progress in Colloid & Polymer Science, Vol. 78
Editors: H.-G. KILIAN; G. LAGALY
Hard cover DM 132,–. US$ 79.00.
ISBN 3-7985-0779-1 (Steinkopff Verlag); ISBN 0-387-91338-6 (Springer-Verlag New York)

The latest advancements and challenges in the rapidly expanding field of the polymeric state of condensed matter are presented in this volume. Comprised of the principal contributions from the 5th International Seminar on Polymer Physics, 1987, Prague, it offers an exhaustive overview by leading researchers from European institutes and universities. Important applications are suggested by findings presented for the superstructure of polymer systems, and the electronic properties of thin polymer films.

Distribution in US and Canada through Springer-Verlag, 175 Fifth Avenue, New York, NY 10010; for other countries, order through your bookseller or directly from Dr. Dietrich Steinkopff Verlag, P. O. Box 11 1442, 6100 Darmstadt, FRG.

Steinkopff Verlag Darmstadt · Springer-Verlag New York

Rheologica Acta

Participating Societies: Australian Society of Rheology · Belgian Group of Rheology · British Society of Rheology · Canadian Rheology Group · Czechoslovak Group of Rheology · Deutsche Rheologische Gesellschaft · Grupo Español de Reologia · Groupe Français de Rhéologie · Società Italiana di Reologia · Sociedad Mexicana de Reologia · Nederlandse Reologische Vereniging · Arbeitsgruppe Rheologie in der Gesellschaft Österreichischer Chemiker · Swedish Society of Rheology

Executive Editor: H. H. Winter, Amherst MA, U.S.A.

Rheologica Acta publishes theoretical and experimental contributions on phenomenological and structural rheology (macro- and micro-rheology), thermo-, electro- and magneto-rheology, rheo-optics and rheometry, as well as applications of rheology to extrusion, spinning, blowing, moulding, calendering, coating, curing, flows through pipes, ducts and porous media, etc., including numerical simulations.
The journal reports investigations in solid and fluid materials, in mono- and multiphase systems as melts, solutions, powders, foams; as well as metals, plastics, rubbers, surfactants, adhesives, building materials, sludges, soils, rocks, biological substances, food, etc.
Rheologica Acta's contributors and readers include engineers, mathematicians, physicists, chemists, biologists, and physicians. Original contributions in English, French, or German are published. The prime consideration in accepting a paper is scientific excellence and specialists working in virtually every part of the world act as referees. Instructions to prospective authors may be found in each issue of Rheologica Acta.
Subscription Information: ISSN 0035-4511
Published bimonthly
1 year: DM 890,– plus postage; US$ 518.00 including postage
Members of all Participating Societies (see above) and subscribers to Colloid and Polymer Science receive 20% discount.

Supplement to "Rheologica Acta," Vol. 27 (1988)

Progress and Trends in Rheology II

Proceedings of the Second Conference of European Rheologists, Prague, June, 1986

H. GIESEKUS, M. F. HIBBERD, Dortmund, FRG (Eds.)
in co-operation with
P. MITSCHKA, P. RIHA and S. SESTAK, Prague, Czechoslovakia
1988. 502 pp.
Subscribers to "Rheologica Acta" receive 20% discount
Cloth DM 280,–. US$ 170.00
ISBN 3-7985-0770-8

Progress and Trends in Rheology II presents a representative survey of current activities in European rheological research. Subjects range from theoretical modelling and prediction on Non-Newtonian flows over rheometry to the rheology and processing of diverse materials, including food and biological substances. An additional 166 abstracts of contributions to the conference are included, many of which contain unpublished results. This volume is indispensable to all those engaged in rheology and its applications in engineering, production, biology, and medicine.

Distribution in US and Canada through Springer-Verlag, 175 Fifth Avenue, New York, NY 10010; for other countries, order through your bookseller or directly from Dr. Dietrich Steinkopff Verlag, P. O. Box 11 1442, 6100 Darmstadt, FRG.

Steinkopff Verlag Darmstadt · Springer-Verlag New York